Control of Energy Storage

Special Issue Editor

William Holderbaum

MDPI • Basel • Beijing • Wuhan • Barcelona • Belgrade

Special Issue Editor
William Holderbaum
University of Reading and Manchester Metropolitan University
UK

Editorial Office
MDPI AG
St. Alban-Anlage 66
Basel, Switzerland

This edition is a reprint of the Special Issue published online in the open access journal *Energies* (ISSN 1996-1073) from 2015–2017 (available at: http://www.mdpi.com/journal/energies/special_issues/control-energy-storage).

For citation purposes, cite each article independently as indicated on the article page online and as indicated below:

Author 1; Author 2. Article title. *Journal Name* **Year**, *Article number*, page range.

First Edition 2017

ISBN 978-3-03842-494-9 (Pbk)
ISBN 978-3-03842-495-6 (PDF)

Table of Contents

About the Special Issue Editor

William Holderbaum received the Ph.D. degree in automatic control from the University of Lille, Lille, France, in 1999. He was a Research Assistant with the University of Glasgow, Glasgow, UK, from 1999 to 2001. He was lecturer (2001–2009), Senior Lecturer (2009–2014) and currently a Professor at the University of Reading, Reading, UK and Manchester Metropolitan University since 2016. His current research interests include control theory and its applications. In particular in the area of energy in smarter grid to help reduce peak demand and increase the stability of the grid, current techniques involve charge incentives for users to cut load at peak periods as well as researching and developing optimal storage devices. He is a member of the IEEE and he has published over 100 papers in leading journals and international conferences.

Preface to "Control of Energy Storage"

Energy storage can provide numerous beneficial services and cost savings within the electricity grid, especially when facing future challenges like renewable and electric vehicle (EV) integration. Public bodies, private companies and individuals are deploying storage facilities for several purposes, including arbitrage, grid support, renewable generation, and demand-side management. Storage deployment can therefore yield benefits like reduced frequency fluctuation, better asset utilisation and more predictable power profiles. Such uses of energy storage can reduce the cost of energy, reduce the strain on the grid, reduce the environmental impact of energy use, and prepare the network for future challenges.

This Special Issue of Energies explore the latest developments in the control of energy storage in support of the wider energy network, and focus on the control of storage rather than the storage technology itself. Specifically, this book encompass:

- Control of energy storage (e.g., for flywheels, batteries or supercapacitors)
- Energy storage systems for transport (e.g., for automotive, shipping and aircraft)
- Energy storage systems for grid support including use with ancillary services
- Intelligent coordination of storage elements in the grid both at micro (i.e., low voltage) and macro (i.e., high voltage) scales
- Monitoring, modelling and other performance assessment methodologies for the control of storage
- Explorations of the future of energy storage systems and associated control problems.

The contributions are based on leading research, as well as cutting-edge exemplars from industrial practice that can be used to encourage sustainable development and performance of control of energy storage systems.

William Holderbaum
Special Issue Editor

Editorial

Control of Energy Storage

Timur Yunusov [1,*], Maximilian J. Zangs [1] and William Holderbaum [2]

[1] Technologies for Sustainable Built Environments Centre, School of Built Environment, University of Reading, Reading RG6 6AF, UK; m.j.zangs@pgr.reading.ac.uk

[2] School of Engineering, Manchester Metropolitan University, Manchester M1 5GD, UK; w.holderbaum@mmu.ac.uk

* Correspondence: t.yunusov@reading.ac.uk

Received: 30 May 2017; Accepted: 10 July 2017; Published: 16 July 2017

1. Introduction

In the attempt to tackle the issue of climate change, governments across the world have agreed to set global carbon reduction targets. For instance, the UK has agreed to reduce the green house gas emissions by 80% of 1990 levels by 2050 [1]. In the pursuit of the carbon reduction, there has been a continuous shift towards the de-carbonisation of major infrastructures such as transport and energy, and an uptake of renewable power generation.

An increasing proportion of renewable energy introduces new challenges for the transmission and distribution system operators. The intermittent nature of the renewable energy resources impacts their power output, causing imbalance in supply and demand across the power system. Since the proportion of inverter-fed generation is also likely to increase, the natural inertia of the system would reduce. This in turn causes the grid's frequency to become less stable and deviate from its target more rapidly than in the present day.

Electrification of major infrastructures will cause an additional demand for electricity, which could potentially coincide with existing demand peaks. Furthering this peak demand imposes additional strain on the distribution network, which pushes both its thermal limits and its voltage constraints.

Energy storage is often viewed as a silver bullet to buffer the differences between the demand and supply. Additionally, it can improve network operation. With advancements in energy storage technologies, today's catalogue of energy storage systems offers a wide range of applications to choose from, where all yield some benefit at different levels throughout the entire network.

The collection of manuscripts in this editorial provides an insight into some of the cutting edge research on the control of energy storage for power systems.

2. Short Review of the Contributions in This Issue

The special issue of the MPDI *Energies* on "Control of Energy Storage" is focused on the control methods of energy storage for a range of applications and degrees of complexity. Specifically, the this special issue addresses the following topics:

- Control of energy storage.
- Energy storage systems for transport.
- Energy storage systems for grid support.
- Intelligent coordination of storage elements in the grid at micro and macro levels.
- Monitoring, modelling, and other performance assessment methodologies for the control of energy storage.
- Explorations of the future of energy storage systems and associated control problems.

The success of energy storage relies on the inclusion of the technical constraints and economic feasibility into the control strategies for the energy storage applications. The research articles included in this special issue cover the full range of aspects for the energy storage applications: from energy storage technology modelling for more predictable performance in real-life applications to micro and macro control strategies for energy storage in power systems of a range of sizes.

2.1. Modelling

Telaretti et al. [2] developed a multi-vector model for energy storage operation taking into account technical, economic, and financial aspects. Along with the model, the paper proposes an energy storage scheduling strategy designed to maximise the profit for the energy storage owner by providing price arbitrage services subject to the technical constraints of the energy storage system (e.g., rating, efficiency, and depth-of-discharge). The performance of the proposed strategy was assessed in a simulation of three energy storage technologies: lithium-ion (Li-ion), sodium-sulfur (NaS), and lead acid.

Looking into more detailed modelling, the performance and lifespan of modern battery chemistries depend on the internal temperature and the voltages on individual cells during the operation. Gao et al. [3] proposed a thermal model and equivalent circuit of a $LiFePO_4$ battery to accurately estimate the state-of-charge and temperature of the battery during operation. The proposed model have been validated on experimental results and shown to have high accuracy cell voltage estimation on a multi-cell $LiFePO_4$ battery.

2.2. Automotive Industry

Energy storage application in automotive industry is presented with unique operational conditions. Bruen et al. [4] presented a study on the effect of vibration on the lifespan and performance of nickel manganese cobalt oxide (NMC) Li-ion batteries, commonly found in electric vehicles (EVs) and plug-in hybrid electric vehicles (PHEV). The results of the study were used to develop an equivalent circuit model for the cells and provide recommendation on the battery management strategies.

Moving on to control strategies for energy storage integrated into power systems, three research articles addressed the application of energy storage for improving the performance and economic efficiency of transport systems. Pietrosanti et al. [5] proposed a power management strategy for the control of a flywheel energy storage system on a rubber tyre gantry crane. A power management strategy was proposed in order to reduce the overall cost of energy that is required to operate the gantry crane. This strategy balances the power demands for container lifting operations, and the recovered energy when lowering the same. Due to the random duration of each such operation, the developed power management strategy was implemented using statistical load distributions. Numerical calculations using MATLAB/Simulink models of the required systems show increased energy savings and reduced peak power demand with respect to current control strategies.

Lin et al. [6] proposed a control strategy for super-capacitor installation to recover breaking energy fromurban rail trains. Introducing variable thresholds for the wayside energy storage system allowed the recuperation to make best use of the train's breaking *V-I* characteristics. Using a dual-loop control method enabled the authors to achieve the best energy-saving effect, which was verified through simulations and an experimental test on the Batong Line of the Beijing subway, using 200 kW wayside supercapacitor energy storage prototypes. Xia et al. [7] proposed a solution for super-capacitor sizing, placement, and control strategy for improving the efficiency of a metro line and improving voltage profile. The proposed solution is based on a novel optimisation method, combining genetic algorithms and simulation platform of an urban rail power system, including network, train, and energy storage system modelling.

2.3. Network Support

Energy storage also has potential to perform energy management and network support in standalone or grid-connected electricity distribution system—microgrids. Zangs et al. [8] proposed an improvement on the additive increase multiplicative decrease (AIMD) algorithm for enabling voltage support services from distributed energy storage devices in a low-voltage distribution network.

The improved algorithm—AIMD+—uses local voltage measurements against location-adjusted thresholds to improve voltage and thermal constraints on the network whilst providing more equal energy storage utilisation.

Nguyen et al. [9] proposed a model predictive control (MPC) system for power control of battery energy storage systems (BESS) in a micro-grid environment. Two variations of MPC—the proposed purely predictive power control and predictive current control with proportional-integral (PI)—are compared against the traditional PI control technique for BESS inverter control. The performance of the control techniques was assessed using MATLAB/Simulink models of a microgrid with a mix of generation sources, two energy storage systems, and a lump load, both in grid-connected and islanded modes. Results showed that MPC-based power control methods are best applied for BESS applications in power import/export control and frequency regulation in a microgrid, and the predictive current with inner PI control loop is more suitable BESS control for smoothing the wind power fluctuations.

Chae et al. [10] highlighted the difference between simulated and actual performance of islanded power systems. Authors presented results from economic feasibility studies of typical island power systems and microgrid island power systems. A representative model of a typical island power system supplied with diesel generators was assessed in a feasibility study tool called HOMER. The results of the study showed that the most economical operational costs remained the same—between 20% and 70% of energy supplied from renewable resources. Study of a planned power system on the test island showed that 91% of the energy will be supplied from the renewable resources, giving an 81% reduction in average fuel consumption. The real operational data showed the 82% of the energy was supplied from renewable resources, achieving fuel consumption savings of 80%. Discussion by the authors highlights the differences between the feasibility study against the actual observations and the effect of microgrid operation on the power quality and operational efficiency of the power system.

2.4. Demand-Side Management

One of the fundamental functions of energy storage is to shift energy usage in time. Demand-side management (DSM) can be viewed as equivalent to energy storage: the energy usage by a controllable load is managed with an aim to minimise the impact on the network (e.g., supply unbalance, frequency regulation, or network support) whilst maintaining the required function of the load for the benefit of the consumer.

Gelazanskas and Gamage [11] proposed a method for scheduling of domestic hot water heaters to compensate for the errors in day-ahead wind generation forecasts. The control system schedules the heating periods every 5 min for the next 12 h to adjust the demand to fill the gap or absorb the excess in supply. An artificial neural network is used to predict the loading of the water heater, allowing the heating periods to be scheduled without causing discomfort to the user. Results showed that the forecasting of energy usage by water heaters combined with scheduling lowers the energy requirement for hot water preparation and reduces the imbalance in supply for wind power generation.

On a larger scale, Kies et al. [12] addressed the issue of demand and supply unbalance in a simplified model of a fully renewable European power system by investigating the impact of DSM on the need for backup generation. Authors use ten years of weather and historical data to perform power flow analysis of several combinations of scenarios for transmission links capacities and distribution of generation capacity across Europe to assess DSM as an energy storage equivalent.

2.5. Frequency Regulation

Imbalance in supply and generation at the grid level causes deviation of frequency from the statutory range. Excess power generation allows the speed of rotating machines (e.g., steam turbines on coal and gas power plants) to increase, which in turn increases the grid frequency. Similarly, lack of supply leads to a decrease in frequency. Significant deviation from the statutory limits could lead to blackouts, as the generation plants and loads will be disconnected from the network by frequency-sensitive relays. Large-scale energy storage devices or coordinated behaviour of multiple small-scale energy storage devices could provide frequency regulation services to assist with maintaining the frequency within the nominal range.

Fu et al. [13] proposed a distributed control algorithm for the coordination of frequency regulation provided by multiple distributed resources. The algorithm uses an agent-based consensus control protocol, where each agent represents a system component capable of providing active power support and, through communication, aims to converge to a new common frequency state. Gatta et al. [14] present an application of $LiFePO_4$ BESS for primary frequency control. Electrical-thermal circuit models were developed for evaluation purposes, taking into account the cycle-life and auxiliary energy consumption. Numerical simulations then showed the trade-off between expected lifetime and overall system efficiency when performing droop controlled frequency control. Yang et al. [15] presented an optimal scheduling algorithm for an energy storage device providing frequency regulation service. The control algorithm uses particle swarm optimisation to compensate for the errors in state of charge estimation and adjust the operation of the energy storage device to maximise profit whilst ensuring availability for the automatic generation control signal.

3. Conclusions

The research articles in the special issue on "Control of Energy Storage" presented contributions from micro to macro scale of energy storage applications. Several works presented models for the prediction of performance and lifespan of the selected energy storage technologies. Control techniques for energy storage applications in transport and microgrid were presented, focusing on improvement of operation efficiency and power quality. On the larger scale, three articles addressed the aspects of frequency regulation provided by energy storage and demand response systems.

The collection of the research articles included in this special issue have demonstrated the wide range applications for energy storage and the role of modelling in delivering effective control systems for energy storage. Energy storage is expected to play an important role in keeping the lights on in the future low-carbon electricity networks. Further integration of renewable generation and low carbon technologies would require greater flexibility from the energy consumers and producers to ensure balance of supply and demand. Energy storage deployed throughout the network levels has the potential to provide the required flexibility and support network operation.

Acknowledgments: The authors are grateful to the MDPI Publisher and the members of the editorial team of "*Energies*" for the invitation to act as guest editors for the special issue.

Author Contributions: Timur Yunusov and Maximilian J. Zangs have reviewed the works included in the MDPI special issue on Control of Energy Storage and wrote the editorial. William Holderbaum is the academic editor for the special issue and have guided and reviewed the writing of the editorial.

Conflicts of Interest: The authors declare no conflict of interest.

References

1. The Stationary Office. *Climate Change Act (c. 27)*; The Stationary Office: London, UK, 2008.
2. Telaretti, E.; Ippolito, M.; Dusonchet, L. A simple operating strategy of small-scale battery energy storages for energy arbitrage under dynamic pricing tariffs. *Energies* **2016**, *9*, 12, doi:10.3390/en9010012.
3. Gao, Z.; Chin, C.S.; Woo, W.L.; Jia, J. Integrated equivalent circuit and thermal model for simulation of temperature-dependent $LiFePO_4$ battery in actual embedded application. *Energies* **2017**, *10*, 85, doi:10.3390/en10010085.
4. Bruen, T.; Hooper, J.M.; Marco, J.; Gama, M.; Chouchelamane, G.H. Analysis of a battery management system (BMS) control strategy for vibration aged nickel manganese cobalt oxide (NMC) lithium-ion 18650 battery cells. *Energies* **2016**, *9*, 255, doi:10.3390/en9040255.
5. Pietrosanti, S.; Holderbaum, W.; Becerra, V.M. Optimal power management strategy for energy storage with stochastic loads. *Energies* **2016**, *9*, 175, doi:10.3390/en9030175.

6. Lin, F.; Li, X.; Zhao, Y.; Yang, Z. Control strategies with dynamic threshold adjustment for supercapacitor energy storage system considering the train and substation characteristics in urban rail transit. *Energies* **2016**, *9*, 257, doi:10.3390/en9040257.
7. Xia, H.; Chen, H.; Yang, Z.; Lin, F.; Wang, B. Optimal energy management, location and size for stationary energy storage system in a metro line based on genetic algorithm. *Energies* **2015**, *8*, 11618–11640.
8. Zangs, M.J.; Adams, P.B.E.; Yunusov, T.; Holderbaum, W.; Potter, B.A. Distributed energy storage control for dynamic load impact mitigation. *Energies* **2016**, *9*, 647, doi:10.3390/en9080647.
9. Nguyen, T.T.; Yoo, H.J.; Kim, H.M. Application of model predictive control to bess for microgrid control. *Energies* **2015**, *8*, 8798–8813.
10. Chae, W.K.; Lee, H.J.; Won, J.N.; Park, J.S.; Kim, J.E. Design and field tests of an inverted based remote microgrid on a Korean Island. *Energies* **2015**, *8*, 8193–8210.
11. Gelazanskas, L.; Gamage, K.A.A. Distributed energy storage using residential hot water heaters. *Energies* **2016**, *9*, 127, doi:10.3390/en9030127.
12. Kies, A.; Schyska, B.U.; Bremen, L.V. The demand side management potential to balance a highly renewable European power system. *Energies* **2016**, *9*, 955, doi:10.3390/en9110955.
13. Fu, R.; Wu, Y.; Wang, H.; Xie, J. A distributed control strategy for frequency regulation in smart grids based on the consensus protocol. *Energies* **2015**, *8*, 7930–7944.
14. Gatta, F.; Geri, A.; Lamedica, R.; Lauria, S.; Maccioni, M.; Palone, F.; Rebolini, M.; Ruvio, A. Application of a $LiFePO_4$ battery energy storage system to primary frequency control: Simulations and experimental results. *Energies* **2016**, *9*, 887, doi:10.3390/en9110887.
15. Yang, J.S.; Choi, J.Y.; An, G.H.; Choi, Y.J.; Kim, M.H.; Won, D.J. Optimal scheduling and real-time state-of-charge management of energy storage system for frequency regulation. *Energies* **2016**, *9*, 1010, doi:10.3390/en9121010.

Article

A Simple Operating Strategy of Small-Scale Battery Energy Storages for Energy Arbitrage under Dynamic Pricing Tariffs

Enrico Telaretti *, Mariano Ippolito and Luigi Dusonchet

Department of Energy, Information Engineering and Mathematical Models, University of Palermo, Viale delle Scienze, 90128 Palermo, Italy; ippolito@dieet.unipa.it (M.I.); dusonchet@dieet.unipa.it (L.D.)
* Correspondence: telaretti@dieet.unipa.it; Tel.: +39-091-238-602-62; Fax: +39-091-488-452

Academic Editor: William Holderbaum
Received: 9 October 2015; Accepted: 16 December 2015; Published: 25 December 2015

Abstract: Price arbitrage involves taking advantage of an electricity price difference, storing electricity during low-prices times, and selling it back to the grid during high-prices periods. This strategy can be exploited by customers in presence of dynamic pricing schemes, such as hourly electricity prices, where the customer electricity cost may vary at any hour of day, and power consumption can be managed in a more flexible and economical manner, taking advantage of the price differential. Instead of modifying their energy consumption, customers can install storage systems to reduce their electricity bill, shifting the energy consumption from on-peak to off-peak hours. This paper develops a detailed storage model linking together technical, economic and electricity market parameters. The proposed operating strategy aims to maximize the profit of the storage owner (electricity customer) under simplifying assumptions, by determining the optimal charge/discharge schedule. The model can be applied to several kinds of storages, although the simulations refer to three kinds of batteries: lead-acid, lithium-ion (Li-ion) and sodium-sulfur (NaS) batteries. Unlike literature reviews, often requiring an estimate of the end-user load profile, the proposed operation strategy is able to properly identify the battery-charging schedule, relying only on the hourly price profile, regardless of the specific facility's consumption, thanks to some simplifying assumptions in the sizing and the operation of the battery. This could be particularly useful when the customer load profile cannot be scheduled with sufficient reliability, because of the uncertainty inherent in load forecasting. The motivation behind this research is that storage devices can help to lower the average electricity prices, increasing flexibility and fostering the integration of renewable sources into the power system.

Keywords: price arbitrage; battery energy storage system; optimal operation; hourly electricity prices; energy management

---

1. Introduction

Electricity customers will face significant challenges in the near future due to the most recent developments in the energy market sector. These changes have been mainly driven by the increasing penetration of renewable and distributed energy sources in the power system, which can positively contribute to a reduction of CO_2 emissions. The diffusion of renewable sources has been made possible thanks to the introduction of support policies, such as those put in place for the photovoltaic (PV) and wind technology [1–4]. Clearly, the transition from the current centralized electricity market structure towards a decentralized market model will require major investments in the electricity grid infrastructure, in order to ensure an adequate level of quality and reliability of the energy supply.

In the spot markets, the electricity price varies stochastically from one day to the next and systematically between seasons. The marginal cost of producing energy has become much more volatile in the last decade, mainly due to the recent moves toward competitive liberalized markets. Indeed, the competition among actors has increased the range of variability in electricity prices, expanding the difference between on-peak and off-peak prices. Normally, electricity users are not exposed to these fluctuations but pay a constant price. In an attempt to reduce demand peaks, several utilities are moving from a conventional fixed-rate pricing scheme to new market-based models, where the electricity cost is free to fluctuate depending on the balance between supply and demand. Such dynamic pricing schemes reflect the prices of the wholesale market and are able to lower demand peaks and the volatility of the wholesale prices [5]. A first example of dynamic pricing tariff is time-of-use (TOU) pricing, which provides two or three periods of different electricity price (generally "on-peak", "mid-peak" and "off-peak" prices), depending on the hour of day. Electricity users are advised in advance about electricity prices that are not normally modified more than once or twice per year. A more flexible electricity-pricing scheme is real-time pricing (*RTP*), for which the retail electricity price closely reflects the wholesale energy price. In this case, customer electricity prices can vary hourly depending on the wholesale market and electricity users can manage their power consumption in a more flexible and economical manner, taking advantage of the price differential. The real-time prices can be notified to electricity customers with different timing, depending on the specific utility's *RTP* program. For example, with Ameren's *RTP* program (an Illinois' Electric Utility), hourly prices for the next day are set the night before and are communicated to customers so they can modify their power consumption in advance. Differently, with ComEd's *RTP* program (another Illinois' Electric Utility), hourly prices are based on the average of the twelve five-minute prices for each hour, and electricity users are notified in real-time, only when the hour has passed. Later on in this article, the *RTP* prices will be considered as day-ahead hourly prices, so electricity customers are advised a day before and can modify their power consumption accordingly.

The highly volatile behavior of the electricity price can be exploited by using an energy storage device in order to capture the price differential. Indeed, if an electricity customer is charged at an hourly-dependent rate, a storage system can be adopted with the aim to shift portions of consumption to different hours than those where they actually occur. The electricity is simply stored when it is inexpensive and resold back to the grid at a higher price [6,7].

The object of this article is to analyze, develop and demonstrate a charge/discharge scheduling method able to maximize the arbitrage benefit of a storage system, subject to technical constraints. The storage system is described by means of its performance parameters, such as the charge and generation capacity, the charge/discharge efficiency, the rated charge/discharge rate, the depth-of-discharge (*DOD*), *etc.*, which are sufficient to evaluate the arbitrage potential of a storage system. The scheduling strategy is based on the definition of an objective function, able to maximize the arbitrage benefit of the storage owner subject to technical constraints, allowing the battery to be charged/discharged at different *DOD*, as further detailed in Section 4. The developed model is valid for any kind of storage, although the simulations refer to a lead-acid, a lithium-ion (Li-ion) and a sodium-sulfur (NaS) battery. Test results show that the proposed operating strategy is effective to maximize the profit for the customer. Unlike the studies reported in the literature, often requiring an estimate of the end-user load profile, the proposed operation strategy is able to properly identify, for each daily period, the charge/discharge hours relying only on the hourly spot market price profile, regardless of the specific facility's consumption. This is made possible thanks to some simplifying assumptions in the sizing and the operation of the battery energy storage system (BESS), as further details in Section 3. This could be particularly useful when the customer load profile cannot be scheduled with sufficient reliability because of the uncertainty inherent in load forecasting. In these cases, identifying a BESS operating strategy that does not depend on the user's power profile can be an important task, since the deviation of the scheduled power profile from the effective one could affect the results obtained using more complete methods. Furthermore, the proposed management strategy requires a low computational burden and can be implemented in simple and available software, for instance in a spreadsheet, representing a friendly but effective instrument to optimize the charge/discharge schedule of a storage device.

The next section summarizes existing literature on the topic of optimal operation of storage systems. In Section 3, the customer energy system used in this paper is briefly described and the basic operational assumptions are outlined. In Section 4, the problem formulation is provided, showing the objective function to be maximized and defining the constraint equations. In Section 5, a case study is presented and the technical and economic parameters for each storage device are provided. Section 6 shows the simulation results and some important remarks about the operating schedule of the storage devices. Finally, Section 7 summarizes the conclusion of the work.

2. Current Literature

Traditionally, most of the studies address the optimal operation of a storage system based on linear programming [8–11], nonlinear programming [12], dynamic programming [13–16] and multipass iteration particle swarm optimization approach [17]. Other charge/discharge strategies are described in [18–25].

2.1. Linear and Nonlinear Programming

In [8], the authors study the optimal operation of an energy storage unit installed in a small power producing facility using a conventional linear programming technique. In [9], the authors determine the optimal charge/discharge schedule by using a linear optimization model of the battery systems (based on Li-ion and lead-acid technology) for arbitrage accommodation. They found that the cost and the efficiency of the storage systems have the highest impact on simulation results. The developed model is linear and can thus be solved without much computational effort. Bradbury *et al.* [10] studied seven real-time US electricity markets and 14 different storage technologies, finding that the optimal profit-maximizing size of a storage device (*i.e.*, hours of energy storage) depends largely on its technological characteristics (round-trip charge/discharge efficiency and self-discharge), rather than the magnitude of market price volatility, which instead increases internal rate of return (IRR). The arbitrage benefit is maximized using a simple linear programming, subject to technical constraints. Graves *et al.* [11] emphasize the fact that using average peak and off-peak prices does not account for the variability in prices and thus leading to significant errors in the optimal management strategy. They also discuss the use of a linear programming for determining the optimal operation strategy.

In [12], the authors present an optimal operation strategy of BESSs to the real-time electricity price in order to achieve maximum profits of the BESS. The algorithm is based on a sequential quadratic programming method as to maximize the profits for the customer. The strategy is promising although operating and maintenance costs of the BESS are not taken into account.

2.2. Dynamic Programming

Linear programming is often considered to be too inflexible, as it typically does not capture the stochastic nature of load profiles. In order to overcome the restriction, dynamic programming methods are employed to capture the uncertainties in load profiles and electricity prices [13]. The algorithm developed in [14] is a multipass dynamic programming that ensures the minimization of the electricity bill for a given battery capacity, while reducing stress on the battery and prolonging battery life. In [15], the authors address the problem of organizing home energy storage purchases as a Markov decision process, showing that there exists a threshold-based stationary cost-minimizing policy. The battery is charged up to the threshold, when the battery level is below the threshold, and discharged when the level is above the threshold. The proposed strategy is interesting, even though the system cost is not considered. In [16], the authors propose a self-learning optimal operating control scheme based on adaptive dynamic programming for the residential energy system with batteries. The algorithm is effective in achieving minimization of the cost through neural network learning. The main feature of the proposed scheme is the ability of the continuous learning and adaptation to improve the performance during real-time operations under uncertain changes in the environment or new system configuration of the residential household.

2.3. *Other BESS Management Strategies*

In [17], a modified particle swarm optimization (PSO) algorithm (called multipass iteration PSO) is used to solve the optimal operating schedule of a BESS for an industrial TOU rate user with wind turbine generators. Thanks to the high computational efficiency, the algorithm can be used to evaluate the optimal operating policy of a BESS in real-time applications, based on the load condition of the user, the energy left in the BESS, and the output of wind turbines. In [18], the authors estimate the benefit of using energy storages for aggregate storage applications, such as energy price arbitrage, TOU energy cost reduction, ancillary services, and transmission upgrade deferral. The maximization of the arbitrage benefit is carried out by maximizing an objective function, under the assumption that the electricity prices are both dependent/independent on the battery operation. In [19], a simple methodology to charge/discharge a residential battery system for energy arbitrage in presence of TOU prices was described. The statistical variability of the household consumption was accounted through a Monte Carlo method. The economic feasibility of the storage system was determined in the context of the Australian retail electricity market, showing that, for various BESSs, the load shifting strategy can be profitable. In [20], the authors present an estimation of the economic feasibility of electricity storage in the west Danish power market, exploiting a simple operation strategy of the BESS in the spot market. The strategy includes two main conditions: (1) the price for buying must be less than the price for selling times the round trip efficiency (in order to ensure positive incomes) and (2) the amount of power bought in a given time period must equal the amount of power sold times the round trip efficiency (in order to ensure the balance of energy). Shcherbakova *et al.* (2014) [21] simulated the operation and resulting profits of small storage batteries (NaS and Li-ion) in South Korea using a charge/discharge strategy based on Hotelling rule. They concluded that neither technology generates a sufficient amount of arbitrage revenue to cover the battery's capital costs. Purvins and Summer [22] presented an optimal battery system management model in distribution grids for lithium-ion battery system used in stationary applications. The proposed approach is based on three management priorities, the first being the maximum utilization of renewable energy sources (RES) energy in distribution grids (preventing situations of reverse power flow at the distribution level), followed by efficient battery utilization (charging at off-peak prices and discharging at peak prices) and residual distribution grid demand smoothing. Finally, in [23,24], the authors evaluate the capacity of storage and active demand side management (DSM) to increase the self-consumed electricity in the residential sector, using a lead–acid battery. The operating strategy is based on self-consumption maximization, reducing the use of the grid and supplying the highest amount of energy from PV generation. In [25], the authors present a home energy management system model that uses a heuristic algorithm to manage and control home appliances based on a combination of energy pricing models including TOU and *RTP* tariffs. The algorithm aims to minimize overall usage and cost of energy without significantly degrading consumer comfort.

3. Energy System Description and Operational Assumptions

The customer energy system consists of a passive user (end-user), interconnected to a storage system through a bidirectional converter, as depicted in Figure 1. The bidirectional converter consists of a rectifier AC/DC (the battery charger) and an inverter DC/AC [26,27]. The battery system is handled in order to ensure an economic benefit for the customer, exploiting a load shifting strategy. Since the system marginal price (SMP) value is available one day ahead and it is defined each hour, the electricity prices are considered as hourly-dependent prices, where each hour of the day has a different electricity price. The reference period used in the study is one day, *i.e.*, the battery operation is defined starting from a vector of 24 elements as input data.

Three different operating modes are considered for the storage system: charging mode, activated when the electricity prices are low; standby mode, in which the power grid supplies directly the end-user without contribution of the storage; and discharging mode, activated when the electricity prices are high, where part of the load is supplied from the battery.

The following assumptions have been made:

- The end-user is allowed to buy the consumed energy at an hourly tariff (*RTP* tariff), defined by the utility on a daily basis. The *RTP* tariffs are assumed to be proportional to the SMP values, by applying a percentage increase to incorporate the benefit for the utility and taxes (electricity tax and value added tax (VAT)).
- The power flow is always directed from the grid to the load. The stored energy can only be used by the customer for load compensation and cannot be sold to the utility.
- The hourly electricity prices are known in advance in a finite horizon setting (daily period) and the use of the storage device does not influence the prices of electricity in the energy market (small price taking storage devices). Predictions about future electricity rates are not part of this work since the aim is to show results based upon the current electricity prices.
- Battery self-discharge is disregarded.
- Battery capacity is assumed constant throughout the battery life, without degradation.
- The common frictions during battery operation are accounted for by incorporating imperfect charging and discharging efficiency;

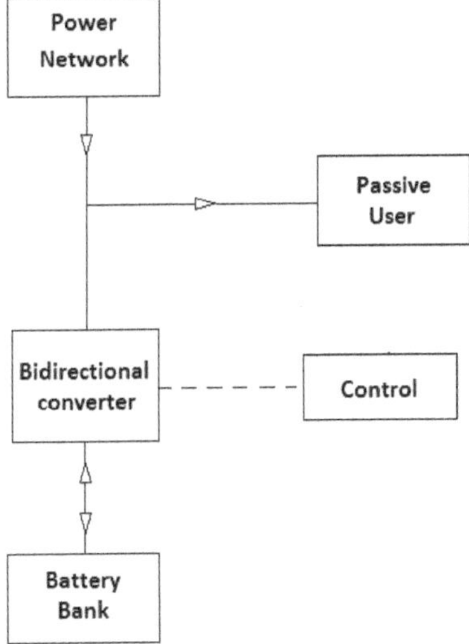

Figure 1. Grid-connected customer energy system operating in parallel with the storage system.

- The charge/discharge rate of the battery is assumed constant and equal to the rated power capacity of the device. Doing so, the storage charge/discharge constraints are automatically satisfied (*i.e.*, the energy charged/discharged into the battery at any time t cannot be more than the rated power capacity of the device). It is worth noting that both the battery capacity and the battery life are influenced by the charging rate. Indeed, at very high rates the capacity cell and the battery life are reduced. Fast charging may also have negative consequences on the battery efficiency [28]. Therefore, the use of a battery at constant charge/discharge rate helps to prolong the battery life, to preserve the rated capacity and to keep the battery efficiency at appropriate values.

- The charging time is assumed equal to the discharging time, in each operating cycle. According to the last two mentioned hypotheses, the battery returns to the initial state-of-charge (SOC) at the end of each operating cycle. Such an operation means that the battery energy constraints are automatically satisfied (*i.e.*, the storage level of the battery cannot be more than the rated energy capacity of the device).
- The *DOD* of the battery can take different discrete states, depending on the value of the objective function.
- The storage capacity is assumed equal to the facilities' energy consumption during peak times (*i.e.*, the hours where electricity prices are the highest) on the day of the year of lowest consumption [29]. In other words, the battery is sized so that it can supply the entire customer load during peak price hours, on the day of the year of lowest consumption, and only a portion of the customer's load on the other days. The choice of the storage capacity is driven by a trade-off between gaining more arbitrage savings during days with relatively high peak loads and wasting idle capacity during days with low peak loads. Among all the possible solutions, the one that ensures the minimum upfront investment cost for the storage owner has been chosen. The aim of this article is to identify a battery operating strategy able to maximize the profit of the storage owner (under the considered assumptions), without attempting to identify the optimal BESS capacity. In other words, the battery has been sized according to a criterion of minimum cost, which is not necessarily the optimal one. As a consequence of this statement, the BESS can be operated regardless of the specific facility's load profile and the power flow is always directed from the grid to the load, without selling to the utility.

4. Problem Formulation

4.1. Preliminary Considerations

The optimal operating strategy of the storage device is able to uniquely determine the daily charge/discharge intervals so as to maximize the economic saving for the customer. Figure 2 shows typical daily profiles of SMP (the national single price of the Italian day-ahead market) for a reference weekly period (from 31 March to 6 April 2014) [30]. The profiles clearly show a first couple of min/max prices in the first semi-daily period and a second couple in the second half of the day. The battery thus will be charged only once a day, twice a day or it will remain idle, depending on the maximization of the objective function. Since the *RTP* tariffs are assumed to be proportional to the SMP values, hereinafter will be referred as *RTP* prices. It is worth noting that weekdays *RTP* values have a first price peak at about 8:00–10:00 a.m. and a second peak at 8:00–9:00 p.m. Differently, Sunday only retains the second peak at 9:00 p.m. As a result, we can expect that the BESS could be charged two times on weekdays (including Saturday), only one time on Sunday.

Since the battery can be charged/discharged at different *DOD*, the algorithm calculates the moving average (*MA*) of *RTP* prices (*MA RTP*) corresponding to each charge/discharge time, *d*, where *d* is a discrete variable denoting the charge/discharge time of the battery (corresponding to different *DOD* values). For example, assuming that the charge/discharge time, *d*, can take *D* different discrete values, the algorithm calculates *D* daily profiles of *MA RTP* prices, for each day of the year, *i*:

$$MA_{RTP_{i,d}}(h) = \sum_{n=h}^{h+d-1} RTP(n)/d \; h = 1, \ldots, 24-d+1; \; d = 1, \ldots, D; \; i = 1, \ldots, 365 \quad (1)$$

where *d* is an index denoting the charge/discharge time of the battery, *i* is an index denoting the day of the year, *h* is an index denoting the hour of the day, *D* is the maximum charge/discharge time of the battery (corresponding to the maximum *DOD*) and $MA_{RTP_{i,d}}(h)$ is the *MA* of *RTP* prices in hour *h*, corresponding to the charge/discharge time *d* in the day *i*. In the following, all equations will be referred to a generic day *i*, and the variability of the index over the year will be omitted.

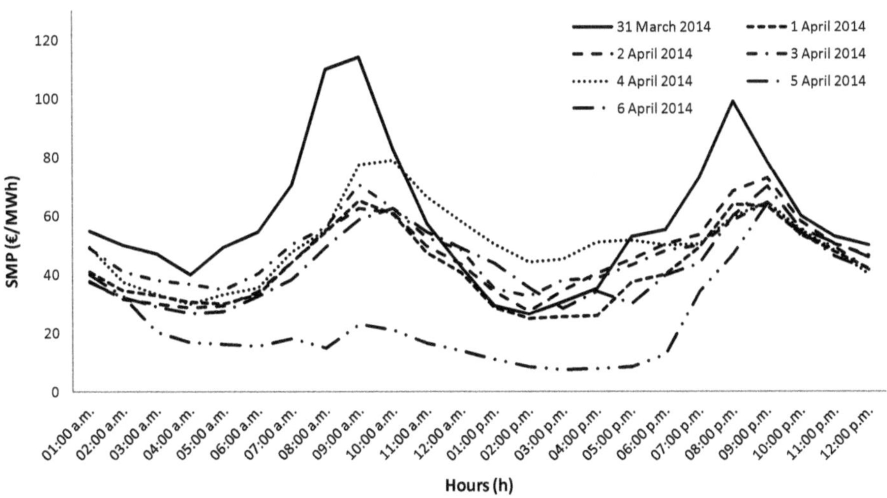

Figure 2. System marginal price (SMP) for the weekly period from 31 March to 6 April 2014.

Since the charge/discharge rate of the battery, P_{BESS}, is assumed constant, the following relation exists between the DOD and the discharged time, d:

$$DOD = \frac{E_{BESS}}{Cap} = \frac{P_{BESS} \cdot d}{P_{BESS} \cdot D_{max}} = \frac{d}{D_{max}} \tag{2}$$

where E_{BESS} is the energy discharged from the storage device, Cap is the rated energy capacity of the BESS, and D_{max} is the maximum theoretical discharging time of the battery, corresponding to a full discharge (this is a theoretical discharging value, since the battery can never fully discharge).

Since the battery can be charged once or twice a day, depending on the maximization of the objective function, the algorithm takes into account two *MA RTP* profiles for each charge/discharge time d, the first referred to a daily period, the second to a semi-daily period. In other words, the algorithm scans both the daily and the semi-daily *MA RTP* profiles, with the aim of verifying whether the maximum of the objective function corresponds to only one cycle or to two cycles per day. Figure 3a,b shows the daily profile of *MA RTP* related to a daily period or to a semi-daily period, together with the daily/semi-daily average value, respectively:

$$Aver_{MA_{i,d}} = \sum_{h=1}^{24-d+1} MA_{RTP_{i,d}}(h) / (24 - d + 1) \; ; \; d = 1, \ldots, D \tag{3}$$

$$\begin{cases} Aver_{MA_{i,d}^{(1)}} = \sum_{h=1}^{12} MA_{RTP_{i,d}}(h)/12 \\ Aver_{MA_{i,d}^{(2)}} = \sum_{h=(12-d+1)}^{(24-d+1)} MA_{RTP_{i,d}}(h)/12 \end{cases} \quad d = 1, \ldots, D \tag{4}$$

where $Aver_{MA_{i,d}}$ is the daily average value of the *MA RTP* profile and $Aver_{MA_{i,d}^{(k)}}$ is the semi-daily average value of the *MA RTP* profile (in the semi-daily period k of the day i, with $k = 1, 2$). Figure 3a,b also shows the min/max values of *MA RTP* profiles in the daily/semi-daily period:

$$MA_{RTP_{i,d,min}}, \; MA_{RTP_{i,d,max}} \; ; \; d = 1, \ldots, D \tag{5}$$

$$MA_{RTP^{(k)}_{i,d,min}}, \; MA_{RTP^{(k)}_{i,d,max}} \quad d = 1, \ldots, D\,;\; k = 1,\,2 \tag{6}$$

where $\left(MA_{RTP_{i,d,min}}, MA_{RTP_{i,d,max}}\right)$ is the couple of min/max *MA RTP* values in a daily period and $\left(MA_{RTP^{(k)}_{i,d,min}}, MA_{RTP^{(k)}_{i,d,max}}\right)$ is the couple of min/max *MA RTP* values in the semi-daily period k of the day i, respectively. The average values and the min/max *MA RTP* values are calculated for each charge/discharge time d and for each day i. The daily profile in Figure 3 corresponds to the *RTP* prices when $d = 1$, to the *MA* of *RTP* prices when $d \neq 1$.

4.2. Optimization Problem Formulation

Since the battery can be charged once or twice a day, depending on the value of the objective function, the algorithm calculates the benefit for the storage owner (electricity customer) in both cases, verifying in which situation the objective function takes the maximum value. In the following sections, the objective function will be defined in both situations, by considering a daily or a semi-daily periodicity, respectively.

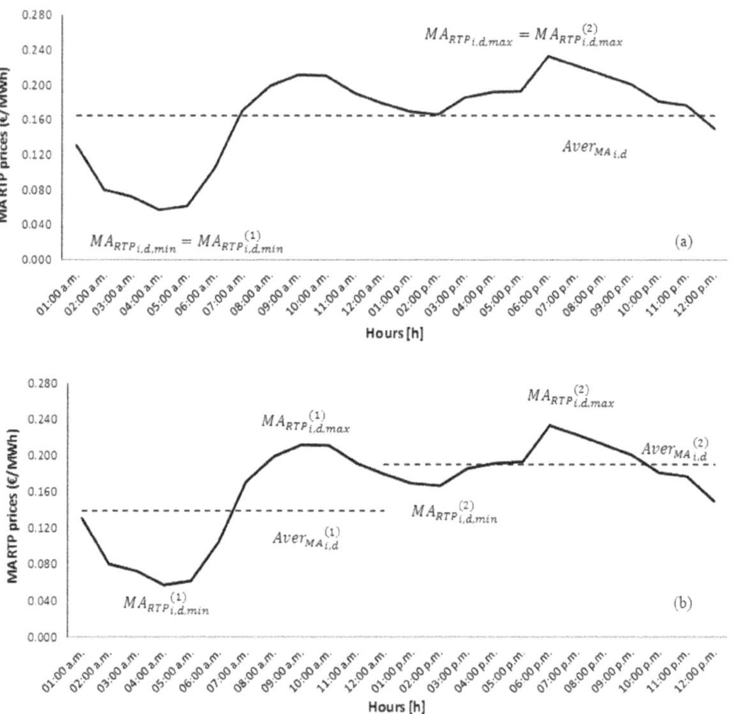

Figure 3. Daily profile of moving average of *RTP* prices (*MA RTP*) with daily average (**a**) and semi-daily average values (**b**).

4.2.1. Semi-Daily Periodicity

Under the assumption of semi-daily periodicity, the storage device will perform two charging cycles per day, according to the *MA RTP* profile shown in Figure 3b. For each battery cycle, the problem comes down to maximizing the following objective function:

$$OF^{(k)}_{i,d} = max\left(S^{(k)}_{BESS,i,d} - C_{BESS_{cycled,d}}\right) \tag{7}$$

where $S_{BESS,i,d}^{(k)}$ is the saving per kWh obtained charging/discharging the BESS over time d, in the semi-daily period k of the day i and $C_{BESS_{cycled,d}}$ is the storage cost per kWh cycled, obtained charging/discharging the BESS over time d.

The saving, $S_{BESS,i,d}^{(k)}$, can be calculated as follows:

$$S_{BESS,i,d}^{(k)} = \frac{E_{BESS,i,d}^{(k)}}{Cap} \cdot \left(MA_{RTP_{i,d,max}^{(k)}} \cdot \mu_d - \frac{MA_{RTP_{i,d,min}^{(k)}}}{\mu_c} \right) = DOD \cdot \left(MA_{RTP_{i,d,max}^{(k)}} \cdot \mu_d - \frac{MA_{RTP_{i,d,min}^{(k)}}}{\mu_c} \right) \quad (8)$$

where $E_{BESS,i,d}^{(k)}$ is the energy discharged from the storage device over time d, and μ_c and μ_d are the charge/discharge efficiencies of the battery, respectively.

The storage cost per kWh cycled can be expressed as:

$$C_{BESS_{cycled,d}} = \frac{C_{TOT_{BESS}}}{Cap \cdot N_{Full\ cycle,d}} \quad (9)$$

where $C_{TOT_{BESS}}$ is the total cost of the storage and $N_{Full\ cycle,d}$ is the number of equivalent full cycles of the battery, corresponding to a charge/discharge time d.

Denoted by $C_{BESS_{kWh}}$, the storage cost per kWh (from Equation (9)) can be expressed as:

$$C_{BESS_{cycled,d}} = \frac{C_{BESS_{kWh}}}{N_{Full\ cycle,d}} \quad (10)$$

The objective function, $OF_{i,d}^{(k)}$, can finally be expressed as:

$$OF_{i,d}^{(k)} = \max \left[DOD \cdot \left(MA_{RTP_{i,d,max}^{(k)}} \cdot \mu_d - \frac{MA_{RTP_{i,d,min}^{(k)}}}{\mu_c} \right) - \frac{C_{BESS_{kWh}}}{N_{Full\ cycle,d}} \right] \quad (11)$$

The only variable that appears in the objective function is the DOD. Indeed, $N_{Full\ cycle,d}$ and $\left(MA_{RTP_{i,d,max}^{(k)}}, MA_{RTP_{i,d,min}^{(k)}} \right)$ are not independent variables, since they are linked to the DOD. The DOD is thus the only variable to be optimized and the search space is the set of all possible charging/discharging times, namely all integers between 1 and D. Ultimately, the maximization of the objective function allows one to obtain the DOD value that maximizes the customer's benefit, for each semi-daily charging/discharging cycle.

4.2.2. Daily Periodicity

In the same manner as was done in the previous section, in presence of a daily periodicity of the *MA RTP* profile, the objective function, $OF_{i,d}$, can be expressed as:

$$OF_{i,d} = \max \left[DOD \cdot \left(MA_{RTP_{i,d,max}} \cdot \mu_d - \frac{MA_{RTP_{i,d,min}}}{\mu_c} \right) - \frac{C_{BESS_{kWh}}}{N_{Full\ cycle,d}} \right] \quad (12)$$

The maximization of the objective function allows one to obtain the DOD value that maximizes the customer's benefit, for each daily charging/discharging cycle.

4.2.3. Constraint Equations

As already stated in Section 3, the battery charging and discharging constraints are automatically satisfied, since the charge/discharge rate of the battery is assumed constant. The storage energy constraints are also satisfied, since the battery returns to the same initial SOC at the end of each charge/discharge cycle (namely the energy discharged is equal to the energy charged, in each battery cycle). Furthermore, charging/discharging periods should not overlap each other. This might happen when the battery performs two operating cycles per day. If this is the case, the charging/discharging period will be reduced accordingly.

The charge/discharge cycle of the battery would only be worth it if the difference between the maximum and minimum values of *MA RTP* is higher than the cost of cycling energy plus the cost of the energy losses in the charge/discharge process. Expressed differently, Equations (11) and (12) must take positive values for the battery operation to be profitable:

$$OF_{i,d}^{(k)} > 0\,,\ OF_{i,d} > 0 \tag{13}$$

If the constraints in Equation (13) are not satisfied, the battery will remain idle, since the arbitrage benefit is not enough to compensate for the cost of cycling energy plus the cost of the energy losses. In the following, the term "eligible" will be used to indicate an objective function whose value is greater than zero.

4.2.4. Selection of the Charging/Discharging Intervals

Once Equations (11) and (12) are calculated, the algorithm checks, for each day of the year, if the summation of the eligible objective functions corresponding to each semi-daily cycle is greater than that corresponding to the daily cycle, namely:

$$\sum_{k=1}^{2} OF_{i,d}^{(k)} \geqslant OF_{i,d}\ d = 1,\ldots,D \tag{14}$$

If Equation (14) is satisfied, the battery is charged in the first half of the day, in the second half or in both, depending on the number of the eligible objective functions, $OF_{i,d}^{(k)}$. The *DOD* for each battery cycle is selected according to Equation (11). If Equation (14) is not satisfied, the battery will make only one cycle per day. The corresponding *DOD* is selected according to Equation (12). Finally, if all the objective functions have negative value (*i.e.*, there are no eligible objective functions), the battery remains idle in the day *i*.

It is worth noting that the proposed operating strategy allows maximizing the customer's benefit under the assumptions described in Section 3. More complex and complete models could lead to higher benefits for the storage owner. Furthermore, the proposed method leads to an effective maximization of the objective function only if the SMP profile is assumed to have a convex form in the charging/discharging intervals, as in most spot electricity markets. If the price profile differs from a convex form, the proposed procedure could lead to suboptimal results, but it was verified that the error margin is narrow.

5. Case Study

The number of equivalent full cycles cannot be estimated directly, as it mainly depends on the energy cycled by the batteries, namely by the *DOD*. For most batteries, manufactures show in their datasheets the curves of number of cycles to failure, $N_{cycle,d}$ *vs.* the *DOD* (for given temperature value), as shown in Figure 4, derived for a lead-acid battery [31].

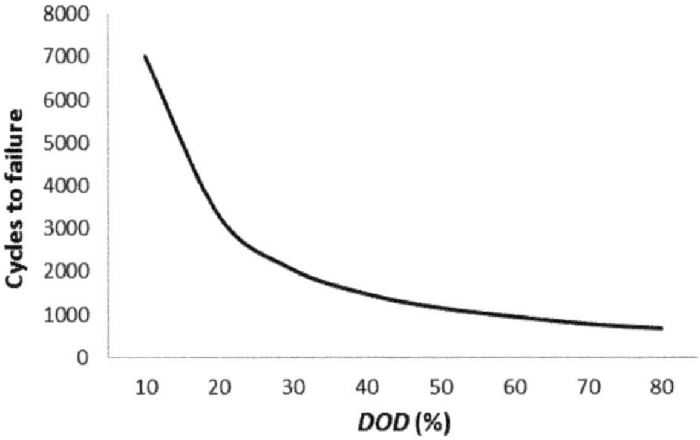

Figure 4. Typical cycles to failure *vs.* depth-of-discharge (*DOD*) curve for lead-acid-batteries.

The number of equivalent full cycles performed by the battery at a given *DOD* can be obtained as [32]:

$$N_{Full\ cycle,d} = DOD \cdot N_{cycle,d} \tag{15}$$

where $N_{cycle,d}$ is the number of cycles to failure, as derived from Figure 4.

For most of electrochemical batteries, the number of equivalent full cycles remains constant (for given operating temperature) and does not depend on the *DOD*. Expressed differently, the total Ah a battery can deliver over its life is approximately constant. However, the relationship deviates for some electrochemistries, especially at low *DOD*. With a view to highlight the changes, Figure 5 shows a comparison of cycles to failure *vs. DOD* curves for three different BESS technologies (lead-acid, Li-ion and NaS battery).

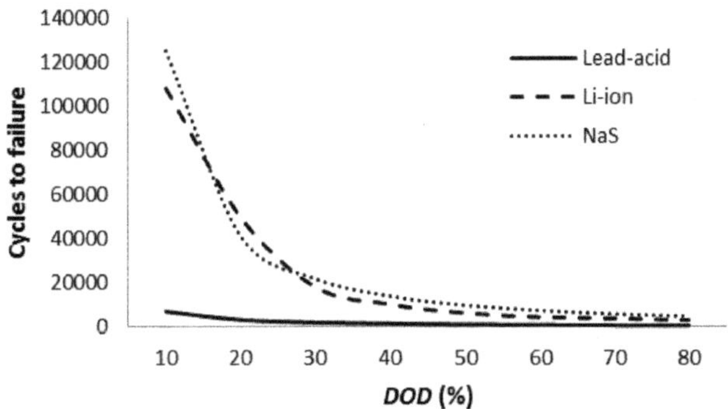

Figure 5. Cycles to failure *vs.* depth-of-discharge (*DOD*) curve for three different battery technologies.

Let us assume $D_{max} = 5\ h$, which corresponds to a discharging time $D = 4\ h$ at a *DOD* = 80%. The number of equivalent full cycles, for each selected *DOD* (ranging from 1 to 4 h), is reported in Table 1, for each of the selected battery technologies. The values were calculated using Equation (11). The number of cycles to failure, N_{cycles_d}, was deduced from the typical cycles to failure *vs. DOD* curve, for each battery option [31,33,34]. Table 1 also shows the percentage increment, $\Delta N_{Full}(\%)$, with respect to the value corresponding to a *DOD* = 80%. It is worth noting that the percentage increment is minimum for lead-acid, maximum for Li-ion battery.

Table 1. Number of equivalent full cycles for each selected *DOD*, for the three battery technologies.

	Lead-Acid Battery		Li-Ion Battery		NaS Battery	
***DOD*(%)**	$N_{Full\ cycles,d}$	ΔN_{Full} (%)	$N_{Full\ cycles,d}$	ΔN_{Full} (%)	$N_{Full\ cycles,d}$	ΔN_{Full} (%)
80%	540		2400		3592	
60%	570	5.56	2640	10	4269	18.85
40%	590	9.26	4040	68.3	5445	51.58
20%	660	22.22	10000	316.7	8253	129.76

The analysis has been carried out by referring to a typical medium-scale public facility (Department of Energy, Information engineering and Mathematical models (DEIM), University of Palermo). For the selected facility, a reference weekly period has been considered, from 31 March to 6 April 2014. The SMP for the reference weekly period have already been reported in Figure 2.

The proposed strategy can be applied to several kinds of storages, but the test results refer to three kind of batteries, lead-acid, Li-ion and NaS, that are, nowadays, the most suitable to be used in residential, commercial or industrial buildings, for load shifting applications. Among the three technologies, Li-ion batteries are the most promising in terms of cost reduction and cycling performance [35]. The technical and economic parameters are reported in Table 2 for each of the selected battery technologies.

Table 2. Technical and economic parameters selected for the three battery technologies.

Components	**Specifications**		
Technology	**Lead-Acid Battery**	**Li-Ion Battery**	**NaS Battery**
Energy capacity (kWh)	20	20	20
Power rating (kW)	5	5	5
Roundtrip efficiency (%)	82	90	81
Operating temperature (°C)	(−20)–(+50)	(−20)–(+45/+60)	300–350
Healthy *DOD* (%)	80	80	NA
Cycles to failure (80% *DOD*)	1100	3000	4500
BESS cost (€/kWh)	171	844	256
PCS cost (€/kW)	172	125	171
BOP cost (€/kW)	70	0	53

The storage cost and the charge/discharge roundtrip efficiency have been selected calculating the arithmetic mean between low and high literature values [36]. In Table 2, the total storage cost has been decomposed as the sum of the power conversion system (PCS) cost, the BESS cost and the balance-of plant (BOP) cost [37]. The operating temperatures and the healthy *DOD* were derived from [29]. The rated energy capacity (equal to 20 kWh for each battery) was selected referring to the facility's energy consumption during peak price hours, on the day of the year of lowest consumption, as already specified in Section 3.

The storage costs per kWh cycled are on average higher than the difference between maximum and minimum electricity prices. Indeed, the average storage costs per kWh cycled are equal to 0.171 €/kWh cycled for lead-acid, 0.103 €/kWh cycled for Li-ion and 0.096 €/kWh cycled for NaS batteries, as against a maximum value of 0.1 €/kWh for the difference between maximum and minimum electricity price. For this reason, a grant equal to 75% of the upfront investment cost is considered in this analysis. The storage costs per kWh cycled have been obtained considering average values of $C_{BESS_{kWh}}$ and $N_{Full\ cycle}$, according to [36].

6. Simulation Results

For each day of the reference period, the algorithm handles the *MA RTP* prices, corresponding to each *DOD*, calculating the value of the objective functions and verifying the fulfillment of condition in Equation (14).

The values of the objective functions together with the charge/discharge time, for the three battery technologies, are reported in Table 3. If Equation (14) is satisfied, Table 3 reports the value of $\sum_{k=1}^{2} OF_{i,d}^{(k)}$ and the column d shows a couple of values, (x,y), denoting the charging/discharging time of the first and the second half day period, respectively. If Equation (14) is not satisfied, the value of the daily objective function, $OF_{i,d}$, is reported and the column d shows a single value denoting the charging/discharging time in the daily period. Finally, if all the objective functions have negative value (*i.e.*, there are no eligible objective functions) the battery remains idle and the corresponding values of the objective function and the charging/discharging times are missing in Table 3.

Table 3. Values of the objective functions in the reference weekly period.

	Lead Acid		Li-ion		NaS	
	OF	*d*	*OF*	*d*	*OF*	*d*
31/03/2014	0.038	4,4	0.036	2,1	0.122	4,4
01/04/2014	-	-	-	-	0.049	4,4
02/04/2014	-	-	0.002	-,1	0.047	4,3
03/04/2014	-	-	-	-	0.018	3,2
04/04/2014	-	-	0.004	1,-	0.042	4,-
05/04/2014	-	-	0.001	1	0.043	4,4
06/04/2014	0.028	-,4	0.01	-,4	0.071	-,4
Weekly *OF*	0.066		0.053		0.392	

The values reported in Table 3 lead to the following fundamental results (valid under the assumption that a subsidy equal to 75% of the upfront investment cost is granted to the storage owner):

- Among the three considered storage options, the use of NaS batteries leads to the maximum benefit for the storage owner (the value of the weekly objective function is around six times the one observed for the lead-acid battery); indeed, although NaS batteries have an acquisition cost higher than lead-acid, the number of cycles to failure is more than three times higher than that of lead-acid battery (see Table 2).
- The lead-acid technology appears to be the least convenient for arbitrage applications, despite its lower cost. This is essentially due to the low number of equivalent full cycles compared to the other battery technologies. The Li-ion technology also has a low profitability for arbitrage applications, essentially because of the high upfront investment cost. However, the situation could rapidly change since Li-ion batteries are the most promising in terms of cost reduction and cycling performance [31].
- Lead-acid battery remains idle during most of the days, since the gap between maximum and minimum electricity price is not enough to compensate for the low number of equivalent full cycles.
- As previously stated in Section 4.1, NaS battery is charged two times per day on weekdays (except on Friday), and only one time on Sunday. This is because weekdays have two price peeks, and the gap between max/min electricity price is high enough to compensate for the cost of cycling energy plus the cost of the energy losses in the charge/discharge process.
- The NaS battery often performs two operating cycles, whereas the Li-ion battery performs two operating cycles only on Monday. This is essentially due to the high upfront investment cost of Li-ion battery compared with NaS technology, and to the lower number of equivalent full cycles.

- On Sunday, the batteries perform only one cycle in the second half of the day, lasting four hours (as previously stated in Section 4.1).

It is worth noting that the battery cycle lasts four hours when the objective function takes a high value, *i.e.*, when the gap between high and low electricity prices is large. Indeed, in this case the first term of the objective function prevails over the second term and the higher *DOD* resulting from the greater discharge duration offsets the number of equivalent full cycles.

Finally, it is possible to assert that, at the current price of storage technologies, the use of batteries for arbitrage applications is not profitable for the storage owner. The battery is charged once a day or twice a day depending on the shape of *RTP* profiles, being the BESS operating cycle dependent on the specific battery technology.

In order to highlight the advantages of the proposed approach compared to other simple methods, a comparison is made with respect to a simple strategy (base case) where the battery is operated in the hours where the gap between the lowest and the highest prices is maximized. The base case differs from the proposed operating strategy since the battery can be operated at different hours, not necessarily uninterrupted, but always regardless of the facility's load profile. Besides, in the base case, the battery is operated always at its maximum *DOD* (4 h), if the discharge duration is compatible with the objective function values, under the fulfillment of constraint conditions.

The values of the objective functions together with the charge/discharge time, in the base case, are shown in Table 4. When the objective functions have negative value, the corresponding values and the charging/discharging times are missing in Table 4.

Table 4. Values of the objective functions for the base case.

	Lead acid		**Li-ion**		**NaS**	
	OF	*d*	*OF*	*d*	*OF*	*d*
31/03/2014	0.038	4,4	0.014	4,4	0.122	4,4
01/04/2014	-	-	-	-	0.049	4,4
02/04/2014	-	-	-	-	0.046	4,4
03/04/2014	-	-	-	-	0.013	4,4
04/04/2014	-	-	-	-	0.042	4,-
05/04/2014	-	-	-	-	0.043	4,4
06/04/2014	0.028	-,4	0.009	-,4	0.071	-,4
Weekly *OF*	0.066		0.023		0.386	
% weekly increase	-		130%		1.5%	

It was found that the percentage increase of the weekly objective function, compared to the base case, is 130% for Li-ion and 1.5% for NaS batteries, as reported in Table 4.

According to the values reported in Table 4, the comparison between the proposed operating strategy and the base case leads to the following considerations:

- For lead acid battery, the values of the objective function are the same (the weekly percentage increase is zero). Indeed, this kind of battery performs the same charging/discharging cycles both in the proposed operating strategy and in the base case.
- For Li-ion battery, the weekly percentage increase of the objective function is large (130%). Indeed, in the base case the Li-ion battery remains idle for most of the days and the value of the objective function on Monday is more than halved compared with the corresponding value reported in Table 3.
- For NaS battery, the weekly percentage increase of the objective function is 1.5%, as a result of an increase of the objective functions on Wednesday and Thursday.

The last conclusion is particularly meaningful since it confirms that operating the battery at low *DOD* can be advantageous for the storage owner when the gap between high and low electricity prices is limited (e.g., when the objective function takes a small value).

Figure 6a,b show the graphic comparison between the objective function values of the two approaches, for NaS and Li-ion battery, respectively.

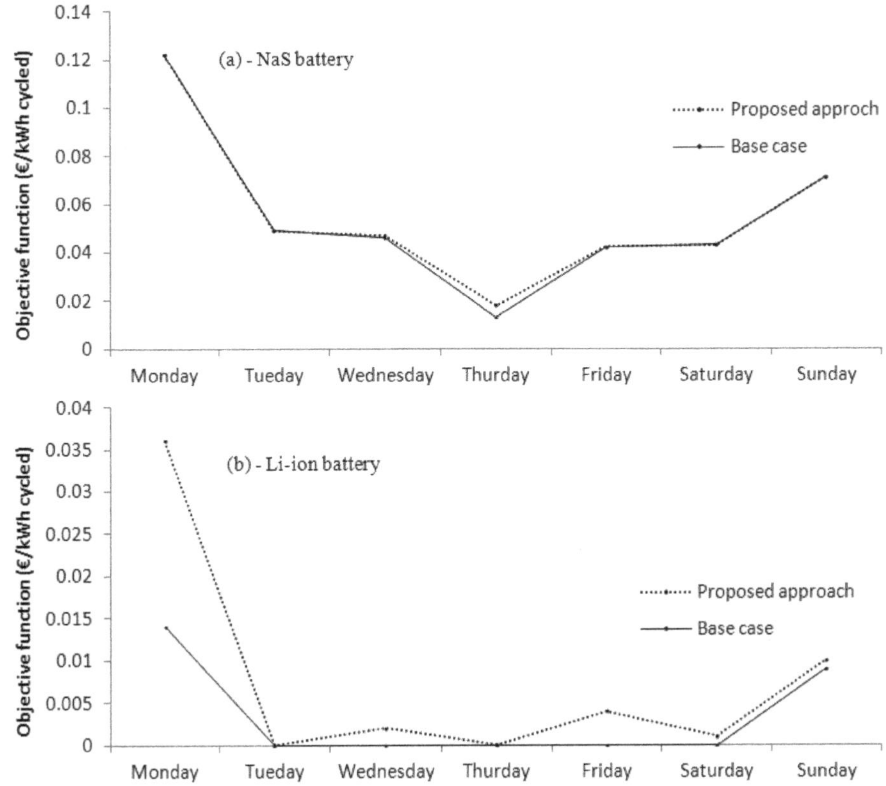

Figure 6. Graphic comparison between the objective function values of the two approaches: (**a**) NaS battery and (**b**) Li-ion battery.

The results obtained from the proposed approach show the effectiveness of the proposed operating strategy compared to the base case.

Finally, the effect of the proposed operating strategy on the daily curve of the energy extracted from the main grid is evaluated. To this aim, the power consumption of the department was registered over a reference period of one week (from 31 March to 6 April 2014).

Figure 7 shows the DEIM power diagram for the reference period, without (Figure 7a) and with (Figure 7b) storage contribution.

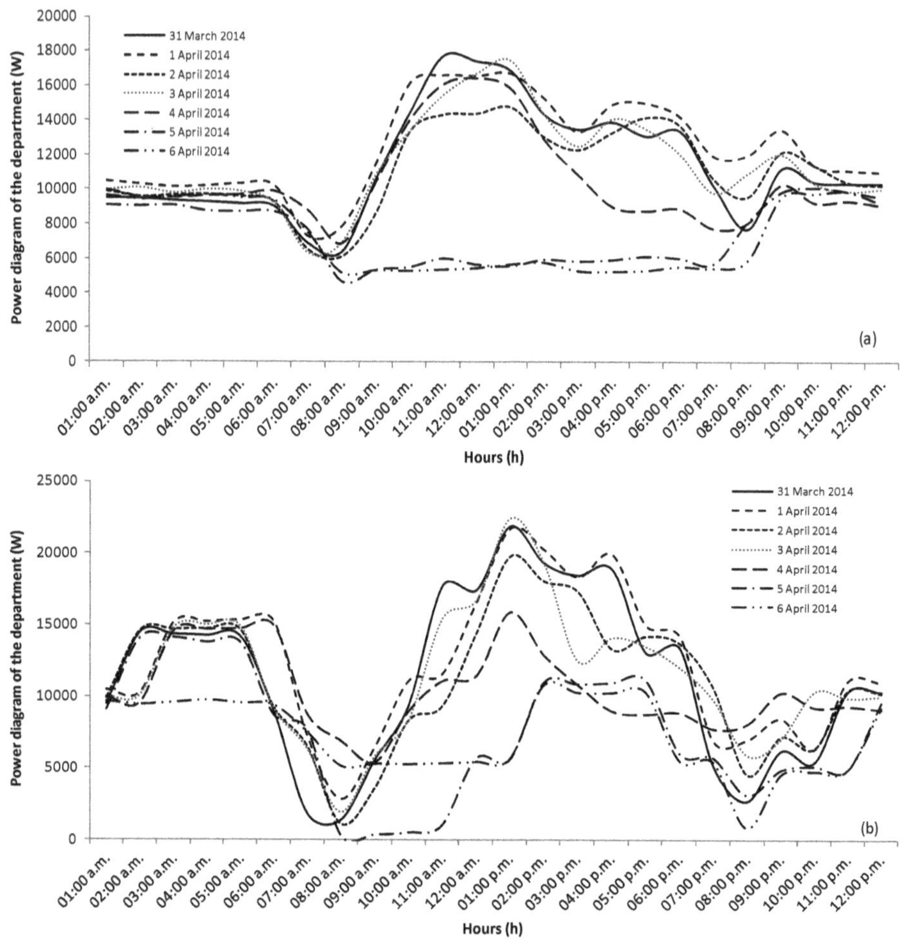

Figure 7. Power diagram of the department without (**a**) and with storage contribution (**b**).

Figure 7b shows power spikes due to the BESS charging/discharging. The maximum weekly peak load is increased at 23 kW (against a value of 18 kW without storage) when the proposed operating strategy is applied. Conversely, the minimum weekly peak load is reduced to zero when the storage is operated (against a value of 5 kW without storage). Therefore, the implementation of the proposed strategy does not lead to a flattening of the power profile but to an increase in the gap between peak and off-peak loads.

7. Conclusions and Future Work

This paper develops a detailed storage model linking together technical, economic and electricity market parameters. The storage system is described by means of its performance parameters, such as the charge and generate capacity, the charge/discharge efficiency, the rated charge/discharge rate, the *DOD*, *etc.*, which are sufficient to evaluate the arbitrage potential of the storage device. The proposed operating strategy aims to maximize the profit of the storage owner (electricity customer) by determining the optimal charge/discharge schedule. Unlike the studies reported in the literature, often requiring an estimate of the end-user load profile, the proposed operating strategy is able to identify the proper charging schedule of the device regardless of the specific facility's consumption. This is made possible since the battery is sized referring to the facilities' energy consumption during peak price hours, on the day of the year of lowest consumption. Under this assumption, the storage will be able to supply the entire customer load during the day of the year of lowest consumption, but only a portion of the customer's load on the other days. This could be particularly useful when the customer load profile cannot be scheduled with sufficient reliability, because of the uncertainty inherent in load forecasting. In these cases, identifying a BESS operating strategy that does not depend on the user's power profile can be an important task, since the deviation of the scheduled power profile from the effective one could affect the results obtained using more complete methods. In order to highlight the advantages of the proposed approach compared to other methods, a comparison is made with respect to a simple strategy (base case) where the battery is charged only one time per day at its maximum *DOD* (equal to four hours). The results obtained from the proposed approach show the effectiveness of the proposed operating strategy. The proposed model can be applied to several kinds of storages but the test results refer to three electrochemical technologies: lead-acid, Li-ion and NaS battery. The simulation results show that the operating schedule of the storage device differs in the various days of the week and it depends on the specific battery used (the most critical parameters being the acquisition cost of the battery bank and the number of cycles to failure). The operating cycle lasts four hours (*i.e.*, the maximum available charge/discharge time) when the objective function takes high values. However, in the days when the objective function has a lower value, the storage device is operated at a lower discharging time. This is because the higher gap between high and low electricity prices and the higher value of equivalent full cycles fully offset the less benefit due to the lower *DOD* (which results in a lower energy discharged). Simulation results show that, at current prices, no BESS technology is cost effective, due to the high upfront investment costs. However, if a subsidy is granted to reduce the initial investment cost, the use of NaS batteries leads to the maximum benefit among the three considered storage options. This is essentially due to the high number of equivalent full cycles (four times higher than that of lead-acid batteries). Conversely, the lead-acid technology appears to be the least convenient for arbitrage applications, despite its lower cost. This is essentially due to the low number of equivalent full cycles compared to the other battery technologies. In addition, the Li-ion technology has a low profitability for arbitrage applications, essentially because of the high upfront investment cost. However, the situation could rapidly change since Li-ion batteries are the most promising in terms of cost reduction and cycling performance.

In a future work, the authors will evaluate the effect of load forecasting uncertainty on the accuracy of storage operating strategies, in order to demonstrate that often the deviation of the scheduled power profile from the effective one could affect the results of more complete methods.

Acknowledgments: This work was supported by the project i-NEXT (Innovation for greeN Energy and eXchange in Transportation), identification code: PON04a2_Hi-NEXT (CUP B71H12000700005).

Author Contributions: This work was conceived by Enrico Telaretti. Preparation of the manuscript has been performed by Enrico Telaretti. Simulation and analysis of the results have been perfomed by Enrico Telaretti. Luigi Dusonchet and Mariano Ippolito supervised the work, giving a final review of the paper. All authors read and agreed to the final article.

Conflicts of Interest: The authors declare no conflict of interest.

Abbreviations

BESS	Battery Energy Storage System
BOP	Balance-of Plant
DEIM	Department of Energy, Information Engineering and Mathematical Models
DOD	Depth-of-Discharge
DSM	Demand Side Management
IRR	Internal Rate of Return
Li-ion	Lithium-Ion
MA	Moving Average
MA RTP	Moving Average of *RTP* Prices
NaS	Sodium-Sulphur
PCS	Power Conversion System
PSO	Particle Swarm Optimization
PV	Photovoltaic
RES	Renewable Energy Sources
RTP	Real-Time Pricing
SMP	System Marginal Price
SOC	State-of-Charge
TOU	Time-of-Use
VAT	Value Added Tax

References

1. Campoccia, A.; Dusonchet, L.; Telaretti, E.; Zizzo, G. Feed-in tariffs for grid-connected PV systems: The situation in the European community. In Proceedings of IEEE Power Tech Conference, Lausanne, Switzerland, 1–5 July 2007; pp. 1981–1986.
2. Campoccia, A.; Dusonchet, L.; Telaretti, E.; Zizzo, Z. Financial Measures for Supporting Wind Power Systems in Europe: A Comparison between Green Tags and Feed'in Tariffs. In Proceedings of IEEE Power Electronics, Electrical Drives, Automation and Motion, Ischia, Italy, 11–13 June 2008; pp. 1149–1154.
3. Sgroi, F.; Tudisca, S.; Di Trapani, A.M.; Testa, R.; Squatrito, R. Efficacy and Efficiency of Italian Energy Policy: The Case of PV Systems in Greenhouse Farms. *Energies* **2014**, *7*, 3985–4001. [CrossRef]
4. Giannini, E.; Moropoulou, A.; Maroulis, Z.; Siouti, G. Penetration of Photovoltaics in Greece. *Energies* **2015**, *8*, 6497–6508. [CrossRef]
5. Borenstein, S. The long-run efficiency of real-time electricity pricing. *Energy J.* **2005**, *26*, 93–116. [CrossRef]
6. Dusonchet, L.; Ippolito, M.G.; Telaretti, E.; Graditi, G. Economic impact of medium-scale battery storage systems in presence of flexible electricity tariffs for end-user applications. In Proceedings of IEEE International Conference on the European Energy Market, Florence, Italy, 10–12 May 2012; pp. 1–5.
7. Dusonchet, L.; Ippolito, M.G.; Telaretti, E.; Zizzo, G.; Graditi, G. An optimal operating strategy for combined RES-based Generators and Electric Storage Systems for load shifting applications. In Proceedings of IEEE International Conference on Power Engineering, Energy and Electrical Drives, Instanbul, Turkey, 13–17 May 2013; pp. 552–557.
8. Youn, L.T.; Cho, S. Optimal operation of energy storage using linear programming technique. In Proceedings of the World Congress on Engineering and Computer Science, San Francisco, CA, USA, 20–22 October 2009; pp. 480–485.
9. Ahlert, K.; Van Dinther, C. Sensitivity analysis of the economic benefits from electricity storage at the end consumer level. In Proceedings of IEEE Power Tech Conference, Bucharest, Romania, 28 June–2 July 2009; pp. 1–8.
10. Bradbury, K.; Pratson, L.; Patino-Echeverri, D. Economic viability of energy storage systems based on price arbitrage potential in real-time U.S. electricity markets. *Appl. Energy* **2014**, *114*, 512–519. [CrossRef]
11. Graves, F.; Jenkin, T.; Murphy, D. Opportunities for Electricity Storage in Deregulating Markets. *Electr. J.* **1999**, *12*, 46–56. [CrossRef]

12. Hu, W.; Chen, Z.; Bak-Jensen, B. Optimal operation strategy of battery energy storage system to real-time electricity price in Denmark. In Proceedings of the IEEE Power and Energy Society General Meeting, Minneapolis, MN, USA, 25–29 July 2010; pp. 1–7.
13. Mokrian, P.; Stephen, M. A stochastic programming framework for the valuation of electricity storage. In Proceedings of 26th USAEE/IAEE North American Conference, Ann Arbor, MI, USA, 24–27 September 2006; pp. 1–34.
14. Maly, D.K.; Kwan, K.S. Optimal battery energy storage system (BESS) charge scheduling with dynamic programming. *IEE Proc. Sci. Meas. Technol.* **1995**, *142*, 454–458. [CrossRef]
15. Van de Ven, P.M.; Hegde, N.; Massoulié, L.; Salonidis, T. Optimal control of residential energy storage under price fluctuations. In Proceedings of International Conference on Smart Grids, Green Communications and IT Energy-aware Technologies, Venice, Italy, 22–27 May 2011; pp. 159–162.
16. Huang, T.; Liu, D. Residential energy system control and management using adaptive dynamic programming. In Proceedings of the International Joint Conference on Neural Networks, San Jose, CA, USA, 31 July–5 August 2011; pp. 119–124.
17. Lee, T.Y. Operating Schedule of Battery Energy Storage System in a Time-of-Use Rate Industrial User With Wind Turbine Generators: A Multipass Iteration Particle Swarm Optimization Approach. *IEEE Trans. Energy Conv.* **2007**, *22*, 774–782. [CrossRef]
18. Abeygunawardana, A.; Ledwich, G. Estimating benefits of energy storage for aggregate storage applications in electricity distribution networks in Queensland. In IEEE Power and Energy Society General Meeting, Vancouver, BC, Canada, 21–25 July 2013; pp. 1–5.
19. Byrne, C.; Verbic, G. Feasibility of Residential Battery Storage for Energy Arbitrage. In Proceedings of Power Engineering Conference (AUPEC), 2013 Australasian Universities, Hobart, TAS, Australia, 29 September–3 October 2013; pp. 1–7.
20. Ekman, C.K.; Jensen, S.H. Prospects for large scale electricity storage in Denmark. *Energy Convers. Manag.* **2010**, *51*, 1140–1147. [CrossRef]
21. Shcherbakova, A.; Kleit, A.; Cho, J. The value of energy storage in South Korea's electricity market: A Hotelling approach. *Appl. Energy* **2014**, *125*, 93–102. [CrossRef]
22. Purvins, A.; Sumner, M. Optimal management of stationary lithium-ion battery system in electricity distribution grids. *J. Power Sources* **2013**, *242*, 742–755. [CrossRef]
23. Castillo-Cagigal, M.; Caamaño-Martín, E.; Matallanas, E.; Masa-Bote, D.; Gutiérrez, A.; Monasterio-Huelin, F.; Jiménez-Leube, J. PV self-consumption optimization with storage and Active DSM for the residential sector. *Sol. Energy* **2011**, *85*, 2338–2348. [CrossRef]
24. Matallanas, E.; Castillo-Cagigal, M.; Gutiérrez, A.; Monasterio-Huelin, F.; Caamaño-Martín, E.; Masa, D.; Jiménez-Leube, J. Neural network controller for Active Demand-Side Management with PV energy in the residential sector. *Appl. Energy* **2012**, *91*, 90–97. [CrossRef]
25. Abushnaf, J.; Rassau, A.; Górnisiewicz, W. Impact of dynamic energy pricing schemes on a novel multi-user home energy management system. *Electr. Power Syst. Res.* **2015**, *125*, 124–132. [CrossRef]
26. Ippolito, M.G.; Telaretti, E.; Zizzo, G.; Graditi, G.; Fiorino, M. A Bidirectional Converter for the Integration of $LiFePO_4$ Batteries with RES-based Generators. Part I: Revising and finalizing design. In Proceedings of 3rd Renewable Power Generation Conference, Naples, Italy, 24–25 September 2014; pp. 1–6.
27. Ippolito, M.G.; Telaretti, E.; Zizzo, G.; Graditi, G.; Fiorino, M. A Bidirectional Converter for the Integration of $LiFePO_4$ Batteries with RES-based Generators. Part II: Laboratory and Field Tests. In Proceedings of 3rd Renew. Power Generation Conference, Naples, Italy, 24–25 September 2014; pp. 1–6.
28. Viera, J.C.; Gonzalez, M.; Liaw, B.Y.; Ferrero, F.J.; Alvarez, J.C.; Campo, J.C.; Blanco, C. Characterization of 109 Ah Ni–MH batteries charging with hydrogen sensing termination. *J. Power Sources* **2007**, *171*, 1040–1045. [CrossRef]
29. Zheng, M.; Meinrenken, C.J.; Lackner, K.S. Agent-based model for electricity consumption and storage to evaluate economic viability of tariff arbitrage for residential sector demand response. *Appl. Energy* **2014**, *126*, 297–306. [CrossRef]
30. GME home page. Available online: http://www.mercatoelettrico.org/it/Default.aspx (accessed on 26 May 2015).
31. Dufo-López, R. Optimisation of size and control of grid-connected storage under real time electricity pricing conditions. *Appl. Energy* **2015**, *140*, 395–408. [CrossRef]

32. Dufo-López, R.; Bernal-Agustin, J.L. Techno-economic analysis of grid-connected battery storage. *Energy Conv. Manag.* **2015**, *91*, 394–404. [CrossRef]
33. The Lithium-Ion Battery. Service Life Parameters. Available online: https://www2.unece.org/wiki/download/attachments/8126481/EVE-06-05e.pdf?api=v2 (accessed on 26 May 2015).
34. Lu, N.; Weimar, M.R.; Makarov, Y.V.; Ma, J.; Viswanathan, V.V. The Wide-Area Energy Storage and Management System–Battery Storage Evaluation. Available online: http://www.pnl.gov/main/publications/external/technical_reports/PNNL-18679.pdf (accessed on 26 May 2015).
35. Divya, K.C.; Østergaard, J. Battery energy storage technology for power systems—An Overview. *Electr. Power Syst. Res.* **2009**, *79*, 511–520. [CrossRef]
36. Battke, B.; Schmidt, T.S.; Grosspietsch, D.; Hoffmann, V.H. A review and probabilistic model of life cycle costs of stationary batteries in multiple applications. *Renew. Sustain. Energy Rev.* **2013**, *25*, 240–250. [CrossRef]
37. Telaretti, E.; Sanseverino, E.R.; Ippolito, M.; Favuzza, S.; Zizzo, G. A novel operating strategy for customer-side energy storages in presence of dynamic electricity prices. *Intell. Ind. Syst.* **2015**, *1*, 233–244. [CrossRef]

Article

Integrated Equivalent Circuit and Thermal Model for Simulation of Temperature-Dependent $LiFePO_4$ Battery in Actual Embedded Application

Zuchang Gao [1], Cheng Siong Chin [2,*], Wai Lok Woo [3] and Junbo Jia [1]

1 School of Engineering, Temasek Polytechnic, 21 Tampines Avenue 1, Singapore 529757, Singapore; jiajunbo@tp.edu.sg (Z.G.); zuchang@tp.edu.sg (J.J.)
2 School of Marine Science and Technology, Newcastle University, Newcastle upon Tyne NE1 7RU, UK
3 School of Electrical and Electronic Engineering, Newcastle University, Newcastle upon Tyne NE1 7RU, UK; Lok.Woo@newcastle.ac.uk
* Correspondence: cheng.chin@newcastle.ac.uk; Tel.: +44-65-6908-6013

Academic Editor: William Holderbaum
Received: 11 October 2016; Accepted: 4 January 2017; Published: 11 January 2017

Abstract: A computational efficient battery pack model with thermal consideration is essential for simulation prototyping before real-time embedded implementation. The proposed model provides a coupled equivalent circuit and convective thermal model to determine the state-of-charge (SOC) and temperature of the $LiFePO_4$ battery working in a real environment. A cell balancing strategy applied to the proposed temperature-dependent battery model balanced the SOC of each cell to increase the lifespan of the battery. The simulation outputs are validated by a set of independent experimental data at a different temperature to ensure the model validity and reliability. The results show a root mean square (RMS) error of 1.5609×10^{-5} for the terminal voltage and the comparison between the simulation and experiment at various temperatures (from 5 °C to 45 °C) shows a maximum RMS error of 7.2078×10^{-5}.

Keywords: lithium-ion battery; battery management system; convective thermal model; cell model; state-of-charge

1. Introduction

In recent years, interest has increased for lithium-ion (Li-ion) batteries [1–3] in power generation and renewable energy applications such as solar energy systems, wave-operated electrical generation systems, wind turbines, battery electric vehicles (BEVs), and portable power storage devices. Compared with other commonly used batteries like lead acid, nickel cadmium (NiCd) and nickel metal hydride (NiMH), $LiFePO_4$ is popular due to its high capacity, low self-discharge current, wide temperature operation range, and long service life that make them better candidates for many applications. However, lithium-ion batteries are sensitive to overcharging or discharging that could deteriorate the performance resulting in a shorter lifetime [4]. The accurate SOC estimation is, therefore, necessary for a properly functioning $LiFePO_4$ battery power system.

Since there is no sensor available to measure SOC directly, it is estimated from physical measurements (such as the current, voltage and temperature) via the battery management system (BMS). Currently, a large variety of methods for battery SOC estimation is proposed in the literature. First, the standard measurement-based estimation approaches, such as the coulomb counting method or ampere-hour (Ah) methods, as well as the open-circuit voltage (OCV) and impedance measurement methods [5–9] give a more intuitive and reliable estimation. Second, the machine learning-based estimation method (also called black-box method), such as artificial neural network (ANN) and fuzzy logic (FL) methods [10–17] require high computational effort for extensive training on a large dataset. Lastly, the state-space model-based estimation using Kalman filter (KF) [18–20] increases the computational load of the BMS. Hence, these methods have its advantages and disadvantages. They mostly focus on estimating the SOC of a single cell and terminal voltage estimation without considering the SOC and temperature differences between the cells in a battery [13]. Few works have been performed to explore the SOC predictions of the battery pack (or multiple cells) system with temperature variations due to the ambient condition and cells. However, the temperature-dependent battery model with the convective heat transfer between the cells are often too complicated to realize for the actual application, and the simulation model is not available.

In this paper, the battery pack model is proposed to simulate the influence of temperature [14,15] between cells and the effect of ambient temperature acting on the cells that are necessary for developing a more reliable SOC estimation and cell balancing algorithm. A total of 12 $LiFePO_4$ cells in a series are used for the modeling and parameter identification process. The parameters are estimated online by a series of lookup tables to provide a good compromise between the high fidelity and computational effort for integrated BMS implementation, where the lookup table uses an array of data to map input values to output values, approximating a mathematical function. If the lookup table encounters an input that does not match any of the table's pre-defined input values, the block interpolates or extrapolates the output values based on nearby table values. Since table lookups and simple estimations can be faster than mathematical function evaluations, using the lookup table method can result in a faster computational time. The SOC estimation algorithm of the battery pack and cell balancing strategy are implemented and validated using the experimental data collected in a laboratory. The battery model under different temperatures is included to improve the battery model. The experiment results show the feasibility of the proposed model for simulation prototyping before the actual implementation.

In summary, the contributions of the paper include a 12-cell temperature-dependent battery model with the convective heat transfer between cells to estimate the SOC of each cell for automatic passive cell balancing. This work also provides a battery simulation prototyping platform to allow different algorithms and battery cells to be simulated and implemented quickly using the Simulink Coder to generate and execute C code from Simulink with less programming needed. The experiments verify the battery model in both near zero and room temperatures (from 5 °C to 45 °C) using the actual duty cycle.

The paper is organized as follows: Section 2 models the battery pack. It is followed by Section 3 that deals with the experimental setup and data acquisition. The simulation model validation using independent experimental data, SOC estimation of the proposed battery pack and cell balancing are presented in Section 4. Finally, Section 5 concludes the work with the future plan.

2. Battery Cell Model Description

The electrochemical model is the most accurate battery model for estimating the SOC. However, the electrochemical models are quite complex and involve partial differential equations [13] to solve in real-time. The black-box models using machine learning have recently been proposed. However, they use much computational effort for training large datasets for real-time embedded applications that slow down the system's output and performance in real-time. The alternative approach is to use the equivalent circuit models (ECM) with a combination of voltage sources, resistors, and capacitors to model the battery behaviors that will provide an interpretable structure for online estimation and implementation.

2.1. Equivalent Circuit Model

The number of RC blocks typically ranges from one to two for various applications. The dynamic voltage responses of 1 RC model and 2 RC model are compared in Figure 1, as well as the experimental data which is from the $LiFePO_4$ battery cell test. It is obvious that a higher number of RC increases the computational resources without significantly improving the model accuracy. Therefore, 1 RC battery cell model is proposed for the embedded applications in this paper, the model structure is shown in Figure 2.

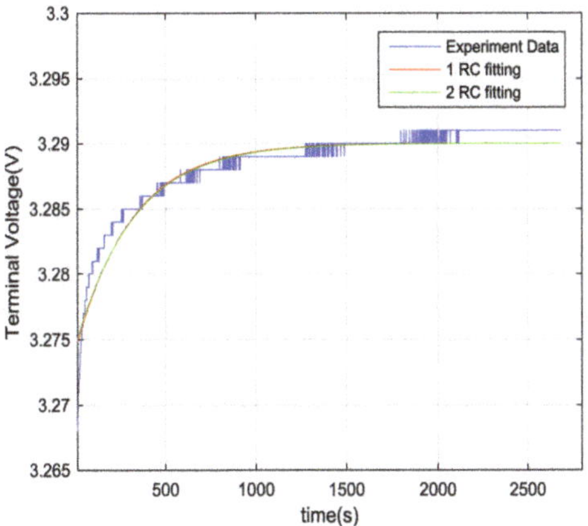

Figure 1. Comparison of 1 RC and 2 RC fitting.

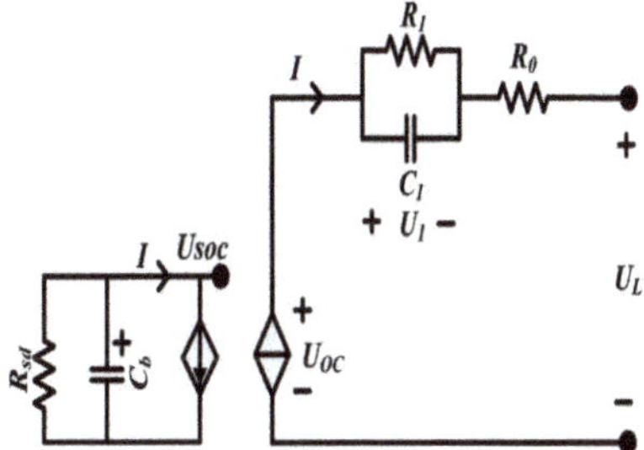

Figure 2. Equivalent circuit model.

The nonlinear mapping from the battery's SOC to the open circuit voltage (OCV) is represented by a voltage-controlled voltage source (VCVS) denoted as U_{oc}. R_0 as the internal resistance. The R_1 and C_1 are polarization resistance and polarization capacitance to simulate the transient response during a charge or discharge process. The whole-charge capacitor is denoted as C_b, its value is the battery capacity in unit (A·s), the self-discharge energy loss due to long time storage is represented by R_{sd}. The voltages across C_1 are denoted as U_1. The terminal voltage and current are denoted as system output U_L and system input I, respectively. Define $I > 0$ when charging; $I < 0$ when discharging. The governing equation of the battery model is as follows.

$$U_L = U_{OC} - I \times R_0 - U_1 \tag{1}$$

As the temperature affects the battery cells' performance, the critical parameters such as OCV, R_0, R_1 and C_1 are function of both SOC and temperature T. The lookup tables are used to establish a direct correlation between electrochemical phenomena inside the cell and the circuit elements. The method can capture nonlinear electrochemical phenomena and yet avoid lengthy electrochemical process calculations to make the model suitable for embedded applications besides the simulation environment.

SOC is one of the most important variables in the BMS to manage the lithium-ion batteries to their optimal performance. It is necessary to monitor the SOC of the battery cell in real-time to prevent the battery cell from either undercharging or over-discharging as any of these conditions could damage the battery cell permanently. In this paper, the Ah method is used to compute the SOC.

$$SOC(k) = SOC(0) - \frac{T}{C_n}\int_0^k (\eta . i(t) - S_d)dt \tag{2}$$

where $SOC(0)$ is the initial SOC, C_n is the nominal capacity of the battery pack, T is the sampling period, $i(t)$ is the load current at time t, η is coulombic efficiency, and S_d is the self-discharging rate. For $LiFePO_4$ battery used in this experiment, $\eta > 0.994$ under room temperature [1]. In this paper, $\eta = 1$, and $S_d = 0$ are assumed.

The series connected cells' capacity is the quantity of electric charge stored in the cells. Theoretically, the series battery capacity is given by the sum of the minimum capacity that can be charged and discharged [2]:

$$C_{series} = \min_{1 \le i \le m} (SOC_i \cdot C_i) + \min_{1 \le i \le m} ((SOC_i - 1) \cdot C_i) \tag{3}$$

where C_{series} is the usable capacity of the series battery pack, C_i is the capacity of the i cell, SOC_i is state of charge of the i cell, m is the number of cells connected in a series.

2.2. Lumped-Capacitance Thermal Model of the Battery Cell

In this paper, the commercial $LiFePO_4$ 26650 cylindrical cells are selected to be the research objects, which are constructed in a multilayer structure in which the radial thermal conductivity is lower than the axial one. Nevertheless, the thermal resistance by the radial conduction is still much less than the convective thermal resistance, as air is used as the coolant (i.e., the Biot number, $\mathrm{Bi} = L_c h_f / k_s < 0.1$). Therefore, a lumped-capacitance thermal model for battery cells assuming a uniform temperature in each cell is sufficient without compromising accuracy of the numerical analysis. The thermal energy balance of the battery cell is modeled by using the first law of thermodynamics:

$$\frac{\mathrm{d}U}{\mathrm{d}t} = Q_{gen}(t) - Q_{loss}(t) \tag{4}$$

where U represents the internal energy and is the total energy contained by a thermodynamic system (in joules). $Q_{gen}(t)$ is the generating heating rate, i.e., the rate of the heat generation occurring in the cell.

On the other hand, U can be determined by the following.

$$\mathrm{d}U = m \times C_P \times \mathrm{d}T_{cell} \tag{5}$$

where m is the mass of the cell (in kilograms), $\mathrm{d}T_{cell}$ is the temperature variation of the cell with time (in kelvin), and C_P is the specific heat capacity of the cell (in J/kg/K).

The volume heat generation rate in a battery body is the sum of numerous local losses such as active heat generation, reaction heat generation, and Ohmic heat generation. In this paper, $Q_{gen}(t)$ is characterized only by ohmic losses because of their simplicity to the model in the embedded applications. Ohmic losses are expressed as follows.

$$Q_{gen}(t) = R_0 \times (I)^2 + R_1 \times (I_1)^2 \tag{6}$$

where I is the battery current, I_1 is the current going through by R_1.

Moreover, $Q_{loss}(t)$ is a value of all the heat transfers as a result of a temperature difference between the cells and the connections of the cells and consists of two parts: convective heat transfer Q_{conv} and conductive heat transfer Q_{cond}.

$$Q_{loss} = Q_{conv} + Q_{cond} \tag{7}$$

1. Convective heat transfer

The convective heat transfer Q_{conv} from the cell to the surrounding is determined by

$$Q_{conv} = h_{conv} S_{area} (T_{cell} - T_{air}) \tag{8}$$

where h_{conv} is the convective heat transfer coefficient, S_{area} is the area of heat exchange, T_{cell} is the cell temperature and T_{air} is the ambient temperature.

2. Conductive heat transfer

The convective heat transfer Q_{cond} represents the thermal diffusion through cell to cell electric connector. It can be modeled by

$$Q_{cond} = \frac{T_{cell_2} - T_{cell_1}}{R_{cond}} \tag{9}$$

where T_{cell_2} and T_{cell_1} are the temperature of battery cell 2 and battery cell 1, respectively. R_{cond} is the thermal resistance of the connection, which includes the top and bottom connection of the battery cell.

In Li-ion battery, the cross-plane thermal conductivity is much smaller than the in-plane thermal conductivity. Heat conduction through the top and bottom of cells are important to the practical system. However, in this study, the experimental battery cells are all brand new, assuming that they are all with good uniformity. The temperature difference $\Delta T = T_{cell_2} - T_{cell_1}$ is ignored. As a result, the conductive heat transfer is also neglected in the model in the paper.

2.3. Coupled Equivalent Circuit Model (ECM) and Thermal Battery Model

A coupled electro-thermal model of the $LiFePO_4$ battery is proposed. In this model, the inputs are the current battery I and the ambient temperature, T_{air}. In the coupled model, both thermal and electrical are considered since the temperature affects the four main parameters (OCV, R_0, R_1 and C_1). As shown in Figure 3, the parameters at different temperatures provide two-dimensional lookup tables for the ECM to compute the terminal voltage and SOC of each cell while the thermal model determines the temperature within the cells due to convection.

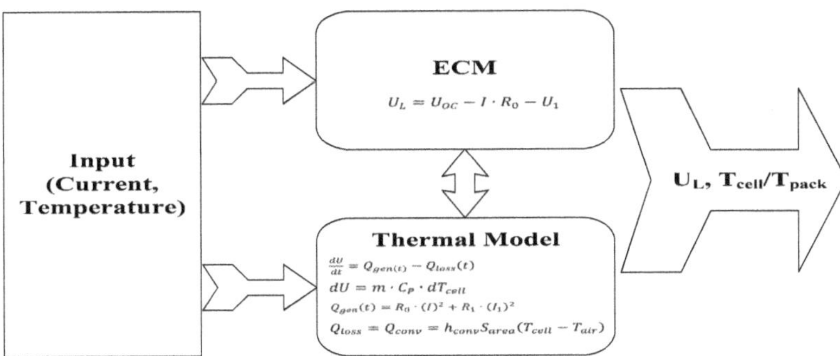

Figure 3. Coupled equivalent circuit model (ECM) and thermal model.

3. Experiment Tests for Battery Characterizations

A test environment for battery characterizations has been built in the laboratory as shown in Figure 4. The commercial $LiFePO_4$ battery cells (ANR26650M1-B from A123 System with Nanophosphate® lithium-ion chemistry) were used in the experiments. The key specification of the battery cell is tabulated in Table 1. Battery cell or battery pack was placed in the temperature chamber as seen in Figure 5 to perform a series of tests under different controlled temperatures. The ambient temperatures 5 °C, 15 °C, 25 °C, 35 °C and 45 °C were used to determine the model parameters of the 12-cell battery. The load current is created using a programmable DC electronic load, and a programmable DC power supply for charging the battery cells. The power supply is utilized as a controlled voltage or current source with the output voltage from 0 to 36 V and current from 0 to 20 A. A current sensor LEM 50-P is used to measure the charge and discharge current. The NTC temperature sensors are utilized to measure the temperatures of the battery cells and the ambient temperature. The National Instruments DAQ device controlled all input and output data. The host PC communicates with the DAQ device to monitor the power supply and charge and discharge status of the battery in real-time. As the data acquisition rate is limited in the embedded system, it is one sample per second. The host PC performs the model simulation and algorithm development using the battery's data received. A custom-designed pulse relaxation that includes the transient part and non-transient part (rather than simple constant current cycles often adopted in the literature) is employed in the SOC estimation as seen in Figure 7.

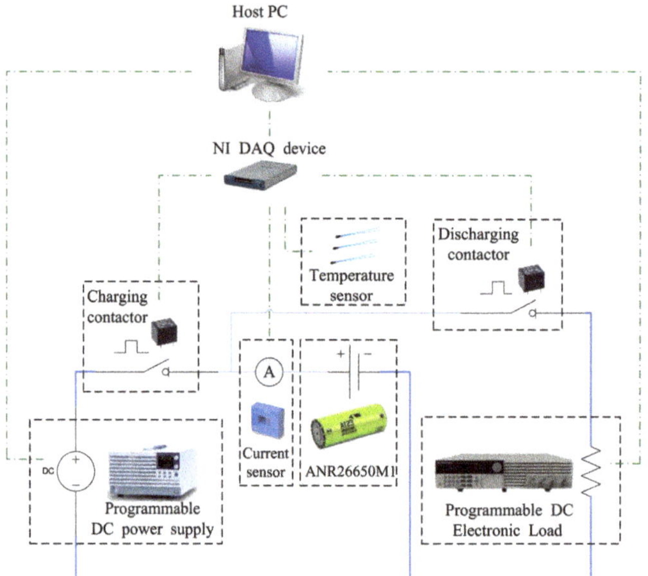

Figure 4. Battery test bench.

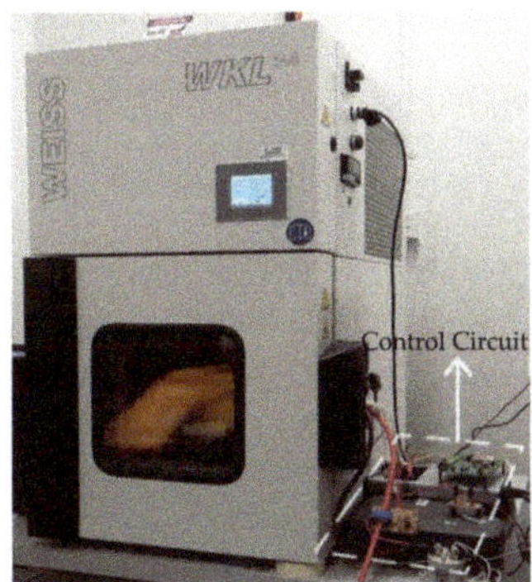

Figure 5. Temperature chamber.

Table 1. Battery cell and thermal specifications.

Cell Dimensions (mm)	Ø 26 × 65
Cell Weight (g)	76
Cell Capacity (nominal/minimum) (0.5 C Rate)	2.5/2.4
Voltage (nominal, V)	3.3
Recommended Standard Charge Method	2.5 A to 3.6 V CCCV for 60 min
Cycle Life at 20 A Discharge, 100% DOD	>1000 cycles
Maximum Continuous Discharge	50 A
Operating Temperature	−30 °C to 55 °C
Storage Temperature	−40 °C to 60 °C
Specific Heat Capacity of the Cell C_p (J/kg/K)	810.53
Convective Heat Transfer Coefficient h_{conv} (W/m^2/K)	5
Surface Area of Heat Exchange S_{area} (m^2)	0.0149
Ambient Temperature T_{air} (°C)	25

3.1. Static Capacity Test

As compared to the nominal capacity, the static capacity of a battery cell varies with the load current and the ambient temperature. The battery capacity testing determines the battery cell capacity in ampere-hours at a constant current (CC) discharge rate. This test provides a baseline for a battery cell for the advanced battery management algorithm development (e.g., SOC, SOH, and cell balancing). The test procedure follows the constant current constant voltage (CCCV) protocol and consists of the following steps.

1. Charge the battery at 0.8 C rate (2 A) to the fully charged state in CCCV mode under the specified temperature. The battery is fully charged to 3.6 V when the current reaches 1 mA.
2. Apply a 15-hour relaxation period before discharging the battery cell.
3. Discharge at a constant current 0.8 C rate until the voltage reaches the battery minimum limit of 2.5 V.
4. Record the data and calculate the static capacity as follows.

$$Q_d = \frac{1}{3600}\int_0^{t_d} I_d(\tau)\mathrm{d}\tau \text{ (Ah)} \tag{10}$$

where Q_d is the static capacity, I_d is the discharge current in ampere, and t_d is the discharge time in second.

3.2. *Pulse Discharge Test*

The pulse discharge test characterizes the battery voltage response (cell dynamics) at various SOCs and temperatures. The test comprises a series of discharge pulses across the full SOC range under specified temperature points. The test procedure is summarized as follows.

1. Charge the battery to a fully charged state, follow step 1 in Section 3.1.
2. Apply a 15-hour relaxation period before discharging the battery cell.
3. Discharge the battery cell at a pulse current 0.8 C rate with 450 s discharging time and 45 min relaxation period, until the terminal voltage reaches the cut-off voltage 2.5 V.
4. Record the data and proceed to model validation and simulation.

3.3. *Cycling Aging Test*

Cycling aging is a major factor that causes the battery to degrade and lose its capacity. When the capacity reduces to 80% of the beginning life capacity, the battery is considered to have reached its end of life (EOL). The static capacity of the battery is a non-linear function of charge-discharge cycling,

$$Q_d = f(N_c) \tag{11}$$

where N_c is the charge-discharge cycling number.

The procedure of the one cycling test is illustrated as follows. The initial state of the battery is assumed to be fully discharged.

1. Charge the battery to a fully charged state, follow step 1 in Section 3.1.
2. Allow the battery to rest for 15 min until its temperature stabilized.
3. Discharge at a constant current 0.8 C rate until the voltage reaches the battery minimum limit of 2.5 V.
4. Record the data and proceed to another cycle after the battery rests for 15 min.

Based on the cycling aging testing designed above and the manufacturer's specifications in Table 1, the test will take around one year to perform. Thus, it is quite time-consuming to conduct such test. Since new battery cells were used in the experiment, the aging effect of the cells is therefore neglected.

4. Battery Model Identification and Results

4.1. *Temperature-Dependent Battery Cell Parameters Identification*

Many parameter identification methods are proposed in the literature. With its limitation in the embedded system resource, the lookup table approach was implemented in the battery pack model to allow more computation time to perform the SOC estimation and cell balancing. The static capacities of the battery cell were identified from the results of the static capacity test. Table 2 and Figure 6 illustrate the results of the static capacity test under the specified ambient temperatures, respectively. As seen in Figure 6, the static capacity of the battery cell increases as the ambient temperature increases. It reaches a steady state value at around 35 °C.

Table 2. Static capacities under specified ambient temperature

Temperature (°C)	**5**	**15**	**25**	**35**	**45**
Static capacity (Ah)	2.2369	2.4474	2.5642	2.5693	2.5706

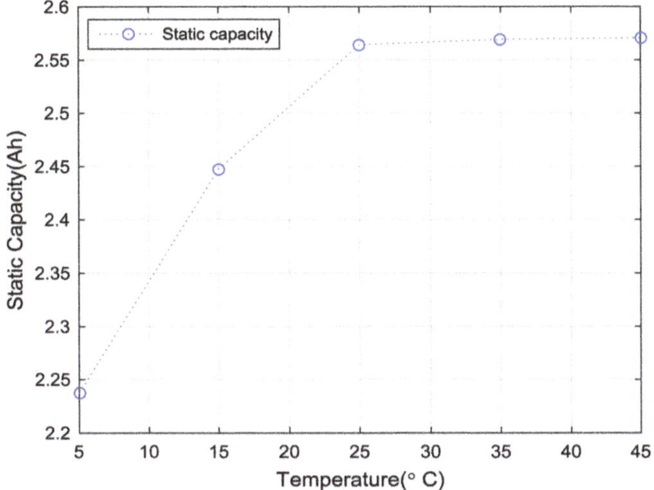

Figure 6. Static capacity under specified ambient temperature.

The value of OCV is identified from the results of the pulse discharge test (PDT) in this paper. The relaxation process example of battery PDT under 25 °C is shown in Figure 7. From the figure, it is obvious that the difference of the terminal voltage between 45 min and 3 h is 0.02 mV (0.06% of the nominal voltage). Hence, to save the experiment time, 45 min was thought to be enough for relaxation due to the small change in the terminal voltage after 45 min for the selected $LiFeO_4$ battery.

Compared with low-rate current charge/discharge method, the proposed PDT method to obtain the OCV at certain SOC intervals (e.g., 10%) can reduce the measurement time by around 90%. The comparison of C/50 low-rate discharge profile and the 10% SOC step incremental OCV curve at 25 °C is shown in Figure 8.

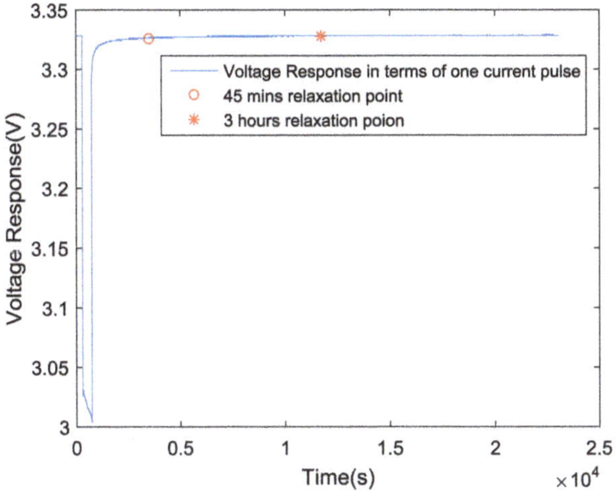

Figure 7. Relaxation process example of battery pulse discharge test (PDT).

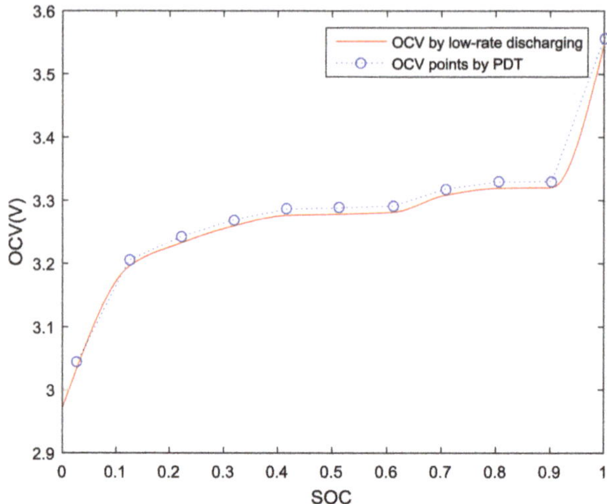

Figure 8. Comparison of C/50 low-rate discharge profile and the 10% state-of-charge (SOC) step incremental open-circuit voltage (OCV) curve (25 °C).

Figure 9 shows a full voltage curve sample of PDT with 45 min relaxation at 25 °C. OCV approximates the terminal voltage of the battery at equilibrium state of every relaxation period. The OCV-SOC relationship curves under different temperatures are shown in Figure 10. As observed in Figure 10a, there is a higher OCV for the SOC value from 0.1 to 0.9. Also reflected in the close view as shown in Figure 10b, is a different OCV between various temperatures. When SOC is 0.2, the maximum OCV is approximately 25 mV with SOC error of 10% which is between 5 °C and 45 °C. Therefore, the OCV cannot be represented by simply a curve fitting method (that is commonly adopted in the literature) to improve the accuracy of the OCV-SOC curve. From the Figure 8, 11 OCV data points can be gained for a full discharge period, which is 0~1 with the 10% SOC intervals. However, they might be insufficient to reflect all electrode features due to the low resolution. Interpolation is a common method to yield additional data. Here, we applied interpolation method for better resolution and as a result of reducing measurement time. Therefore, a lookup table with interpolation techniques is applied to obtain the real-time OCV under different temperatures. The lookup table is created and stored in the embedded microcontroller. The curves of the OCV are illustrated in Figure 11. With various SOCs and temperatures, the corresponding OCV can be obtained from the lookup table.

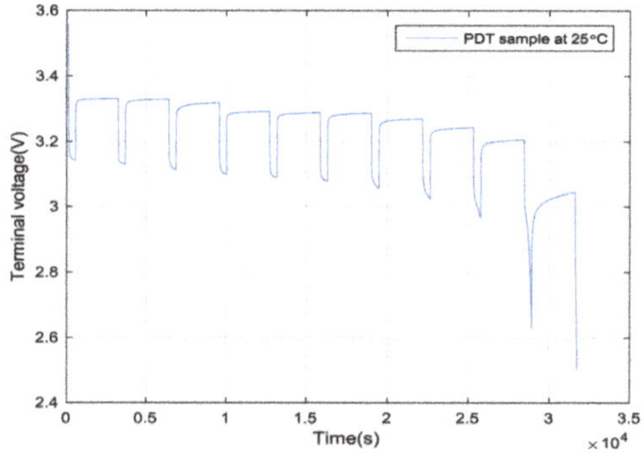

Figure 9. Terminal voltage of pulse discharges test.

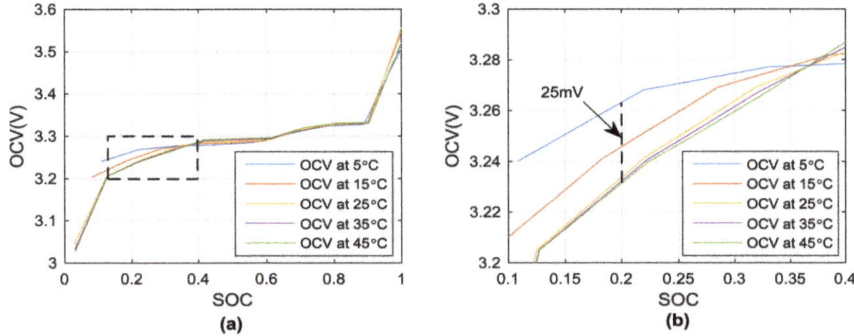

Figure 10. (**a**) OCV-SOC relationship curves under different temperatures; (**b**) detailed view from SOC 0.1 to 0.4.

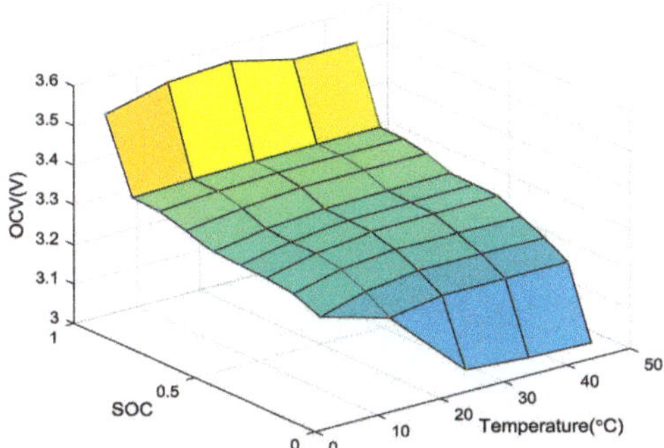

Figure 11. OCV-SOC value curves.

In this Section, R_1, C_1 and R_0 are identified from the results of PDT. Figure 12 is the relaxation cycle used to identify these parameters. The DC internal resistance R_0 is calculated from the instantaneous rise of voltage using the following equation.

$$R_{0,n} = \frac{U_n - U_{n-1}}{I} \tag{12}$$

where U_n is the terminal voltage of sample n, and I is the discharge current.

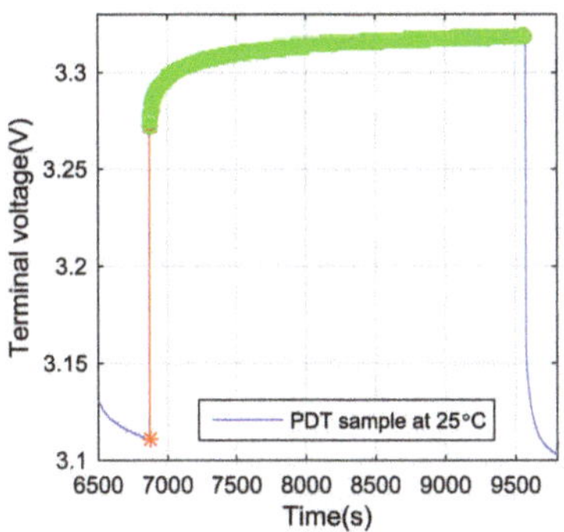

Figure 12. Relaxation period after the discharge pulses.

The abovementioned R_0 is identified using the segment marked in red color shown in Figure 12. R_1 and C_1 represent the transient response of the battery voltage during the relaxation period. The identification process starts in the segment marked in green color as shown in Figure 12. The values of each parameter of the RC networks can be identified. The identified values will be tabulated in the 2-D lookup tables as shown in Figures 13–15. The experiments were conducted at the following temperatures 5 °C, 15 °C, 25 °C and 45 °C in order to include the influence of the ambient temperature to parameters R_0, R_1 and C_1. In this paper, a simplified lookup table with interpolation technique is applied to obtain the real-time OCV, R_1 and C_1, which is a highly efficient method for the microcontroller in the embedded applications. Compared with other models with a complex identified process such as adaptive least square (ALS) and extended Kalman filter (EKF), this method reduces the burden on the processor greatly without large deficiency in performance. Table 3 shows the comparison result of R_0 identification, which obtains the R_0 value by the ALS method, EKF method and lookup table method implemented in the MATLAB environment, respectively. As shown in Table 3, the lookup table method can save much identification time without big differences in RMS.

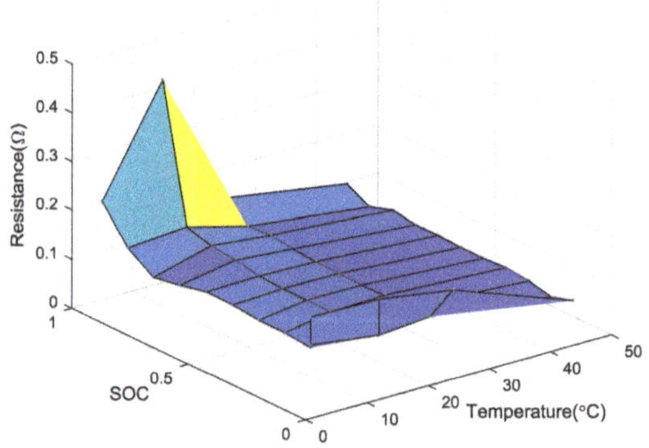

Figure 13. R_0 value curves.

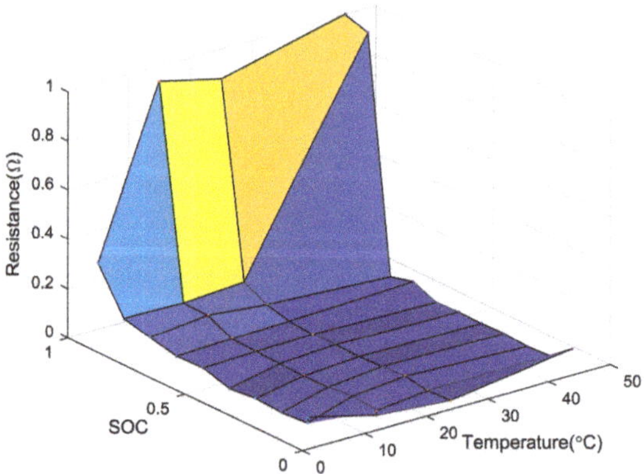

Figure 14. R_1 value curves.

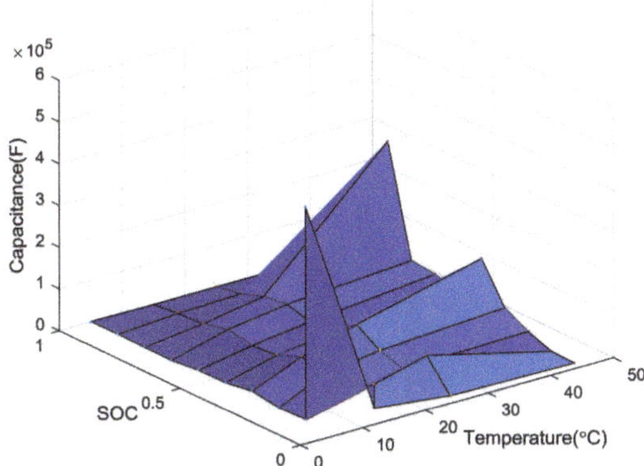

Figure 15. C_1 value curves.

Table 3. Comparison results between adaptive least square (ALS), extended Kalman filter (EKF) and lookup table methods.

R_0	ALS Method	EKF Method	Lookup Table
RMS	0.0055	0.0042	0.0058
Computation time	1.35 s	1.25 s	0.021 s

4.2. Temperature-Dependent Battery Cell Parameters Validation

The model output terminal voltage was compared with the measured terminal voltage at the similar current loads to validate the equivalent circuit battery model. As shown in Figure 16, a cell battery model is implemented in the MATLAB/Simulink platform using the model identified in Section 4.1. The comparisons of the battery terminal voltages for both the experimental data and the simulation outputs under different temperatures such as 5 °C, 15 °C, 25 °C and 45 °C are shown in Figures 17–20, respectively. As seen in the individual error plots across the simulation time, it is evident that the model outputs follow the experimental data closely with a small error throughout the simulation time.

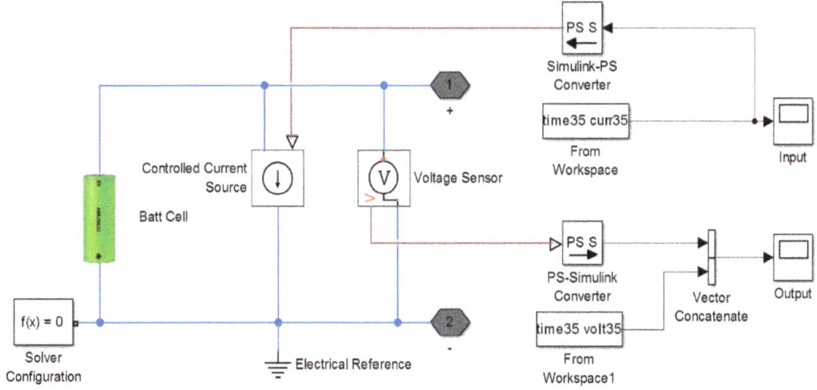

Figure 16. Battery cell model for validation.

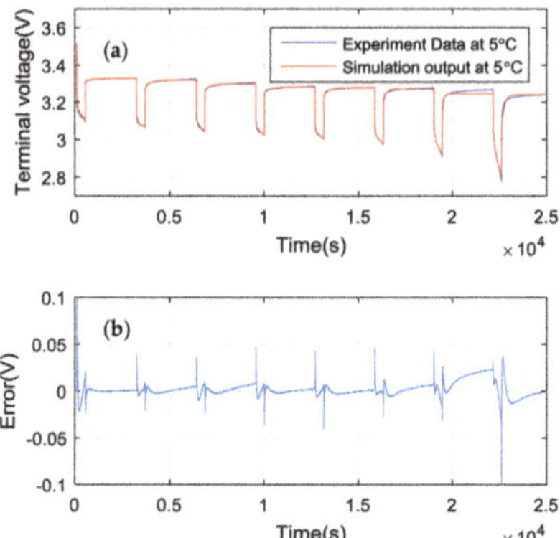

Figure 17. (**a**) Model validation at 5 °C; (**b**) Error between model output and experimental data.

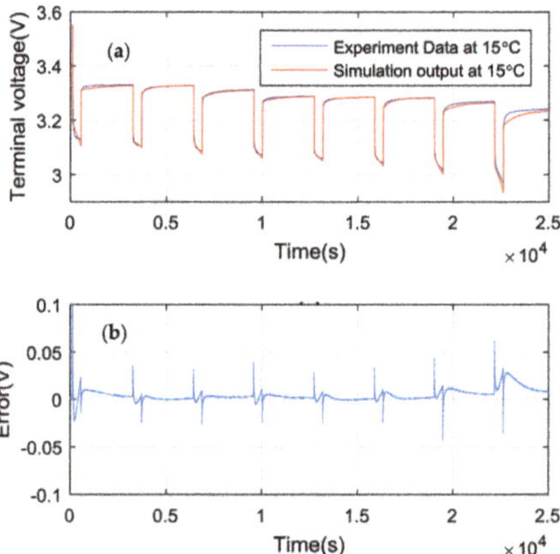

Figure 18. (**a**) Model validation at 15 °C; (**b**) Error between model output and experimental data.

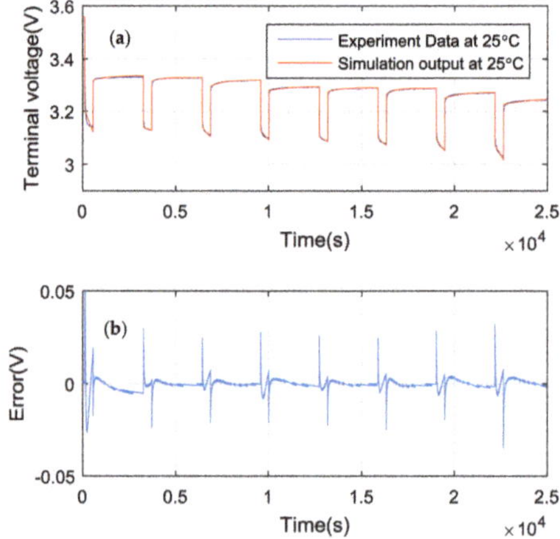

Figure 19. (**a**) Model validation at 25 °C; (**b**) Error between model output and experimental data.

For comparison purposes, the root means square errors of the terminal voltage between the simulation and experimental results shown in Figures 17–20 at different temperatures are tabulated in Table 4. The comparison between the simulation and experiment at various temperatures shows a maximum RMS error of 7.2078 $\times$ 10^{-5}. It shows the battery cell indeed operating quite poorly at a lower temperature (a common characteristic of a battery cell). From the figures, it is evident that the terminal voltage errors due to the suddenly changed current can be converged to around 0 quickly (e.g., within 1.2 $\times$ 10^{-5} s); this means the model has a certain degree of robustness, which is relevant to the further study of the advanced algorithms. To test the robustness of the model under different ambient temperatures, a set of experimental data under 35 °C (not used for the parameter identification) was compared with the simulation model. The results in Figure 21 show that the RMS error of the terminal voltage between the simulation and experiment is approximately 1.5609 $\times$ 10^{-5}. It indicates that the temperature-dependent battery model output can estimate the terminal voltage at a different ambient temperature with an acceptable error for the embedded applications.

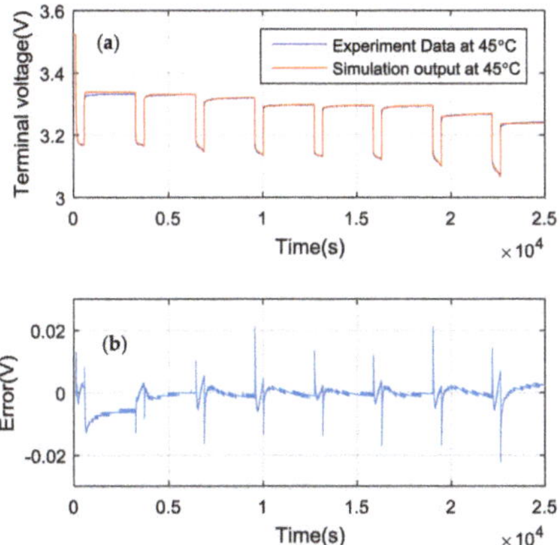

Figure 20. (**a**) Model validation at 45 °C; (**b**) Error between model output and experimental data.

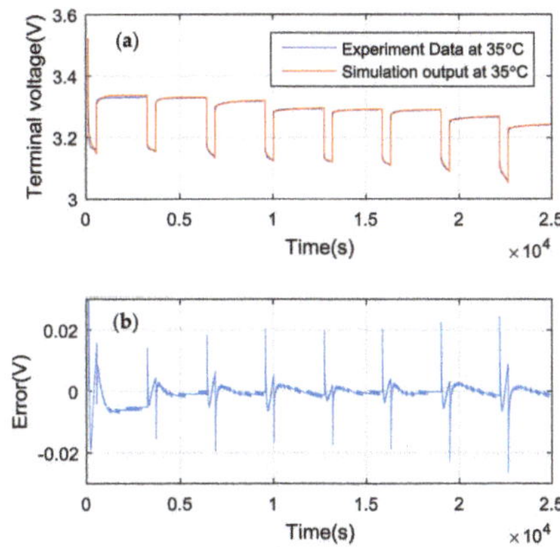

Figure 21. (**a**) Model validation at 35 °C; (**b**) Error between model output and experimental data.

Table 4. Root mean square (RMS) error of terminal voltage between simulation and experimental results.

Temperature (°C)	5	15	25	45
RMS error	6.8736×10^{-5}	7.2078×10^{-5}	2.2671×10^{-5}	9.5907×10^{-6}

4.3. Temperature-Dependent 12-Cell Battery Model with Convective Heat Transfer Simulation

With the cell battery model validated, a 12-cell battery pack is used for SOC estimation and subsequent cell balancing. Figure 22 shows the power and thermal connection of the battery pack model in the series. The link of the battery cells is physically arranged serially as illustrated in Figure 23. The SOC estimation of the battery pack is different from a single battery cell. Figure 24 is the top level simulation environment for SOC estimation and cell balancing for the battery pack. The SOC estimation block computes the SOC of the 12-cell using the data collected from the experiment. Some researchers considered the whole pack as a single cell without taking into account the differences between the cells. However, the non-uniformity of the cells in the battery pack cannot be neglected for the embedded BMS development. Hence, a different SOC value of each cell is required to be estimated. In this case, the Ah method as shown in Figure 25 is used.

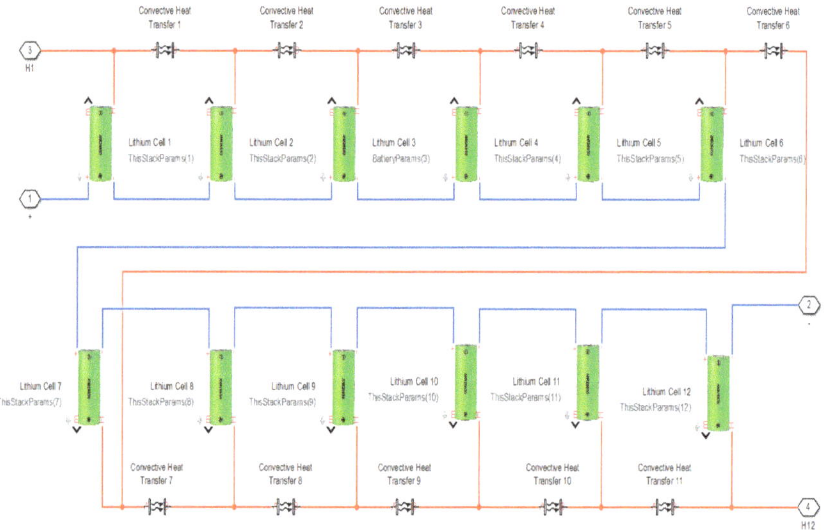

Figure 22. Battery cells' power connection and convective heat transfer in a battery pack model.

Figure 23. Current battery pack's physical connection (**Left**); and proposed final compact enclosure design (**Right**).

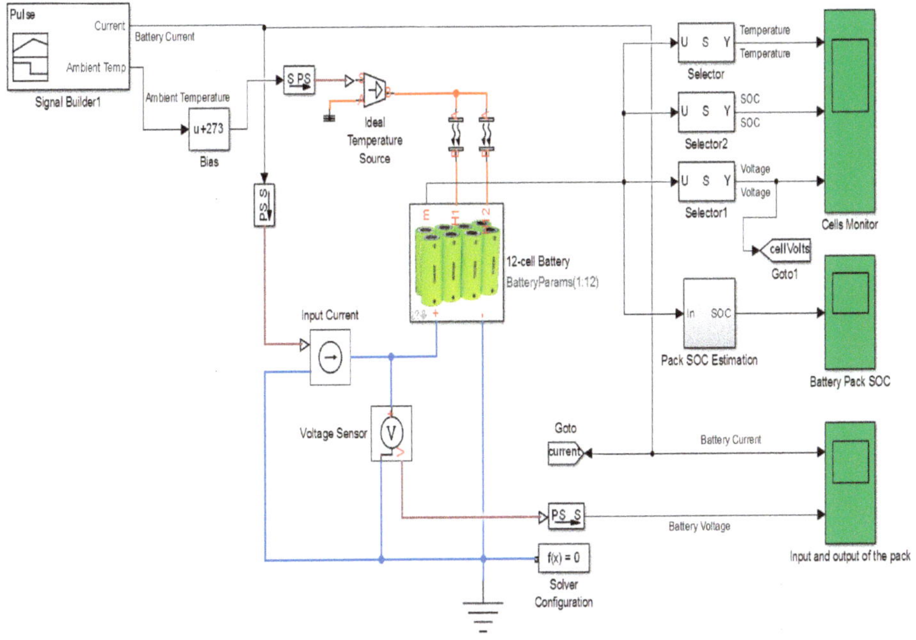

Figure 24. Battery pack validation set up in Simulink.

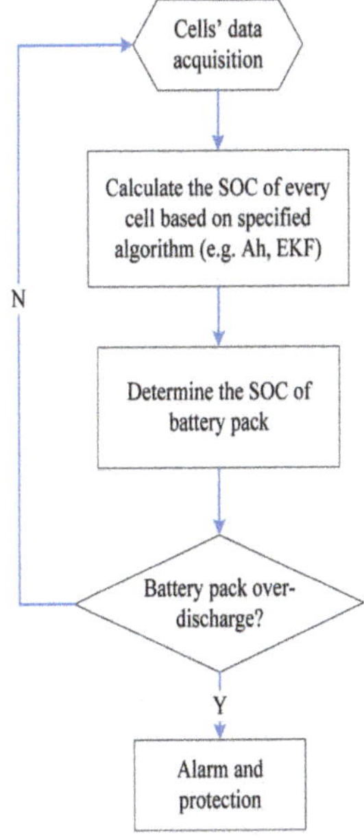

Figure 25. Flow chart for SOC estimation.

The SOC estimation algorithm is shown in Figure 26. The SOC computation is grouped into two main parts. First, the SOC of each cell is determined. The Pack SOC #1 and Pack SOC #2 determine the minimum SOC and the average SOC, respectively. The battery cell is checked for any over-charging before it sends out an alarm or warning signal. The SOC estimation algorithm also gives additional SOC information for the subsequent cell balancing algorithm to prioritize the cell that needs to balance first. In this case, it should be the cell with the lowest SOC.

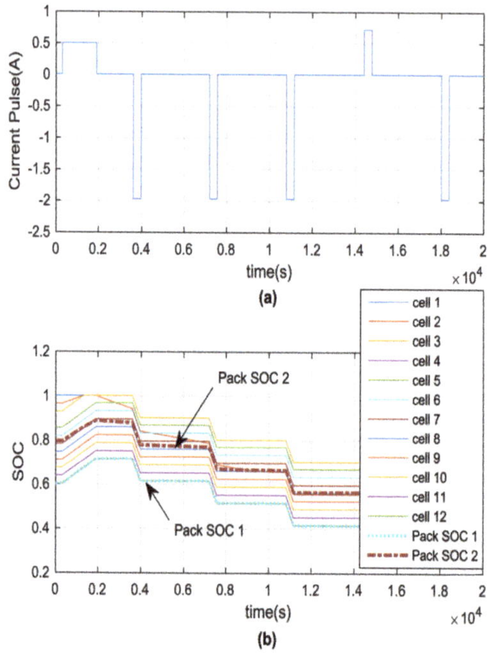

Figure 26. (**a**) Current pulse; (**b**) simulation battery pack SOC result for battery pack SOC.

Since the SOC in each cell can be different, the cells need to be checked for over-charging and balanced to operate for a longer endurance in embedded applications. Hence, the cell balancing becomes an indispensable feature for real embedded BMS as it affects the lifespan and eventual safety of the battery power system. Different types of model-based cell balancing algorithms can be developed and validated in the MATLAB/Simulink environment using the battery pack model. For clarity, only the battery cells #1 to #4 of the battery pack are used for comparison. Cells #1 and #2 are employed for the passive balancing due to its simplicity and reliable performance. Cells #3 and #4 are not used for any balancing function as shown in Figure 27. Instead, it is used to compare the cell balancing results with cells #1 and #2 (that are not balanced). The balancing scheme is solely based on average voltage. If the cell is not equal to the mean voltage, the cell balancing will begin to increase the SOC. To do that, the SOC in each cell provides an input variable to the balancing decision block (named "cell balancing block"). The initial SOC values of cells #1 to #4 are pre-set to 100%, 96%, 92% and 81%, respectively, to show different initial SOV values. The simulation result (without the cell balancing) is provided in Figure 28. The results indicate that battery cells with different initial states will lead to different terminal voltages, SOC distributions, and thermal behaviors. As shown in Figure 29, cells #1 and #2 after cell balancing can maintain the SOC across each cell, and the SOC across all the cells have improved by approximately 60% as compared to the one without the cell balancing.

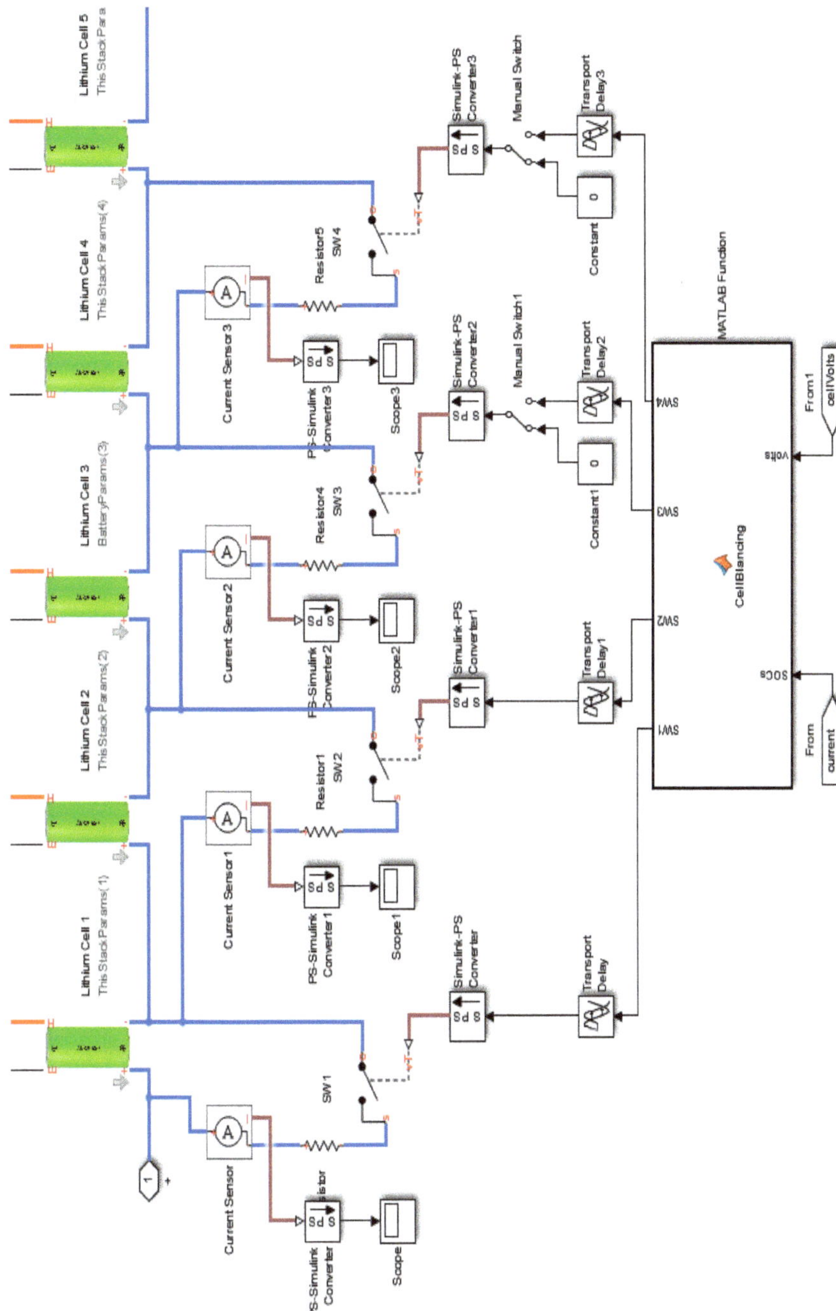

Figure 27. Cell balancing algorithm development in Simulink (for clarity only show 5-cell).

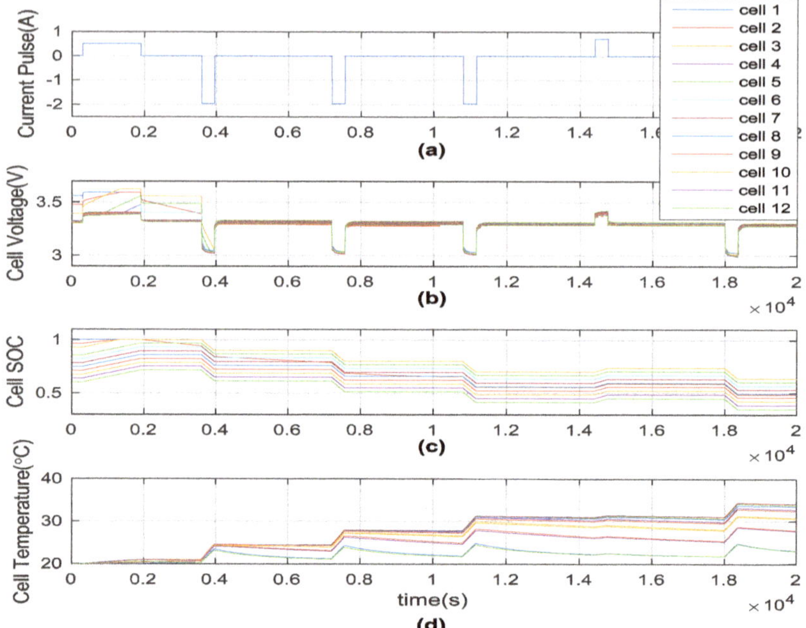

Figure 28. Battery pack simulation result without cell balancing: (**a**) current pulse (active when charging and negative when discharging); (**b**) voltage response of each cell; (**c**) SOC of each cell; (**d**) temperature of each cell.

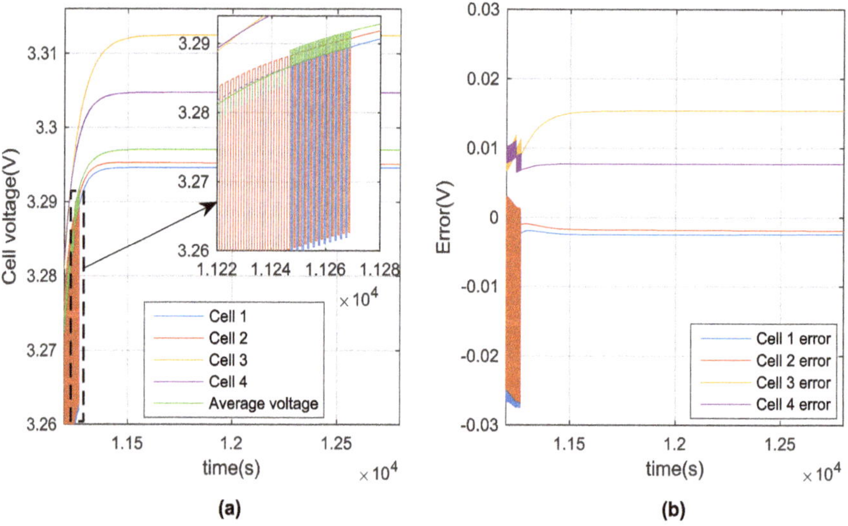

Figure 29. Simulation result with cell balancing: (**a**) comparison of cell #1 to #4; (**b**) error between cell voltage and average voltage of cell #1 to #4 after balancing.

In summary, the proposed battery pack model can estimate the SOC of each cell and temperature between the cells. The passive cell balancing scheme was applied on the temperature-dependent battery model. Although the active balancing system has attracted more attention as of late, it is quite costly, possesses a sophisticated control structure and requires higher power consumption than passive cell balancing.

5. Conclusions

In this paper, a simple but effective battery model was proposed, which was suitable to be implemented in a microcontroller with limited resources for the embedded applications. Simplified lookup table with interpolation technique is applied to obtain the real-time open-circuit voltage (OCV), R_1 and C_1, which is a highly efficient method for the microcontroller implementation in the embedded applications. Furthermore, based on the proposed cell model, a 12-cell series connected battery pack is systematically modeled, simulated and validated by actual experimental results. This paper mainly focused on the 12-cell $LiFePO_4$ battery pack for a more realistic simulation instead of a single battery cell. As a trade-off between the high fidelity and computation effort, the conductive thermal transfer is neglected in this paper. Instead of using the temperature as an external disturbance acting on the battery power system, the thermal influence due to convective heat transfer of each cell was included as parameters to couple both the equivalent circuit model (ECM) and the thermal model. Also, the temperature-dependent battery model was included to estimate the SOC that was balanced by an automatic cell balancing scheme. As compared with the experimental results, there exists a minimal root mean square error of the terminal voltage at a different ambient temperature (from 5 °C to 45 °C). The proposed simulation model allows SOC and temperature estimation of the battery cells for the embedded implementation. It can be used to develop and validate any advanced algorithms using the proposed battery cell/pack model.

For future works, the high current rate and effects of aging will be included. More experimental works will be conducted. The fault diagnosis approach will be performed on the final battery model. The mechanical enclosure will be used to hold the battery pack.

Acknowledgments: This work was supported by Singapore Maritime Institute (SMI) through the project (SMI-2013-MA-05). The authors would like to thank Newcastle University, Mahesh Menon and John McCann from Soil Machine Dynamics (SMD) Ltd and Cham Yew Thean from Temasek Polytechnic for providing research supports.

Author Contributions: Zuchang Gao is credited with the majority of the theoretical formulation, simulation and experimental works performed in this paper for his Ph.D. study and project involvement under SMI-2013-MA-05. Cheng Siong Chin who is the main supervisor defined the flow of the paper and participated in the battery power system simulation and verification that drove this research with Wai Lok Woo and Junbo Jia to test its relevance for industrial applications. Wai Lok Woo and Junbo Jia provided the co-supervision and academic support during the project. All authors discussed and provided comments on the results at all stages to the final proofreading.

Conflicts of Interest: The authors declare no conflict of interest.

References

1. Chang, M.-H.; Huang, H.-P.; Chang, S.-W. A new state of charge estimation method for $LiFePO_4$ battery packs used in robots. *Energies* **2013**, *6*, 2007–2030. [CrossRef]
2. Alhanouti, M.; Gießler, M.; Blank, T.; Gauterin, F. New electro-thermal battery pack model of an electric vehicle. *Energies* **2016**, *9*. [CrossRef]
3. Zhang, C.; Li, K.; Pei, L.; Zhu, C. An integrated approach for real-time model-based state-of-charge estimation of lithium-ion batteries. *J. Power Sources* **2015**, *283*, 24–36. [CrossRef]
4. Awadallah, M.A.; Venkatesh, B. Accuracy improvement of Soc estimation in lithium-ion batteries. *J. Energy Storage* **2016**, *6*, 95–104. [CrossRef]
5. Hansen, T.; Wang, C.-J. Support vector based battery state of charge estimator. *J. Power Sources* **2005**, *141*, 351–358. [CrossRef]
6. Chaoui, H.; Golbon, N.; Hmouz, I.; Souissi, R.; Tahar, S. Lyapunov-based adaptive state of charge and state of health estimation for lithium-ion batteries. *IEEE Trans. Ind. Electron.* **2015**, *62*, 1610–1618. [CrossRef]

7. Gandolfo, D.; Brandão, A.; Patiño, D.; Molina, M. Dynamic model of lithium polymer battery—Load resistor method for electric parameters identification. *J. Energy Inst.* **2015**, *88*, 470–479. [CrossRef]
8. Lee, J.L.; Chemistruck, A.; Plett, G.L. Discrete-time realization of transcendental impedance models, with application to modeling spherical solid diffusion. *J. Power Sources* **2012**, *206*, 367–377. [CrossRef]
9. Coleman, M.; Lee, C.K.; Zhu, C.; Hurley, W.G. State-of-charge determination from emf voltage estimation: Using impedance, terminal voltage, and current for lead-acid and lithium-ion batteries. *IEEE Trans. Ind. Electron.* **2007**, *54*, 2550–2557. [CrossRef]
10. Hussein, A.A. Capacity fade estimation in electric vehicle Li-ion batteries using artificial neural networks. *IEEE Trans. Ind. Appl.* **2015**, *51*, 2321–2330. [CrossRef]
11. Li, I.H.; Wang, W.Y.; Su, S.F.; Lee, Y.S. A merged fuzzy neural network and its applications in battery state-of-charge estimation. *IEEE Trans. Energy Convers.* **2007**, *22*, 697–708. [CrossRef]
12. Charkhgard, M.; Farrokhi, M. State-of-charge estimation for lithium-ion batteries using neural networks and ekf. *IEEE Trans. Ind. Electron.* **2010**, *57*, 4178–4187. [CrossRef]
13. Gao, Z.; Chin, C.S.; Woo, W.L.; Jia, J.; Toh, W.D. Lithium-ion battery modeling and validation for smart power system. In Proceedings of the 2015 International Conference on Computer, Communications, and Control Technology (I4CT), Kuching, Malaysia, 21–23 April 2015; pp. 269–274.
14. Liu, S.; Jiang, J.; Shi, W.; Ma, Z.; Wang, L.Y.; Guo, H. Butler–volmer-equation-based electrical model for high-power lithium titanate batteries used in electric vehicles. *IEEE Trans. Ind. Electron.* **2015**, *62*, 7557–7568. [CrossRef]
15. Melin, P.; Castillo, O. Intelligent control of complex electrochemical systems with a neuro-fuzzy-genetic approach. *IEEE Trans. Ind. Electron.* **2001**, *48*, 951–955. [CrossRef]
16. Lin, H.T.; Liang, T.J.; Chen, S.M. Estimation of battery state of health using the probabilistic neural network. *IEEE Trans. Ind. Inform.* **2013**, *9*, 679–685. [CrossRef]
17. Wang, S.C.; Liu, Y.H. A pso-based fuzzy-controlled searching for the optimal charge pattern of Li-ion batteries. *IEEE Trans. Ind. Electron.* **2015**, *62*, 2983–2993. [CrossRef]
18. Kim, J.; Cho, B.H. State-of-charge estimation and state-of-health prediction of a Li-ion degraded battery based on an ekf combined with a per-unit system. *IEEE Trans. Veh. Technol.* **2011**, *60*, 4249–4260. [CrossRef]
19. He, W.; Williard, N.; Chen, C.; Pecht, M. State of charge estimation for Li-ion batteries using neural network modeling and unscented Kalman filter-based error cancellation. *Int. J. Electr. Power Energy Syst.* **2014**, *62*, 783–791. [CrossRef]
20. Partovibakhsh, M.; Liu, G. An adaptive unscented Kalman filtering approach for online estimation of model parameters and state-of-charge of lithium-ion batteries for autonomous mobile robots. *IEEE Trans. Control Syst. Technol.* **2015**, *23*, 357–363. [CrossRef]

Article

Optimal Power Management Strategy for Energy Storage with Stochastic Loads

Stefano Pietrosanti [1,*], William Holderbaum [1] and Victor M. Becerra [2]

[1] School of Systems Engineering, University of Reading, Whiteknights, Reading RG6 6AY, UK; w.holderbaum@reading.ac.uk

[2] School of Engineering, University of Portsmouth, Anglesea Road, Portsmouth PO1 3DJ, UK; victor.becerra@port.ac.uk

* Correspondence: s.pietrosanti@pgr.reading.ac.uk; Tel.: +44-0118-378-6086

Academic Editor: K. T. Chau
Received: 29 January 2016; Accepted: 3 March 2016; Published: 9 March 2016

Abstract: In this paper, a power management strategy (PMS) has been developed for the control of energy storage in a system subjected to loads of random duration. The PMS minimises the costs associated with the energy consumption of specific systems powered by a primary energy source and equipped with energy storage, under the assumption that the statistical distribution of load durations is known. By including the variability of the load in the cost function, it was possible to define the optimality criteria for the power flow of the storage. Numerical calculations have been performed obtaining the control strategies associated with the global minimum in energy costs, for a wide range of initial conditions of the system. The results of the calculations have been tested on a MATLAB/Simulink model of a rubber tyre gantry (RTG) crane equipped with a flywheel energy storage system (FESS) and subjected to a test cycle, which corresponds to the real operation of a crane in the Port of Felixstowe. The results of the model show increased energy savings and reduced peak power demand with respect to existing control strategies, indicating considerable potential savings for port operators in terms of energy and maintenance costs.

Keywords: energy storage; power management; optimization; stochastic loads; flywheel; RTG crane

1. Introduction

Energy storage is beneficial in situations where power production is intermittent or the load varies in intensity, as the objective of the storage is to mitigate variability in generation and demand by acting as a buffer. Economical feasibility limits the size and power rating of the storage system, resulting in the need to optimise the power flow in order to maximise the efficacy of the limited available resources. In the case of a known load profile, it is possible to define a strategy that results in the optimal solution for a given storage system, placing the focus on finding the most suited technology for the single application. Slow processes get the most benefit from using batteries or other forms of high-capacity storage systems (compressed air, pumped hydro, *etc.*), while fast loads characterised by short and high power demand are paired with flywheels, supercapacitors or other technologies capable of reacting in a very short time, outputting relatively high power [1–3]. Usually, the variability of the load also depends on the time constant of the application: the power demand in regional power network fluctuations is periodical with peaks occurring at around the same time of the day and of the year, leading to optimal solutions for power management that account for the deviation from the typical daily or yearly profile [4–8]. When loads are limited to a short period of time (tens of seconds or less), the variability tends to be defined by three main factors: when the demand occurs, its intensity and duration. As an example, electric vehicles show this sort of unpredictability, as there is no prior knowledge of the acceleration that the driver requires when driving, and therefore, the stochastic behaviour of the demand is taken into account when developing control strategies [9–11].

The focus of this paper is to develop an optimal control strategy to be applied to energy storage in rubber tyre gantry (RTG) cranes, which are found in container ports and whose task is to stack containers in the yard area. Most of the energy consumption comes from hoisting containers (weighing up to 54 tons) for a typical full height of 15 m at vertical speeds that can reach 0.85 m/s, resulting in short intense loads of limited durations. In order to reduce the stress on the primary source, which can be either a diesel generator or the port's electrical network, energy storage can be used for peak shaving during the lifting phase and to recover potential energy during the lowering phase [12–14]. Although other authors proposed RTG cranes equipped with batteries [13], these are best suited for reducing idle power consumption due to their low power density. Flywheels and supercapacitors are more suited for the high power flows deriving from the hoist motor (both when motoring and generating), and they have been the subject of multiple studies [12,14–17]. Flywheel energy storage systems (FESSs) in particular have been found particularly suited for this task, as they show similar performance to supercapacitors, while being characterised by excellent ageing characteristics, which are independent of the charge rate or depth of discharge [1,18], allowing their lifetime to match that of the portal frame. Their disadvantage is high standing losses, which are particularly evident in more resilient designs (such as the use of normal ball bearings instead of magnetic bearings), as are the ones used on cranes; however, this does not affect the use for short power loads, and it only requires the storage to be charged shortly before use. Energy storage then can be beneficial for the reduction of energy demand and also peak power demand; for this reason, it is critical to develop a power management strategy (PMS) that takes full advantage of the storage capacity, while keeping its cost at the minimum. An effective control strategy focused on storage in RTG cranes could have major benefits globally, as they are present in all major container ports and are a key element in the export and import processes. Their activity is energy and power intense; nonetheless, it involves spending a large fraction of time idling before the next lift cycle. In a typical cycle, a container lift is preceded by a lowering of the headblock mechanism (which securely locks to the container) and a locking sequence, giving a forewarning of the incoming lifting cycle. The weight of the container that will need to be lifted is known, as cranes need to measure the weight for safety reasons, giving a good estimate of the power demand of the following lift. The only remaining variability is the duration of the load; it is usually unknown (apart from a few ports with advanced terminal operating systems) with the duration being proportional to the height that the container needs to reach in a specific lift cycle, and it depends on the configuration of the stack in the precise location.

This paper introduces a novel PMS, which optimises the use of storage under uncertainties on the duration of power loads, unlike previous works that assume full knowledge of the load profile [11,13,19,20]. The aforementioned PMS is tailored for load characterised by known intensity and random duration, representing the power consumption caused by the lifting of containers; nonetheless, it could also be applied to any hybrid system characterised by loads of unknown duration. The paper is organised as follows: Section 2 describes the topology of the power system. Section 3 presents the optimal control problem and the proposed PMS. The RTG model and the simulation results are presented and discussed in Section 4, with Section 5 presenting a summary of the work.

2. System Topology

The RTG crane under analysis, shown in Figure 1, is manufactured by Shanghai Zhenhua Heavy Industries (ZPMC), and it is currently used at the Port of Felixstowe. It is equipped with a diesel generator, but it is has also been retrofitted with a connection to the terminal power network through a conductor bar running along the stack, allowing it to be powered by the electric grid without using the diesel generator. The simplified diagram in Figure 2 shows the primary energy source (either generator or grid) connected to a diode rectifier, which powers a DC bus. The main motors (hoist, gantry, trolley, *etc.*) are all connected to the shared DC bus, as well as the brake resistors, which engage automatically when the DC bus voltage raises above 750 V. The energy storage can be connected to the bus, drawing current when the voltage raises above a threshold (regeneration phase) and supplying power according to a PMS. The energy storage system consists of a motor drive, a switched reluctance (SR) motor and a flywheel coupled to its axis. The primary energy source supplies all of the power that is required for the DC bus, which is the power consumption of the motors $p_L(t)$ minus the power supplied by the storage $p_s(t)$. As in [19,21], the power system can be represented as shown in Figure 3, where the load, the generated power and the power from the storage are respectively $p_L(t)$, $p_g(t)$ and $p_s(t)$, resulting in the following equation:

$$p_g(t) = p_L(t) - p_s(t),\ t \in \mathbb{R}^+ \tag{1}$$

Figure 1. A rubber tyre gantry crane.

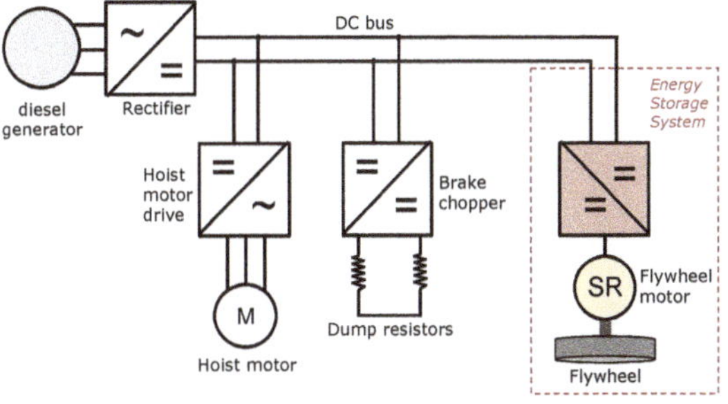

Figure 2. Simplified diagram of the main electric components of an RTG crane.

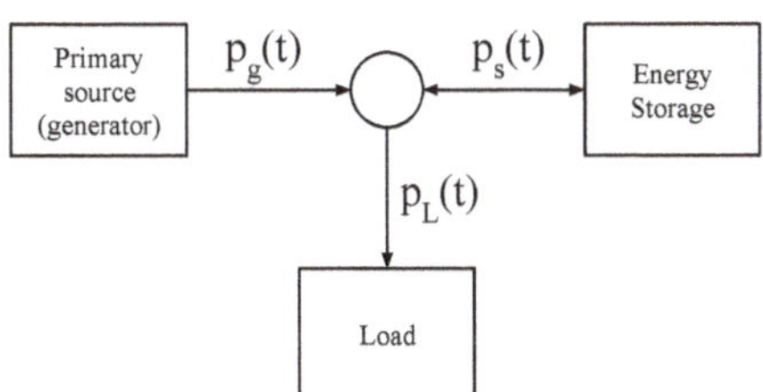

Figure 3. Topology of the power system.

2.1. The Primary Source

The main power supply provides all of the power demanded by the crane motors, and it can be represented as an infinite source of power and energy, as both the diesel generator and the power network supply are rated above the maximum possible load of the crane. It is unidirectional, as power cannot be converted back to fuel or regenerated into the grid (due to the absence of an active front-end) and has a cost associated with every unit of energy delivered. This cost is represented by the positive definite function $D(p_g(t))$, and it encompasses fuel consumption, efficiency and aggregated fixed costs (higher power consumptions have a relatively higher cost due to the need for larger generators). The objective of the proposed PMS is to use the stored energy to minimise the total cost of the energy production for the duration of a lift, which is:

$$D_{tot} = \int_0^T D(p_g(t))\, dt \tag{2}$$

with T indicating the maximum possible duration of the lift cycle.

2.2. The Load

During a lift, the power required by the hoist motor is assumed known and constant with value P_L. Nonetheless, the value of $p_L(t)$ is not known for every instant, as the duration of the lift is unknown. The load profile can then be simplified as follows:

$$p_L(t) = \begin{cases} P_L, & \text{if } 0 \leq t \leq t_f \\ 0, & \text{otherwise} \end{cases} \tag{3}$$

where t_f represents the final time of the lift and is a random variable modelled by a distribution, which is assumed known. It would be possible to calculate the optimal $p_s(t)$ assuming a deterministic and well-known load profile [19,20], but in the case discussed in this paper, the load behaviour is stochastic, creating a new challenge.

2.3. The Energy Storage

An FESS stores recovered energy obtained from lowering containers in the form of a rotating mass, whose angular speed gives an exact value of the stored energy $W_s(t)$:

$$W_s(t) = \frac{1}{2} J \omega^2(t) \tag{4}$$

where J is the flywheel inertia and ω is the angular speed. To limit torque output and wear, the flywheel will have an actual range of speeds limited by a lower limit ω_{min} and an upper limit ω_{max}, so the usable energy at time t will be the following:

$$W_s(t) = \frac{1}{2} J (\omega^2(t) - \omega^2_{min}) \tag{5}$$

and it will also be bounded:

$$0 \leq W_s(t) \leq W_{max} \tag{6}$$

with W_{max} the energy corresponding to the maximum speed ω_{max}.

The flywheel is powered by an electric motor, which is assumed equipped with a control system that is able to follow with negligible delay the instantaneous power command issued by the PMS. Given the high standing losses of a flywheel storage system, it is necessary to model a system that loses energy over time with the following approximation:

$$\dot{W}_s(t) = -\eta_1 W_s(t) - \eta_2 - p_s(t) \tag{7}$$

where η_1 is a constant value that links losses to the stored energy (friction and windage losses), η_2 is a constant power loss, which is independent of the quantity of stored energy (e.g., power supply, cooling), and finally, $p_s(t)$ is the power exchanged with the system. The values for η_1 and η_2 can be measured or estimated from the characteristics of the storage.

3. Optimal Power Management Strategy

The objective is to minimise the total cost D_{tot} expressed in Equation (2), which is associated with the generated energy. The optimal controller should then be designed to find the storage output that minimises Equation (2), which, given Equation (1), is equal to:

$$D_{tot} = \int_0^T D(p_L(t) - p_s(t))\,\mathrm{d}t \tag{8}$$

The cost function for generated power is assumed to result in no cost when no power is demanded (*i.e.*, $D(0) = 0$). Therefore, knowing from Equation (3) that $p_L(t) = 0$ when $t > t_f$, we can reduce the limits of the integral in Equation (8), as the integrand will be zero if we choose the trivial solution $p_s(t) = 0\ \forall t > t_f$ (corresponding to no storage output when there is no load):

$$D_{tot} = \int_0^{t_f} D(p_L(t) - p_s(t))\,\mathrm{d}t \tag{9}$$

The total cost value in Equation (9) is calculated for a specific t_f, which, in reality, is a random value whose distribution L is known. To account for the stochastic behaviour, it is then necessary to calculate the expected value of the total cost, by considering a single value of t_f weighted by the probability of its occurrence. By defining $f_L(t_f)$ as the probability that a certain t_f occurs, the expected value of the cost D_E is the following:

$$D_E = \int_0^T f_L(t_f)\left(\int_0^{t_f} D(p_L(t) - p_s(t))\,\mathrm{d}t\right)\mathrm{d}t_f \tag{10}$$

where T is the maximum possible value of t_f. We can assume that $f_L(t_f) = 0\ \forall t < 0$, as the container lifts have positive duration, and also $f_L(t_f) = 0\,\forall t > T,\ T < \infty$, as the duration is finite. Equation (10) spans the whole range of possible values of t_f and calculates the cost of applying a certain control strategy $p_s(t)$ in all possible scenarios, weighting the cost with the probability of that scenario to occur. An optimal control strategy $p_s^*(t)$ is then one that satisfies:

$$p_s^*(t) = \arg\min_{p_s(t)} \int_0^T f_L(t_f)\left(\int_0^{t_f} D(p_L(t) - p_s(t))\,\mathrm{d}t\right)\mathrm{d}t_f \tag{11}$$

By defining $F_L(t)$, the cumulative distribution function (CDF) [22] of the probability density function $f_L(t_f)$, we have:

$$F_L(t) = \int_{-\infty}^{t} f_L(t_f)\,\mathrm{d}t_f = \int_{-\infty}^{0} f_L(t_f)\,\mathrm{d}t_f + \int_0^t f_L(t_f)\,\mathrm{d}t_f = \int_0^t f_L(t_f)\,\mathrm{d}t_f \tag{12}$$

and $F_L(t)$ has the property that the final value is one:

$$\lim_{t\to\infty} F_L(t) = F_L(T) = \int_0^T f_L(t_f)\,\mathrm{d}t_f = 1 \tag{13}$$

The integrand $D(p_L(t) - p_s(t))$ represents finite quantities, and it can be assumed that:

$$\int_0^\infty |D(p_L(t) - p_s(t))|\,\mathrm{d}t < \infty \tag{14}$$

as the load energy is limited (from Equation (3)), as well as the stored energy. When taking into account Equations (12) and (14), the expression in Equation (11) is equivalent to:

$$p_s^*(t) = \arg\min_{p_s(t)} \int_0^T (1 - F_L(t))D(p_L(t) - p_s(t))\,\mathrm{d}t \tag{15}$$

which is a more manageable form of minimisation, as it involves a single integrand when the CDF is known. The proof of Equation (15) is located in Appendix A.

3.1. Constraints

The problem in Equation (15) has the trivial solution of setting $p_s(t) = p_L(t)$ if the available stored energy is infinite. In the real world, the energy capacity will be limited, and this is reflected by adding the constraint Equation (6), which will have the practical effect of limiting the amount of energy that can be provided by the storage, creating a limited fuel problem:

$$\int_0^T p_s(t) \le W_s(0) \tag{16}$$

Furthermore, the dynamics of the storage system with its losses need to be taken into account; the expression in (7) dictates how the system loses energy over time; hence, Equation (16) is too approximative and overestimates the available energy, so it needs to be replaced by Equation (17).

$$\int_0^T (\eta_1 W_s(t) + \eta_2 + p_s(t))\,\mathrm{d}t \le W_s(0) \tag{17}$$

The last constraint is the power rating of the storage, which cannot exceed the maximum rated value, and it is assumed to be the same, in absolute value, when motoring and generating:

$$-P_s \le p_s(t) \le P_s \tag{18}$$

This type of optimal control problem has not yet been solved analytically [23–25], but it can be solved numerically by dynamic programming, accurately reducing the number of combinations to iterate through.

3.2. Numerical Calculation

The non-convex problem defined in Equation (15), with the domain defined by the constraints in Equations (17) and (18), has been discretised in order to perform the numerical minimisation. This is because the numerical calculation requires the quantisation of the instantaneous control value $p_s(t)$, subject to constraints that are difficult to include in the minimisation process.

$$p_s^*(k) = \arg\min_{p_s(k)} \sum_{k=0}^{N} [(1 - F_L(k))D(p_L(k) - p_s(k))]\,T_s \tag{19}$$

where T_s is the chosen sampling time. To represent the cost, $D(p_g(k))$, it has been chosen as $p_g^2(k)$, as it indicates the higher costs associated with higher power demands to the generator and the network. The constraint expressed in Equation (17) is discretised, as well:

$$\sum_{k=0}^{N} [\eta_1 W_s(k) + \eta_2 + p_s(k)]\,T_s \le W_s(0). \tag{20}$$

The search space for the minimisation is reduced by parametrising the control function $p_s(k)$ using an interpolation method that maintains the monotonicity of the function. The method chosen is piecewise cubic Hermite interpolating polynomial (PCHIP) [26,27], which is a variant of cubic

Hermite interpolation, which, unlike methods like *spline* and*Bessel* , preserves monotonicity, avoiding "bumps" and overshoots in the resulting signal. A PCHIP interpolant is continuously differentiable, and its extrema are located at the extremal points. An interpolant function $P(k)$ is generated from a finite number N of data points (t_j, y_j), $j = 1 \dots N$, which are then the values of the control signal. The number of values N define the complexity of the interpolant and also the computing time; N must be chosen as a trade-off between the time required by the calculation and the resolution of the signal. In order to further reduce the search space, it is assumed that the optimal $p_s(k)$ will be monotonically decreasing for $k \in [0, T]$ due to the monotonically increasing characteristics of the scaling factor $(1 - F_L(k))$.

3.3. The Output

The outcome of the minimisation is $p_s^*(k)$, which is specific for a set of parameters: the durations distribution, the initial conditions of the storage, the load intensity and the storage dynamics. In case of distinct possible initial conditions, it is necessary to calculate the optimal strategy for each scenario. For example, if the energy stored at the beginning is not a constant, the controller must account for the possible range of initial storage levels and produce an output that is appropriate for that particular initial condition.

4. Simulations and Results

The optimal PMS proposed in this paper was tested on a model of an RTG crane equipped with a flywheel storage system. The optimal control is tailored to the particular storage system used, whose parameters are presented in Table 1. The RTG and FESS models are described in more detail in Section 4.3.

Table 1. Parameters of the flywheel energy storage system (FESS) used in the RTG model.

Parameter	Value
P_s	150 kW
W_{max}	3.6 MJ
η_1	1%
η_2	1 kW

4.1. Numerical Calculation of Optimal Values

The number of calculations to be performed is too large for a single system; for this reason, an HTCondor cluster [28] composed of over 300 nodes has been used to calculate the optimal strategy when varying the values of $W_s(0)$ and $p_L(t) = P_L$. The nodes work concurrently on different initial conditions to find the minimum cost by searching in the $\mathbb{R}^2$ parameter space defined by the points (t_j, y_j), which can vary from $[0, -P_s]$ (corresponding to the initial time and minimum power output) to $[T, P_s]$. The time distribution of these points has been selected with the objective of maximising the variability of the interpolant when the changes in power output are more expected; that is, when the probability $f_L(t)$ is higher. The points need to be adequately distributed, as well, so the choice for the values t_j is the following: given τ_j a series of N points linearly distributed in the interval $[0, 1]$ and given a CDF $F_L(t)$, the values t_j are:

$$t_j = F_L^{-1}(\tau_j)\ \forall j = 1 \dots N \tag{21}$$

resulting in a distribution of t_j, which tends to concentrate the values where the CDF changes more rapidly, which in turn corresponds to the instants when the probability $f_L(t)$ is higher, as is visible in Figure 4.

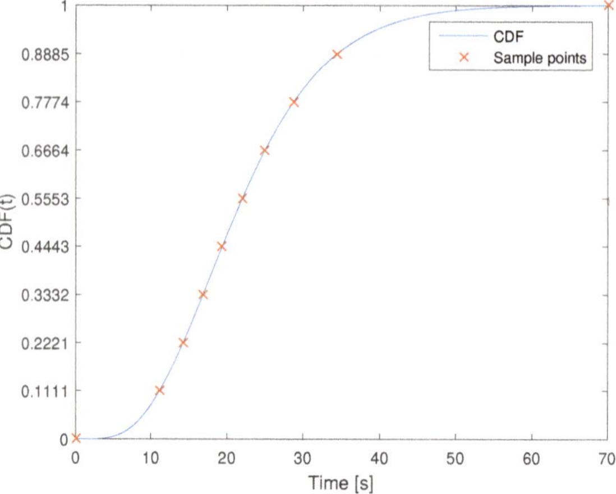

Figure 4. Temporal distribution of 10 sample points of a piecewise cubic Hermite interpolating polynomial (PCHIP) interpolant superimposed on the CDF. The vertical position of the PCHIP points indicates the distribution in the $[0, 1]$ region.

The output of the calculation is a matrix of parameters linked to an individual initial condition of the storage and value of the load. Given a particular scenario, the control system reads the optimal values calculated off-line and generates the optimal power output $p_s^*(k)$ in real time via PCHIP interpolation. Figure 5a shows an example of the output of the calculation, as well as the PCHIP interpolation used as a reference for the control system. The parameters used in the minimisation (including discretisation and load ranges) are presented in Table 2. ΔP_L and $\Delta W_s(0)$ indicate, respectively, the resolution used for the load power and the initial stored energy. A sample of the results of the calculations is shown in Figure 5b.

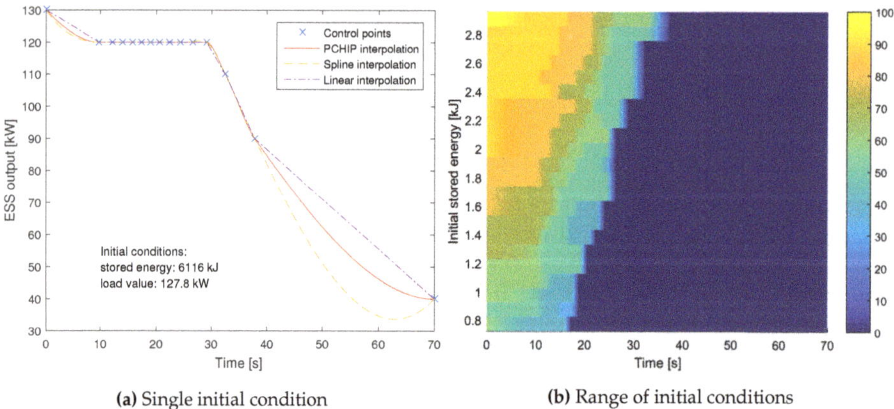

(a) Single initial condition

(b) Range of initial conditions

Figure 5. Examples of control points calculated in the minimisation. (**a**) The control points calculated for a single pair of initial conditions. Three different interpolations are also shown (PCHIP, spline and linear). Notice how the spline interpolation does not maintain monotonicity. (**b**) A range of optimal control strategies calculated for 0.72 MJ $< W_s(0) < 3.00$ MJ and $Pl = 100$ kW. The colour bar on the right shows the power output of the storage expressed in kW.

Table 2. Parameters of the numerical minimisation.

Parameter	Value
P_s	150 kW
T	70 s
ΔP_L	10kW
$W_s(0)$ (range)	[720, 3470] kJ
$\Delta W_s(0)$	101.8 kJ

4.2. Distribution of Lift Durations

Measurements of the activity of an RTG crane were recorded at the Port of Felixstowe, including the duration of container lifts measured as the interval between the start of a lift (*i.e.*, the hoist motor speed becomes positive) and its end when reaching zero speed. The data were collected for a period of six days, after which they were analysed and fit to a Gamma distribution, which has been chosen because it results in the best fit to the data and is defined by the constants α and β. The probability density of a Gammadistribution L is described by the following equation:

$$f_L(t) = \frac{t^{\alpha-1}e^{-\frac{t}{\beta}}}{\beta^{\alpha}\Gamma(\alpha)} \quad \text{for } t > 0 \text{ and } \alpha, \beta > 0 \tag{22}$$

where $\Gamma(\alpha)$ is the Gamma function evaluated at α, and it is a constant:

$$\Gamma(\alpha) = \int_0^{\infty} x^{\alpha-1}e^{-x}\, dx \tag{23}$$

The actual random variable t has a realistic upper limit T, as the duration of the load is limited in time, so the distribution used in the calculation has been truncated at $t = T$. The parameters α and β have been found by minimising the squared error and are presented in Table 3. Figure 6 shows the histogram of lift durations superimposed on the Gamma distribution.

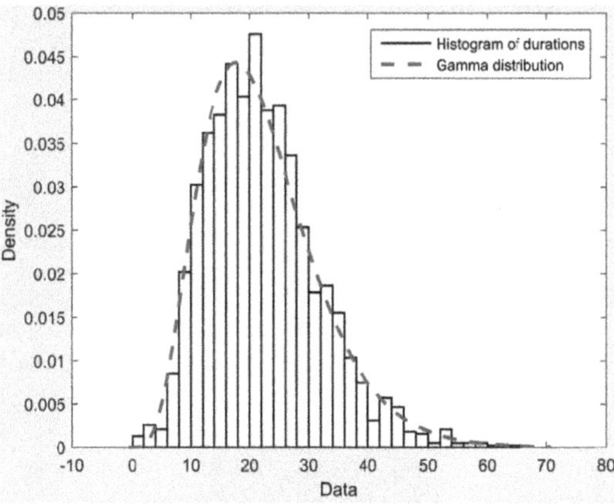

Figure 6. Histogram of the lift durations superimposed on the Gamma distribution that fits the data.

Table 3. Parameters of the Gamma distribution that fit the lift duration data.

Parameter	Value
α	5.0292
β	4.3923

The CDF $F_L(t)$ can be easily pre-calculated off-line for each instant by integration or by using the following equation:

$$F_L(t) = \int_0^t f(u)\,du = \frac{\gamma\left(\alpha, \frac{t}{\beta}\right)}{\Gamma(\alpha)} \tag{24}$$

where $\gamma\left(\alpha, \frac{t}{\beta}\right)$ is the lower incomplete gamma function and equal to:

$$\gamma(s,x) = \int_0^x t^{s-1}\, e^{-t}\, \mathrm{d}t \tag{25}$$

4.3. *Model of the RTG Crane*

The main electrical and mechanical components of an RTG crane have been modelled in MATLAB/Simulink using the SimPowerSystems toolbox. This model was originally developed to study the operations of RTG cranes at the Port of Felixstowe [29,30], but it has been extended to be used to test the PMS by adding a model of an FESS connected to the DC bus of the crane, as is visible in Figure 7. The model is composed of three main elements: a primary source, the hoist motor and the FESS.

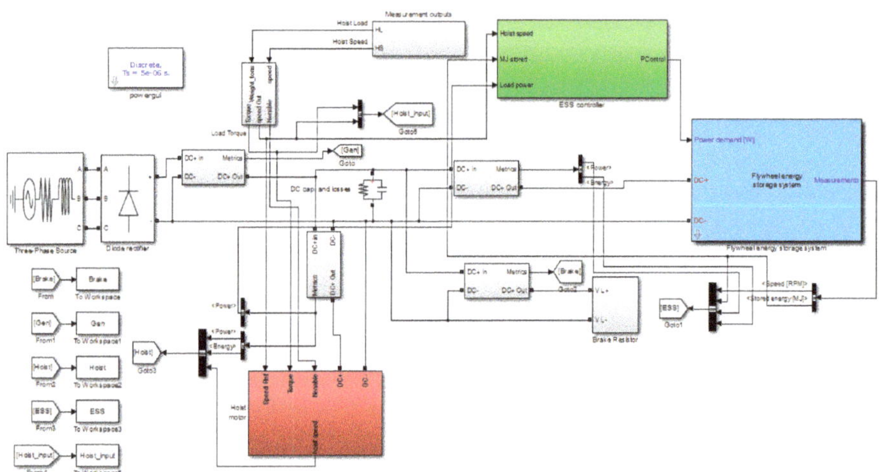

Figure 7. Simulink model of the RTG crane used for the simulations.

4.3.1. Primary Source

In the model, the primary source is an ideal three-phase source connected directly to a diode rectifier, which powers the DC bus. Measurements are taken on the three-phase side to measure the energy consumed by the crane. The benefits of the storage and its control strategy will be assessed by analysing the energy consumed by the primary source. A good control strategy will need, as a primary objective, to reduce the energy consumption and also limit the peak power demand.

4.3.2. The Hoist Motor

An induction motor rated at 200 kW is connected to the DC bus. It is powered by a drive, which controls the motor speed following a reference value extracted directly from measurements taken at the Port of Felixstowe. It draws power from the DC bus when raising a container and then generates power back into the bus when lowering. The regenerated energy is collected by the energy storage until it reaches maximum capacity, then the remaining energy is dissipated into brake resistors.

4.3.3. The Flywheel Energy Storage System

The storage system model has been developed to test the PMS proposed in this paper. The model, shown in Figure 8, is based on a prototype powered by a 150-kW, 12/10 pole switched reluctance motor whose model and low-level control system has been provided by the manufacturer Nidec SR Drives (Harrogate, UK); the motor is coupled to a flywheel whose inertia has been measured to be 3.0447 kg/m^2. The maximum rotational speed is 15,000 RPM, and the minimum speed is set to 5000 RPM; the total capacity is then 3.34 MJ, which equates to 0.927 kWh. The low-level control system is designed to provide any output power up to the rated value of $\pm$150 kW with a maximum delay of 0.6 ms; this is fast compared to the dynamics of the hoist motor, meaning that the low-level control is fully transparent to the PMS.

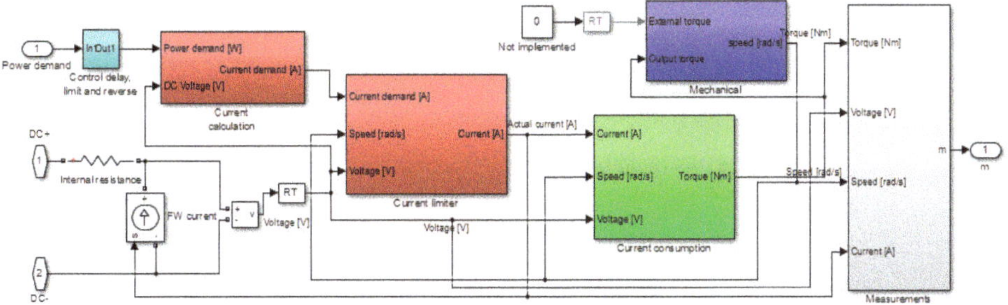

Figure 8. Simulink model of the flywheel energy storage system.

4.4. *Test Cycle*

The activity of a diesel-powered RTG crane was measured for the duration of one hour during a typical day. Among the recorded signals were the speed of the hoist motor and the container weight, which have been used in the RTG model to simulate the operation of the crane. The characteristics of this hour of crane operations are presented in Table 4 and represent the typical activity of an RTG crane, consisting mainly of the combination of four basic hoist movements and their accompanied energy flows:

1. The empty headblock is lowered over a container: a small amount of energy is regenerated;
2. The load is hoisted to a height decided by the crane operator: a large amount of energy is consumed;
3. The load is lowered in place: a large amount of energy is regenerated;
4. The headblock is hoisted back into the starting position: a small amount of energy is consumed.

During the simulation, in the first and third movements, the energy is either dissipated as heat or, when a storage system is present, is stored in the ESS. The crane then uses the stored energy, if available, in the second and fourth movements according to the control system implemented.

The simulation was repeated with different ESS scenarios as follows:

1. *No ESS*: In this scenario, no storage is installed, and all of the recovered energy is dissipated through the brake resistors;
2. *Constant power*: The ESS uses a set-point control strategy where the ESS output is limited to a value that is the average load power, *i.e.*, $p_{set}(t) = \max\{72\text{ kW}, P_L\}$;
3. *Proposed PMS*: An ESS with the optimal control strategy proposed in this paper;
4. *Infinite capacity*: An ideal ESS with unlimited energy capacity and set to absorb or generate energy with a power limit of 150 kW with no time limitations, similarly to the second scenario, but with no capacity constraints and with the highest power limit.

In all of the scenarios, the ESS is set to charge only using recovered energy from lowered containers (no trickle charge, as the storage is never charged directly by the primary source). Scenario 2 is a simple and robust control strategy, which has already been implemented [14] and replicates most power-sharing control strategies. Scenario 4 extends 2 by increasing the power upper limit to the maximum rating of the storage and also increases the capacity to an unlimited value, in order to represent a simple and ideal scenario.

Table 4. Characteristics of the test cycle.

Duration	1 h
Number of lifts (container and empty headblock)	89
Energy consumed	18.24 kWh
Average load weight (container plus headblock)	19.09 t
Average hoist power (when lifting)	72.74 kW

4.5. Results of the Simulation and Analysis

The simulations produced different results depending on the scenario, with the presence of the storage resulting in reduced energy consumption, as shown in Figure 9. With no storage, the crane wasted a significant amount of energy in the brake resistors, as all of the potential energy recovered from lowering the container is not being stored. The addition of the storage greatly increases the efficiency of the system, as it enables the reutilisation of regenerated energy, and the benefits are evident in Figure 9a. The three different control strategies used in Scenarios 2, 3 and 4 produced significantly different results. In particular, the set-point control with constant power output reduced the energy consumption by 35.9%; in this particular test cycle, this is the best achievable outcome for this kind of control, as the power limit is set by knowing in advance the average power consumption. The ideal energy storage with infinite capacity reached a reduction of 39.0%, and this is the upper bound for a system with a power rating of 150 kW. The proposed PMS is only slightly worse than the ideal case with a 38.47% reduction (see Figure 9b).

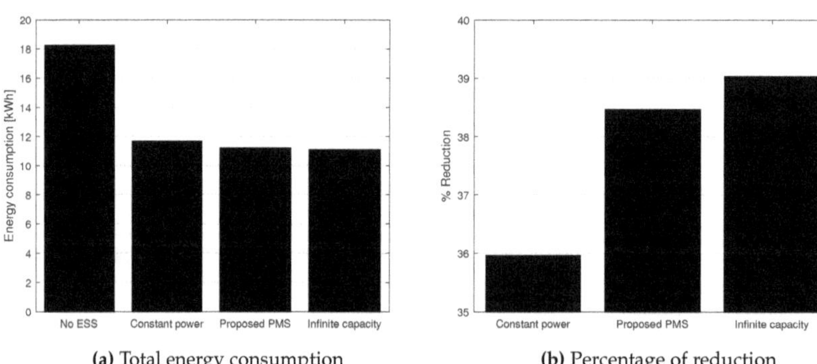

(a) Total energy consumption

(b) Percentage of reduction

Figure 9. Results from the test cycle. (**a**) The total energy from the primary source for the four scenarios; (**b**) the percentage of reduction of energy consumption for the three storage scenarios with respect to the first scenario.

The minimisation criteria used in the PMS also originate a significant reduction in peak power demand, since the quadratic cost function penalises large quantities of primary supply power. This is reflected in Table 5, where the percentage of time that the primary source is outputting more than 150 and 200 kW is shown. The proposed PMS has promising results with respect to the other control systems, including the scenario with infinite storage, as it effectively reduces the stress on the primary source during peak demand. The proposed PMS limits the peak better than the infinite capacity scenario due to the fact that the latter uses all of the recovered energy at the beginning of the lift, causing the storage to be depleted prematurely and forcing the generator to take over for the rest of the duration of the lift. The reduction in peak power consumption achieved by the proposed PMS reduces the stress on the primary source and allows for the downsizing of the diesel generator or the substation feeding the hybrid cranes; this could potentially be an opportunity for further reductions in costs for terminal operators (in addition to the energy cost savings).

Table 5. Percentage of time that the primary source power output is over 150 and 200 kW. PMS, power management strategy.

Scenario	Percentage of Time over 150 kW	Percentage of Time over 200 kW
No ESS	3.997%	0.0437%
Constant power	0.902%	0.0167%
Proposed PMS	1.365%	0.0028%
Infinite capacity	1.856%	0.0139%

An example of the behaviour of the energy storage is visible in Figure 10, which shows a slice of the simulation when the hoist motor performs a lowering and a lifting. In the example, the energy storage starts with around 0.4 MJ of stored energy (Figure 10b), and it recovers the regenerated power (shown in Figure 10a as the negative power). At the end of the lowering, the ESS reaches almost 1.1 MJ of stored energy (relatively low, as it was only lowering an empty headblock), with the flywheel spinning at around 9500 RPM. The stored energy is used in the second part, during the lift of a container, reducing the power demand on the primary source. At the end of the cycle, the ESS ends with very little remaining energy.

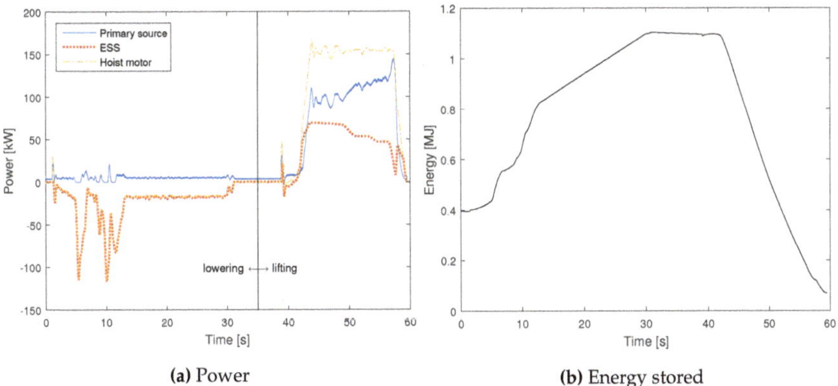

(**a**) Power (**b**) Energy stored

Figure 10. Example extracted from the simulation. (**a**) The power flows of the three main elements in the model; when lowering, the ESS power is equal to the hoist power; (**b**) the profile of the energy stored in the ESS.

The intensity of the activity of the hybrid crane has an impact on the energy savings: low utilisation causes lower savings due to the larger times between lifts (and consequent high standing losses of the storage). Nevertheless, the proposed PMS optimisation process is limited to the single lift and is not dependent on their frequency, and it is only sensible to variations in the distribution of lift durations. Major changes in the container terminal operations could cause, for example, a reduction or an increase in average stack heights, leading to shorter or longer lift durations. This will need to be addressed by recalculating the optimal strategy with the new distribution parameters.

5. Conclusions

A power management strategy has been developed that minimises the energy costs associated with systems subjected to stochastic loads with a random duration. An optimal control problem has been established where the cost function takes into account the variability of the load, penalising the energy cost depending on the probability that a certain condition occurs. After accurately reducing the search space, an HTCondor cluster has been used to perform the numerical calculations aimed at obtaining the global minimum. A set of control strategies has been calculated for a range of possible initial conditions, and the resolution has been increased by interpolating the results.

The calculated PMS has been implemented in a MATLAB/Simulink model of a rubber tyre gantry crane equipped with a flywheel energy storage system; the results of the simulation show that the proposed power management strategy performs better than existing control strategies and very close to the ideal case. In particular, the results show that energy storage with optimal control reduces energy consumption and peak power demand, resulting in an efficient utilisation of the limited capacity of the storage. As the simulations show, using the proposed strategy for energy storage could reduce the energy consumption in container ports by a significant amount, as RTG cranes account for a large portion of the port's total demand. The added benefit of peak power reduction could potentially minimise the maintenance costs for diesel generators and/or reduce the stress on the electrical infrastructure of the port.

Acknowledgments: This work was supported by Climate-KIC (Knowledge and Innovation Community) through the project Delivering sustainable energy solutions for ports (SUSPORTS). The authors are grateful for the help given by the Port of Felixstowe in collecting data on RTG cranes and providing knowledge about the crane operations.

Author Contributions: All of the authors contributed to the editing and improvement of the manuscript. Stefano Pietrosanti developed the control strategy, the model and analysed the results.William Holderbaum and Victor M. Becerra provided the methodology, contributed to the analysis of results and supervised all of the processes.

Conflicts of Interest: The authors declare no conflict of interest.

Abbreviations

The following abbreviations are used in this manuscript:

PMS: Power Management Strategy
RTG: Rubber tyre gantry, a type of container crane
ESS: Energy storage system
FESS: Flywheel energy storage system
CDF: Cumulative distribution function
PCHIP: Piecewise cubic Hermite interpolating polynomial
SR: Switched reluctance (electric motor)

Nomenclature

$p_L(t)$	Power demand from the load
$p_g(t)$	Power from the primary source (e.g., diesel generator)
$p_s(t)$	Power from the storage system
P_s	Power rating of the storage system
D_{tot}	Total cost associated with the energy production
$D(cdot)$	Cost function associated with energy production

P_L	Constant power demand of the load
$W_s(t)$	Energy stored in the storage system at time t
W_{max}	Energy capacity of the storage system
η_1, η_2	Constants defining the dynamic properties of the storage system
$f_L(t)$	Probability that an event occurs at time t when defined by a distribution L
$F_L(t)$	Cumulative distribution function associated with the distribution L
α, β	constant parameters that define a Gamma distribution

Appendix A

Theorem A1. *Given a distribution L whose known probability density function is $f(x)$, with $x \in [0, T]$, and whose CDF is $F(t)$, and given continuous functions $g : \mathbb{R} \to \mathbb{R}$ and $u : \mathbb{R} \to \mathbb{R}$ that satisfy the following:*

$$\int_0^{\infty} |g(u(t))| \, dt < \infty \tag{A1}$$

then the following is true:

$$\arg\min_{u(t)} \int_0^T f(t_f) \left(\int_0^{t_f} g(u(t)) \, dt \right) dt_f = \arg\min_{u(t)} \int_0^T (1 - F(t)) g(u(t)) \, dt \tag{A2}$$

Proof. A sufficient condition for the equality in Equation (A2) is that Equation (A3) is true given the assumptions stated in the Theorem.

$$\int_0^T f(x) \int_0^x g(u(t)) \, dt \, dx = \int_0^T (1 - F(t)) g(u(t)) \, dt \tag{A3}$$

The CDF $F(t)$ is the integral of $f(x)$ over x:

$$F(t) = \int_0^t f(x) \, dx \tag{A4}$$

and, given that the maximum value for x is T, it satisfies the following:

$$F(T) = \int_0^T f(x) \, dx = 1 \tag{A5}$$

Given that $f(x)$ does not depend on t, we can rewrite the left integral of Equation (A3) as follows:

$$\int_0^T f(x) \int_0^x g(u(t)) \, dt \, dx = \int_0^T \int_0^x f(x) g(u(t)) \, dt \, dx \tag{A6}$$

The domain of t is $[0, x]$, while the domain of x is $[0, T]$. By using Fubini's theorem, we can invert the order of integration (given the assumption in Equation (A1)), resulting in the following domains: $t \in [0, T]$, $x \in [t, T]$. Equation (A6) is then equal to:

$$\int_0^T \int_t^T f(x) g(u(t)) \, dx \, dt \tag{A7}$$

The function $g(u(t))$ does not depend on x and can be moved outside the inner integral:

$$\int_0^T g(u(t)) \left(\int_t^T f(x) \, dx \right) dt \tag{A8}$$

From Equations (A4) and (A5), we know that:

$$\int_t^T f(x)\,\mathrm{d}x = \int_0^T f(x)\,\mathrm{d}x - \int_0^t f(x)\,\mathrm{d}x = 1 - \int_0^t f(x)\,\mathrm{d}x = 1 - F(t) \tag{A9}$$

which, when inserted into Equation (A8), results in:

$$\int_0^T g(u(t))(1 - F(t))\,\mathrm{d}t \tag{A10}$$

proving Equation (A3). □

References

1. Ter-Gazarian, A.G. *Energy Storage for Power Systems*; The Institution of Engineering and Technology (IET): Stevenage, UK, 2011.
2. Zeng, Y.; Cai, Y.; Huang, G.; Dai, J. A review on optimization modeling of energy systems planning and GHG emission mitigation under uncertainty. *Energies* **2011**, *4*, 1624–1656.
3. Chen, H.; Cong, T.N.; Yang, W.; Tan, C.; Li, Y.; Ding, Y. Progress in electrical energy storage system: A critical review. *Prog. Nat. Sci.* **2009**, *19*, 291–312.
4. Rowe, M.; Yunusov, T.; Haben, S.; Singleton, C.; Holderbaum, W.; Potter, B. A peak reduction scheduling algorithm for storage devices on the low voltage network. *IEEE Trans. Smart Grid* **2014**, *5*, 2115–2124.
5. Haben, S.; Ward, J.; Vukadinovic Greetham, D.; Singleton, C.; Grindrod, P. A new error measure for forecasts of household-level, high resolution electrical energy consumption. *Int. J. Forecast.* **2014**, *30*, 246–256.
6. Rowe, M.; Yunusov, T.; Haben, S.; Holderbaum, W.; Potter, B. The real-time optimisation of DNO owned storage devices on the LV network for peak reduction. *Energies* **2014**, *7*, 3537–3560.
7. Zhang, Y.; Gatsis, N.; Giannakis, G.B. Robust energy management for microgrids with high-penetration renewables. *IEEE Trans. Sustain. Energy* **2013**, *4*, 944–953.
8. Liang, H.; Zhuang, W. Stochastic modeling and optimization in a microgrid: A survey. *Energies* **2014**, *7*, 2027–2050.
9. Brahma, A.; Guezennec, Y.; Rizzoni, G. Optimal energy management in series hybrid electric vehicles. In Proceedings of the 2000 American Control Conference, Chicago, IL, USA, 28–30 June 2000; Volume 1, pp. 60–64.
10. Lin, W.S.; Zheng, C.H. Energy management of a fuel cell/ultracapacitor hybrid power system using an adaptive optimal-control method. *J. Power Sources* **2011**, *196*, 3280–3289.
11. Romaus, C. Optimal energy management for a hybrid energy storage system for electric vehicles based on stochastic dynamic programming. In Proceedings of the 2010 IEEE Vehicle Power and Propulsion Conference (VPPC), Lille, France, 1–3 September 2010.
12. Flynn, M.; Mcmullen, P.; Solis, O. Saving energy using flywheels. *IEEE Ind. Appl. Mag.* **2008**, *14*, 69–76.
13. Baalbergen, F.; Bauer, P.; Ferreira, J. Energy storage and power management for typical 4Q-load. *IEEE Trans. Ind. Electron.* **2009**, *56*, 1485–1498.
14. Iannuzzi, D.; Piegari, L.; Tricoli, P. Use of supercapacitors for energy saving in overhead travelling crane drives. In Proceedings of the 2009 International Conference on Clean Electrical Power, Capri, Italy, 9–11 June 2009; pp. 562–568.
15. Xu, J.; Yang, J.; Gao, J. An integrated kinetic energy recovery system for peak power transfer in 3-DOF mobile crane robot. In Proceedings of the 2011 IEEE/SICE International Symposium on System Integration (SII), Kyoto, Japan, 20–22 December 2011; pp. 330–335.
16. Kim, S.-M.; Sul, S.-K. Control of rubber tyred gantry crane with energy storage based on supercapacitor bank. *IEEE Trans. Power Electron.* **2006**, *21*, 262–268.
17. Hellendoorn, H.; Mulder, S.; de Schutter, B. Hybrid control of container cranes. In Proceedings of the 18th IFAC World Congress, Milan, Italy, 28 August–2 September 2011; Volume 19, pp. 9697–9702.
18. Hedlund, M.; Lundin, J.; de Santiago, J.; Abrahamsson, J.; Bernhoff, H. Flywheel energy storage for automotive applications. *Energies* **2015**, *8*, 10636–10663.

19. Levron, Y.; Shmilovitz, D. Optimal power management in fueled systems with finite storage capacity. *IEEE Trans. Circuits Syst. I Regul. Pap.* **2010**, *57*, 2221–2231.
20. Levron, Y.; Shmilovitz, D. Power systems' optimal peak-shaving applying secondary storage. *Electr. Power Syst. Res.* **2012**, *89*, 80–84.
21. Singer, S. Canonical approach to energy processing network synthesis. *IEEE Trans. Circuits Syst.* **1986**, *33*, 767–774.
22. Montgomery, D.; Runger, G. *Applied Statistics and Probability for Engineers*; John Wiley & Sons: Hoboken, NJ, USA, 2010.
23. Driessen, B.J. On-off minimum-time control with limited fuel usage: Near global optima via linear programming. *Optim. Control Appl. Methods* **2006**, *27*, 161–168.
24. Singhose, W.; Singh, T.; Seering, W. On-off control with specified fuel usage. *J. Dyn. Syst. Meas. Control* **1999**, *121*, 206–212.
25. Kirk, D.E. *Optimal Control Theory: An Introduction*; Courier Corporation: Mineola, NY, USA, 2004; p. 452.
26. Fritsch, F.N.; Carlson, R.E. Monotone piecewise cubic interpolation. *SIAM J. Numer. Anal.* **1980**, *17*, 238–246.
27. Kahaner, D.; Moler, C.; Nash, S. *Numerical Methods and Software*; Prentice-Hall: Englewood Cliffs, NJ, USA, 1989; p. 495.
28. Thain, D.; Tannenbaum, T.; Livny, M. Distributed computing in practice: The Condor experience. *Concurr. Pract. Exp.* **2005**, *17*, 323–356.
29. Knight, C.; Becerra, V.; Holderbaum, W.; Mayer, R. A consumption and emissions model of an RTG crane diesel generator. In Proceedings of the TSBE EngD Conference, TSBE Centre, Whiteknights, UK, 5 July 2011.
30. Knight, C.; Becerra, V.; Holderbaum, W.; Mayer, R. Modelling and simulating the operation of RTG container cranes. In Proceedings of the 6th IET International Conference on Power Electronics, Machines and Drives (PEMD 2012), Bristol, UK, 27–29 March 2012; pp. F24–F24.

Article

Control Strategies with Dynamic Threshold Adjustment for Supercapacitor Energy Storage System Considering the Train and Substation Characteristics in Urban Rail Transit

Fei Lin *, Xuyang Li, Yajie Zhao and Zhongping Yang

School of Electrical Engineering, Beijing Jiaotong University, No.3 Shangyuancun, Beijing 100044, China; 14121438@bjtu.edu.cn (X.L.); 12121590@bjtu.edu.cn (Y.Z.); zhpyang@bjtu.edu.cn (Z.Y.)

* Correspondence: flin@bjtu.edu.cn; Tel.: +86-10-51687064

Academic Editor: William Holderbaum

Received: 30 January 2016; Accepted: 22 March 2016; Published: 31 March 2016

Abstract: Recuperation of braking energy offers great potential for reducing energy consumption in urban rail transit systems. The present paper develops a new control strategy with variable threshold for wayside energy storage systems (ESSs), which uses the supercapacitor as the energy storage device. First, the paper analyzes the braking curve of the train and the *V-I* characteristics of the substation. Then, the current-voltage dual-loop control method is used for ESSs. Next, in order to achieve the best energy-saving effect, the paper discusses the selection principle of the charge and discharge threshold. This paper proposes a control strategy for wayside supercapacitors integrated with dynamic threshold adjustment control on the basis of avoiding the onboard braking chopper's operation. The proposed control strategy is very useful for obtaining good performance, while not wasting any energy in the braking resistor. Therefore, the control strategy has been verified through simulations, and experimental tests, have been implemented on the Batong Line of Beijing subway using the 200 kW wayside supercapacitor energy storage prototype. The experimental results show that the proposed control is capable of saving energy and considerably reducing energy consumption in the braking resistor during train braking.

Keywords: energy storage system (ESS); supercapacitor; control strategy; train braking characteristics; traction substation; charge and discharge threshold

1. Introduction

With the continuous economic development in China in recent years, urban rail transit has also undergone rapid development. From 2003 to 2013, the operating mileage of China urban rail transit increased from 290.4 km to 2326.0 km, the highest in the world [1]. In the urban rail transit system, braking energy of the train is commonly fed back to the catenary through regenerative braking. However, due to the 24-pulse diode rectifier unit used in the traction substation, surplus regenerating energy cannot provide feedback to the medium-voltage power grid. When a train is braking, if there are no adjacent traction trains or energy storage devices that can absorb the regenerative energy, then the pantograph voltage would exceed the normal range, thus leading to the onboard braking chopper operating, *i.e.*, the braking energy is wasted by the resistor [2,3]. Even worse, regeneration cancellation may occur. Therefore, in order to maximize the use of electric braking energy, while reducing mechanical braking and resistor braking of urban rail trains, currently two main options are energy storage and energy feedback [4]. At present, the main storage devices available are batteries, supercapacitors and flywheels.

Flywheels present relatively high overall efficiencies and elevated energy and power densities. However, they have a potential risk of explosive shattering in case of catastrophic failure and they have higher mechanical structure requirements [5–7].

The batteries present high energy density with discharge times ranging from tens of minutes to hours, which can provide power for the vehicle running to the safe place if the power supply is cut off because of trouble. Besides, the batteries can be used in the urban rail transit to absorb the braking energy and reduce the voltage fluctuation of the DC bus. Their cycle life is shorter and the power density is relatively low.

The supercapacitors have higher power density. They are suitable for supplying power peaks and absorbing the braking power peaks. In addition, they have longer cycle life. In this paper, a cycle is defined as the supercapacitor is charged to the maximum and discharged to the minimum value, which is considered one cycle. However, there are many definitions for cycle life, and it is related to the control strategy [8,9]. Another added value is that, unlike batteries, which require complex algorithms to estimate the state of charge (SOC), the determination of supercapacitor SOC is easily obtained by measuring their terminal voltage [10,11]. The urban rail transit operation has frequent starts and stops, and voltage peaks obviously fluctuate; supercapacitors match the operational characteristics of urban rail transit.

Depending on the placement of the supercapacitor energy storage system (ESS), the ESSs can be divided into the two types: onboard and off-board. The onboard type ESSs can absorb one train's braking energy and decrease the transmission loss during the process of energy flow. There are many applications with ESSs onboard in the trams [12,13]. While installed, ESSs will increase the weight of vehicles. Besides, the vehicles need more space to install the ESSs, so it might not be suitable for the metro trains to install onboard ESSs.

The off-board ESSs are able to absorb the braking energy from all the vehicles linked to the contact lines and feed the energy back into the contact lines for subsequent accelerations. Stationary ESSs are usually placed in the traction station or along the contact lines. Due to this, the ESSs can reduce the volume requirement. Therefore, off-board ESSs are usually applied in subway systems [14–16].

According to the system function, the station type and line type can be used to describe the ESSs. The station type supercapacitor ESS is typically placed in the traction substation, as shown in Figure 1, mainly for the recovery of regenerative braking energy. The line type is set in the middle of the line, primarily to reduce the voltage drop [17,18].

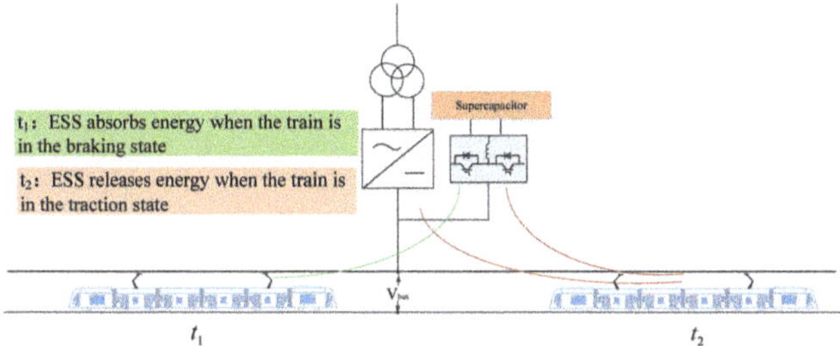

Figure 1. Operating principle of wayside supercapacitor energy storage system (ESS).

Station type supercapacitor ESSs usually use voltage and current bicyclic control method to achieve charging and discharging operations [19–21]. In [19], when the drive load switches from positive to negative (from motor mode to generator mode). The ultracapacitor begins to be charged until its voltage reaches the maximum reference U_{C0max}. The dc-bus voltage V_{BUS} increases until it reaches the reference V_{BUSmax}. The magnitude of the current is adjusted by the cascaded controllers $G_{vBUSmax}$ and G_{uC0} at this level, so as to maintain the dc-bus voltage constant. While the main is interrupted, the dc-bus voltage begins to decrease until it reaches the minimum reference V_{BUSmin}.

This allows deeper discharge of the ultracapacitor and regulation of the dc-bus voltage at the minimum V_{BUSmin}. However, the charging and discharging thresholds are all constants in [19–21], as they never change after setting.

However, these control methods do not analyze how to set the appropriate threshold, yet, charge and discharge threshold settings have extremely important impacts on the energy saving effect of the ESS [22–24]. This paper aims to acquire a control strategy of wayside ESSs, which is oriented to the optimization of the energy saving and reduction of the braking resistor's operation. The control is mainly based on the actual train braking characteristic, and takes the 24-pulse rectifier unit output characteristics into account. Due to the above characteristics analysis, a threshold setting study has been undertaken with the aim of better energy savings.

The organization of this work is as follows: The urban rail transit characteristics are analyzed in Section 2. Then, the wayside supercapacitor ESS compositions and its control strategy are introduced in Section 3. Next, Section 4 further analyzes the threshold selection strategy of the ESSs and a real-time adjustment of the threshold method is put forward. The simulation results are obtained in Section 5. Then, the experimental tests in the Beijing subway fully confirm the correctness of theoretical analysis, namely the threshold setting is closely related to the energy savings. Finally, Section 6 is the conclusion.

2. Analysis of Urban Rail Transit Characteristics

2.1. Braking Characteristics of the Trains

The braking curve of the induction traction motor in metro trains can be divided into three regions: constant torque, constant power and natural characteristics [25], shown as Equation (1) and Figure 2.

$$\begin{cases} F_{\mathrm{t}}(v) = C_1 \\ F_{\mathrm{t}}(v) \cdot V = C_2 \\ F_{\mathrm{t}}(v) \cdot V^2 = C_3 \end{cases} \tag{1}$$

where C_1, C_2, and C_3 are constants, $F_{\mathrm{t}}(v)$ is the traction effort and V is the speed of the train.

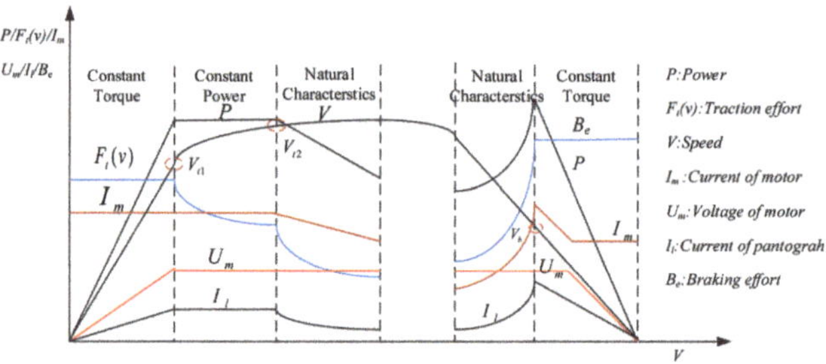

Figure 2. Traction/braking characteristics of metro train.

Taking trains from the Batong Line of Beijing subway in China as an example, the train brakes using air braking and electric braking, which includes regenerative braking and resistor braking. A brake control unit usually consists of M and T cars, and according to the train braking demand, electric braking acts first. Air braking force will compensate or substitute the electric braking when the electric braking force is insufficient or invalid.

When the electric braking operates, in order to prevent the DC voltage being beyond the permissible range, the controller controls the brake chopper throughout the voltage, across the filter capacitor of traction converters, to distribute the regenerative braking and resistor braking power. The working area of resistor braking is the shaded portion in Figure 3. Each inverter unit is equipped with a braking resistor, and the resistance is 1.203 Ω under room temperature. The regenerative energy will give feedback to the grid first when the train is under electric braking, until the absorption capacity of the grid is insufficient. In addition to this, after the grid voltage is increased to 900 V, the resistor chopper will be activated. As the grid voltage gradually increases, the chopper working power gradually increases. The vehicle braking resistor operates at full power until reaching the maximum voltage of 1000 V.

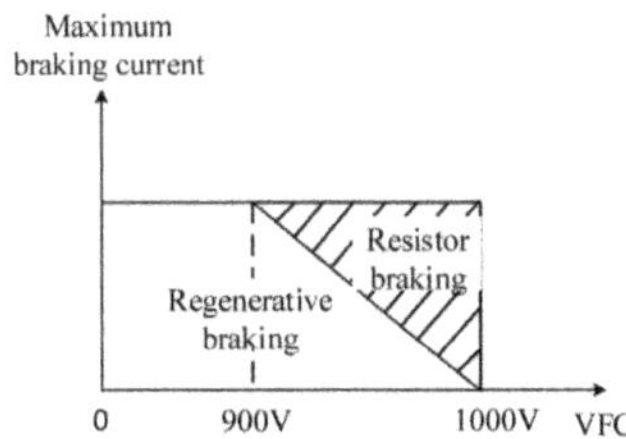

Figure 3. Mixed brake schematic of Batong train.

2.2. *Characteristics of the Equivalent 24-Pulse Rectifier*

The equivalent 24-pulse rectifier unit consists of two 12-pulse transformers and rectifiers, and the windings of the two rectifier transformers are moved to +7.5° and −7.5°, respectively, as shown in Figure 4.

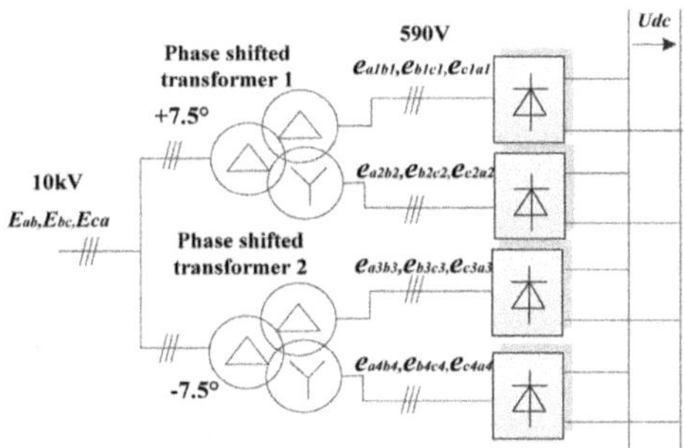

Figure 4. 24-pulse rectifier of the 750 V metro power supply system.

In multi-pulse rectifier technology, the output voltage U_{dc0} of the rectifier unit is proportional to the grid-side no-load voltage [26].

$$\begin{cases} U_{d0} = \frac{P}{2\pi}\int_{\frac{\pi}{P}-\alpha}^{\frac{\pi}{P}+\alpha} \sqrt{2}U_2\cos\theta d\theta = \frac{\sqrt{2}PU_2}{\pi}\sin\frac{\pi}{P}\cos\alpha \\ U_2 = NU_1 \\ U_1 = U_{1N}(1+\delta\%) \end{cases} \quad (2)$$

In Equation (2), U_{d0} is the rectifier output no-load voltage, P is the number of pulses, U_2 is the voltage of the valve side, U_1 is the line voltage of medium-voltage grid, N is the turns ratio of transformer primary to secondary, U_{1N} is the rating line voltage of medium voltage network and δ is the fluctuation ratio. The output voltage of the 24-pulse rectifier without load can be derived from Equation (3) as follows:

$$U_{dc0} \approx 1.41NU_1 \quad (3)$$

As the load current increases, the output DC voltage of the equivalent 24-pulse rectifier unit reduces accordingly.

The external characteristic is mainly related to the impedance of the rectifier transformer, the topology of the rectifier circuit, the impedance of the AC power system, the operation status of the rectifier, and so on. According to engineering experience, the DC output voltage U_{dc} of the 24-pulse rectifier unit can be calculated using the following equation [27]:

$$U_{dc} = U_o - \frac{k_r U_d}{100} \times \frac{U_n^2}{0.9nS_T} \times I_{sub} \tag{4}$$

where U_n represents the rating voltage of the DC side (kV), U_d is the short-circuit voltage percentile of the transformer, S_T is the capacity of the transformer (MVA), n is the number of 24-pulse rectifier, k_r is the coefficient resistance and 0.9 is matching coefficient between the transformer and rectifier.

Based on the output voltage and current data measured at the Beijing subway traction substation, the output characteristics can be plotted as shown in Figure 5.

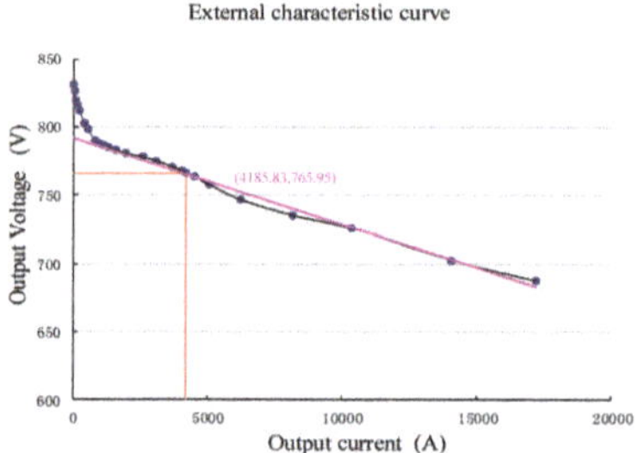

Figure 5. The output character of the 24-pulse rectifier.

When the output current is approximately 4185.8 A, the rated output voltage of traction substations is 765.9 V, at which time the load rate of traction substation is 100%. The output voltage of traction substation decreases as the output current increases, and the slope of the characteristic curve changes in pace with the output current increase.

3. Wayside Supercapacitor Energy Storage System and Control Strategy

3.1. System Components

The station type supercapacitor ESS consists of a bi-directional DC/DC converter and supercapacitor, as shown in Figure 6. The bi-directional DC/DC converter is the key component of the whole system, undertaking the tasks of system voltage level shift and energy management.

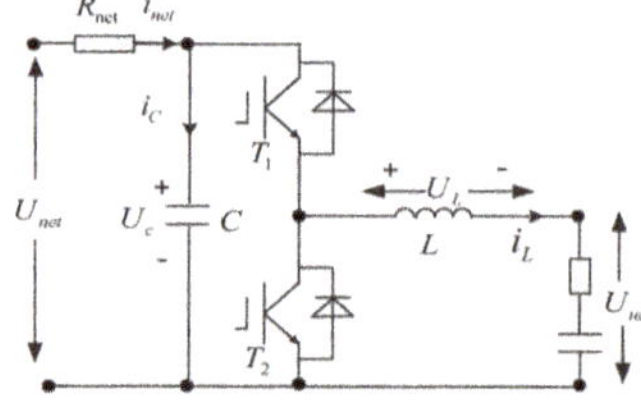

Figure 6. Supercapacitor storage system based on half-bridge topology.

Depending on different conditions of the traction substation, the supercapacitor ESS will operate when in charging or discharging status, the bidirectional DC/DC converter through control T_1 and T_2

switching to realize supercapacitor's charging or discharging, thus achieving the different directions of chopping inductor current i_L. Operating in either the charging or discharging status, the supercapacitor ESS can be represented using the unified model shown in Figure 7, where R_{on} is the on-state resistance of the IGBT, R_L is the equivalent resistance of the chopping inductor, R_{net} is the equivalent resistance of the input side, and R_{uc} is the ESR of the supercapacitor.

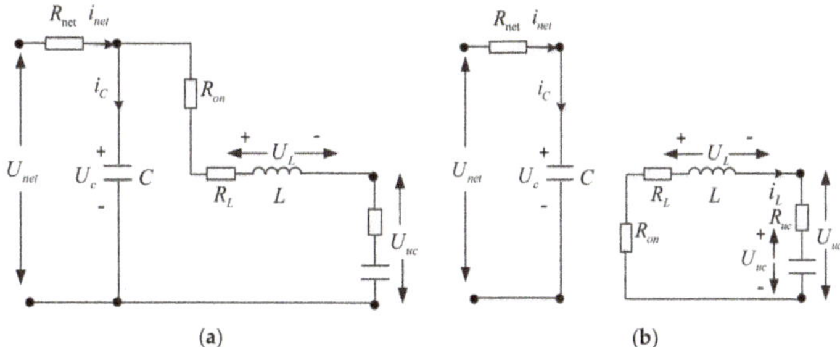

Figure 7. Equivalent model of supercapacitor storage system in switching period. (**a**) $0 < t < DT_s$; (**b**) $DT_s < t < T_s$.

3.2. System Control Strategy

The control module of the ESS can be divided into four parts, the overall control block diagram of which is shown in Figure 8.

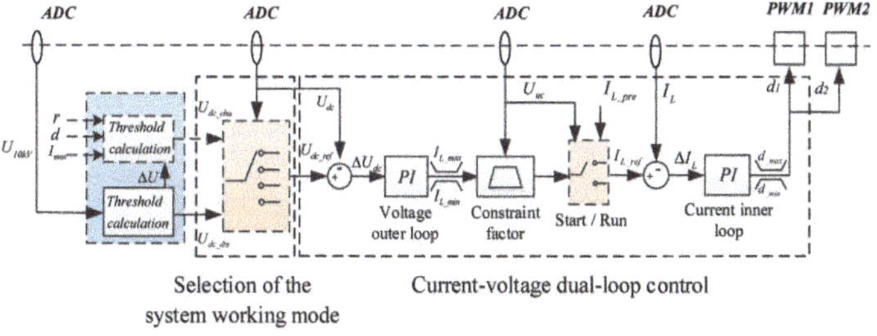

Figure 8. Control block diagram of wayside supercapacitor storage system.

3.2.1. Current-Voltage Dual-Loop Control

As shown in Figure 8, the system uses the cascade control strategy to control the DC bus voltage loop and supercapacitor current loop, and the controller uses the traditional PI control. During the controller design, the small signal analysis model of the supercapacitor ESS can be modeled through the state-space averaging method [28]. Each part of the system transfer function is shown as follows:

$$G_{id}(s)=\frac{\hat{i}_L}{\hat{d}}=\frac{(CRR_pU_{net}+CDR^2U_{uc})s+(R_p-D^2R)U_{net}}{CLR(D^2+R_p)s^2+[CD^2R^2R_p+(LD^2+CR_p{}^2+LR_p)]s+(D^4R^2+2DR_m+R_p{}^2)} \quad (5)$$

$$G_{ud}(s)=\frac{\hat{u}_C}{\hat{d}}=\frac{LR_L(U_{uc}-DU_{net})s+[(D^2R_L{}^2+R_n+R_z)U_{uc}-2D(R_n+R_z)U_{net}]}{CLR(D^2+R_p)s^2+[CD^2R^2R_p+(LD^2+CR_p{}^2+LR_p)]s+(D^4R^2+2DR_m+R_p{}^2)} \quad (6)$$

where G_{id} is the transfer function of the supercapacitor current to the duty cycle, and G_{ud} denotes the transfer function of the bus voltage to the duty cycle. $R = R_{net}$, $R_p = R_{on} + R_L$, $R_m = R_{on}\cdot R_{net}$, $R_n = R_{net}\cdot R_L$, $R_z = R_{on}\cdot R_L$.

3.2.2. Selection of the System Working Mode

In view of the different operating states of the train and achieving the multi-modal switchover, the working area of the supercapacitor ESS must be selected to reasonably realize recycling braking energy. Therefore, based on the voltage state of the traction power supply, the workspace is divided as shown in Figure 9.

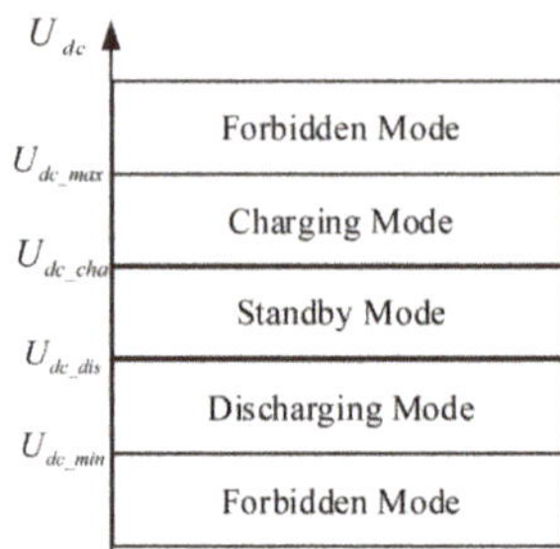

Figure 9. Principle of mode-transition.

On the one hand, the discharge and charge threshold are the switching symbols of different modes, while on the other hand they are also the target regulation values of outer voltage loop of the traction grid. Therefore, the design and selection of the threshold determine whether the ESS can work, and even whether it can work in the best possible state.

4. Threshold Selection of Wayside Supercapacitor Energy Storage System

4.1. Analysis of the Discharge Threshold's Impact and Setting Methods

In the discharge mode, in order to analyze the whole system, including traction substations, ESSs and train loading system, a model is constructed (Figure 10).

The relationship among the parameters in Figure 10 can be obtained as follows:

$$i_{\mathrm{LOAD}} = i_{\mathrm{SUB}} + i_{\mathrm{ESS}} \tag{7}$$

$$V_{\mathrm{REC}} = V_{\mathrm{REC0}} - i_{\mathrm{SUB}} \cdot R_{\mathrm{REC}} \tag{8}$$

Parameter i_{LOAD} is the load current of the train, i_{SUB} is the output current of the traction substation, i_{ESS} is the output current of supercapacitor ESS in the high voltage side, V_{REC0} is the output no load voltage of the rectifier unit, and V_{REC} is the output voltage of rectifier unit.

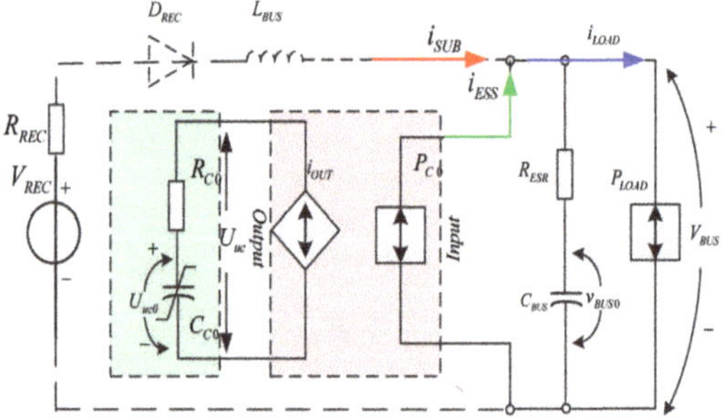

Figure 10. System structure during the discharging mode.

Under the condition that the power of the storage system matches with the braking power of the train, the relationship between the discharge threshold of the ESS and the output voltage of the traction substation can be shown as Equation (9):

$$V_{REC} = V_{REC0} - i_{SUB} \cdot R_{REC} = U_{dc\_dis} \tag{9}$$

The energy storage device output coefficients α is defined as Equation (10), which is used to signify the output situation of the energy storage device and the traction substation:

$$\alpha = \frac{i_{ESS}}{i_{LOAD}} \tag{10}$$

Based on Equations (7)–(10), under different values of i_{LOAD}, how the setting of discharge threshold affects the energy storage device output coefficients is shown in Figure 11.

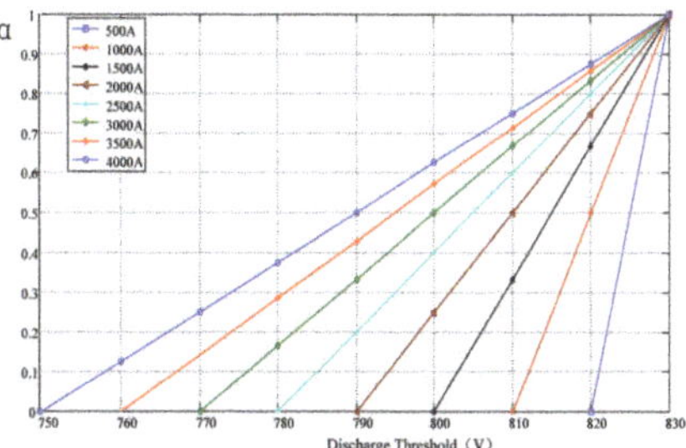

Figure 11. Influence of U_{dc_dis} to α.

From Figure 11, we take the current curve as an example when the load current of train is 2000 A. In the case of the discharge threshold being 830 V (output voltage of the rectifier when there is no load), the output coefficient of the storage system is 1. Nevertheless, the output coefficient decreases linearly as the discharge threshold lessens. When set to 805 V, the output factor of energy storage device is 0.5, *i.e.*, the energy storage device and traction substation bear equal load currents. The worst situation occurs when the discharge threshold is set at 780 V or lower, and the output coefficient of the storage device is 0, indicating that the load current is entirely borne by the traction substation and the storage device is no longer bearing the load current, namely the storage energy devices cannot effectively release the energy stored.

In order to achieve recycling the regenerative braking energy, the energy storage device shall ensure an adequate energy margin, thus considering the requirement that the energy storage device releases energy effectively, and thus the control strategy discharge threshold should preset as Equation (11):

$$U_{dc\_dis} = V_{REC0} \tag{11}$$

4.2. Analysis of the Charge Threshold and Its Impacts

The charge threshold is not only the flag that the storage system runs into charging mode, but also the value of the voltage regulator of the outer loop. In the analysis of the train braking characteristics in Section 2.1, it is shown that when the DC bus voltage rises above 900 V, the braking chopper of the train will be put into operation to curb the bus voltage rising. The operation of the braking chopper means that the regenerative braking energy is almost entirely consumed in the braking resistor, which is unfavorable for saving energy. Therefore, the value of the charge threshold must be reasonable in order to make the voltage at the pantograph be no more than 900 V throughout, therefore allowing the braking resistor's operation to be avoided. This also means that the ESS maximizes the absorption of regenerative braking energy.

The metro power supply system can be divided into traction substation and step-down substation, and a supercapacitor ESS is typically installed in the traction substation. Traction substation and step-down substation are usually set on different metro stops, and the distance between the train braking point and the location of the supercapacitor can be divided into two conditions, as shown in Figure 12. The first is that the braking point is near the metro stop that has a traction substation, which also means that supercapacitor energy storage is performed, while the second is that the braking point is near the metro stop which has a step-down substation, meaning that it has a greater distance with energy storage device than the first circumstance.

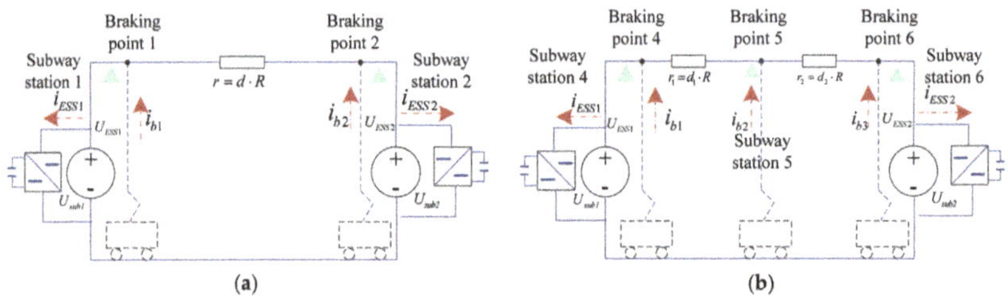

Figure 12. Metro power supply system with supercapacitor storage system: (**a**) Two subway stations with two braking points; (**b**) Three subway stations with three braking points.

In the first circumstance there are no subway stations contained in the middle of two subway stations, with train brakes on the adjacent sides of both stops, as shown in Figure 12a. Another case is that the train has at least three braking points, in which two cases are adjacent to the ESS and one situation is at a greater range, as shown in Figure 12b.

Figure 12a illustrates that the train braking point is in the vicinity of the ESS, thus the voltage drop caused by the line impedance is negligible. Unlike Figure 12a, in Figure 12b the train braking point is not only close to the supercapacitor system, it is also in the middle of the subway stations. The moment at which the voltage drop caused by the line impedance should be considered is shown in Figure 13. If the power and capacity of the ESS are sufficient to absorb electric braking energy, then the bus voltage of ESS side can be stabilized in charge threshold $U_{\text{dc\_cha}}$, and also satisfies the following equation:

$$U_{\text{dc\_dis}} = V_{\text{bus}} - r \cdot i_{\text{LOAD}} = V_{\text{bus}} - d \cdot R \cdot i_{\text{LOAD}} \tag{12}$$

where V_{bus} is the voltage of train pantograph, d is the distance between the braking point and the location of ESS, R is the line impedance and i_{LOAD} is the supercapacitor current regenerated by train braking.

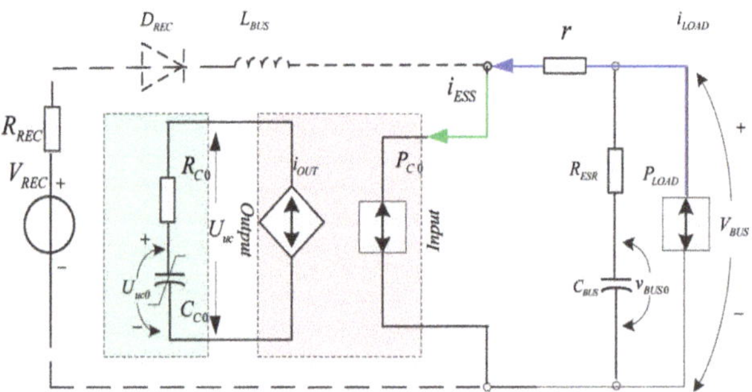

Figure 13. System structure during charging mode.

Defining the train braking energy conversion factor is β, which is used to characterize the electric train braking power conversion efficiency (Equation (13)):

$$\beta = \frac{i_{LOAD}}{i_{REG}} \tag{13}$$

The charge threshold's impact on the conversion factor is shown in Figure 14.

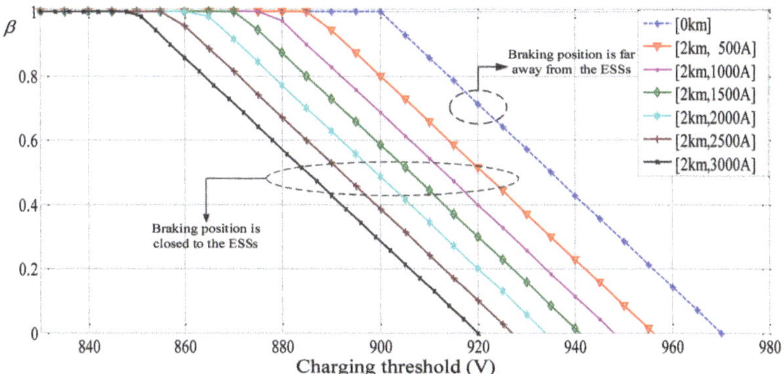

Figure 14. Influence of U_{dc_cha} to the β.

The following conclusions can be obtained through Figure 14.

(a) If the train braking points are in the vicinity of the traction substation, which means being close to the supercapacitor storage system as well, then the line impedance is negligible. The charging threshold is between 830 V and 900 V, due to the fact that charging threshold is smaller than the threshold of 900 V, which is the brake chopper's start symbol. Based on this, the braking resistor does not operate, the electric braking energy is completely absorbed by the supercapacitor system, and the conversion factor of the electric braking power is constantly 1. When the charging threshold is set between 900 V and 970 V, β gradually decreases as the charge threshold increases, *i.e.*, the power of the braking resistor is growing, while the regenerative energy absorbed by the ESS becomes lower and lower. When the charging threshold is set at 970 V, β is 0, indicating that all of the regenerative energy of the train is consumed by resistor braking and mechanical braking.

(b) In the case that the braking position and energy storage device has a far distance, which is assumed as 2 km, the line impedance of steel and aluminum contact rail unit is 0.0069 Ω/km. Taking 2000 A as the feedback current of the train as an example, the charge threshold value is set between 830 V and 860 V, the conversion factor of electric braking power is constantly 1. When the charging threshold is set between 860 V and 930 V, β gradually decreases as the charging threshold increases. The worst situation is when set at 930 V, and β is 0, indicating that all of the braking energy is consumed by the braking resistor and mechanical braking.

On the premise of ensuring that the power of energy storage device matches the braking power of the train, the regenerative braking energy of the train acquire get the feedback to the greatest extent, thus avoiding energy waste. The train braking point is not only close to the supercapacitor system, it is also in the middle of the subway stations. Taking the Figure 12b for example, there are three braking points. The distance between the two traction substations is defined as d. The distance between the braking position and ESS is from 0 to d. If the distance is chosen as d, when the vehicles brake near the ESSs, the charging threshold is too low. Considering the better effect of energy recovery in different braking points, the equivalent distance between the braking position and ESS can be chosen as $d/2$. No matter the vehicle brakes near or keep away the ESSs, the charging threshold is an eclectic selection.

I_{max} is the maximum feedback current in the process of braking, which is calculated through traction calculation. Though the current I is chosen at the maximum situation, the charging threshold keeps a relatively small value to ensure the ESSs can recycle the regenerative braking energy every time.

Therefore, considering the sufficient margin, the charge threshold value should be set as follows:

$$U_{dc\_cha} = U_{dc\_lim} - d_{equ} \cdot R \cdot I_{max} \tag{14}$$

U_{dc_lim} should be set as the start voltage threshold of the train braking chopper, and d_{equ} is the equivalent distance between the braking position and ESS, which is chosen as $d/2$. R is the impedance of contact rail, which is a constant determined by the contact rail material.

4.3. *Dynamic Threshold Adjustment Strategy*

The charging and discharging threshold effects and settings are discussed in Sections 4.1 and 4.2 respectively, and the above conclusions are analyzed under the ideal substation output voltage of 830 V without loads. Considering the fact that the traction substation takes non-ideal output characteristics into account, the control strategy of the wayside supercapacitor ESS is proposed based on dynamic threshold adjustment, shown in Figure 15, and it is the threshold calculation part of Figure 8.

To identify the output voltage of traction substation when there is no load, using the transformer grid side (10 kV) voltage as a reference variable, real-time load voltage value V_{REC0} can be calculated by Equation (15), then discharge threshold U_{dc_dis} can also be obtained:

$$U_{dc\_dis} = V_{REC0} = m \cdot n \cdot U_{10kV} \tag{15}$$

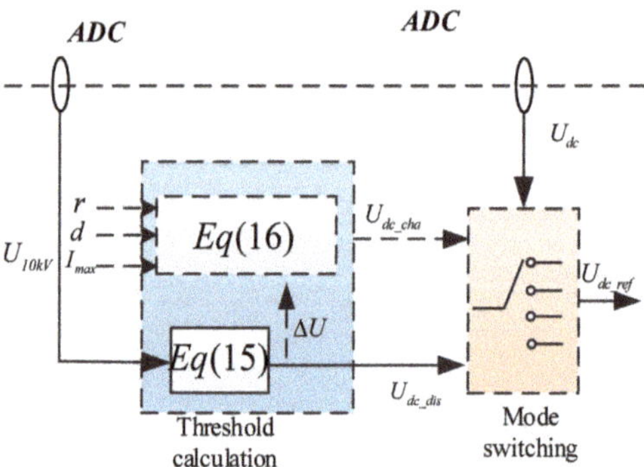

Figure 15. Control strategy based on dynamic threshold.

In Equation (15), m is the coefficient between the 24-pulse rectifier transformer valve side to the DC output voltage of the rectifier, n is the turns ratio coefficient of the primary side of the transformer to the valve side, and U_{10kV} is the line voltage RMS of the transformer primary side. In the case of output voltage of the traction substation has fluctuations, in order to ensure the reliability of the modal switch in preventing accidental charging, the settings of the charging threshold in Equation (14) should be improved, shown as follows:

$$U_{dc\_cha} = \begin{cases} U_{dc\_lim} - d_{equ} \cdot R \cdot I_{max}; & (U_{dc\_lim} - d_{equ} \cdot R \cdot I_{max} \geqslant V_{REC0} + U_{\Delta}) \\ V_{REC0} + U_{\Delta} = m \cdot n \cdot U_{10kV} + U_{\Delta}; & (U_{dc\_lim} - d_{equ} \cdot R \cdot I_{max} < V_{REC0} + U_{\Delta}) \end{cases} \tag{16}$$

Most of the time, the charging threshold is determined by the equation $U_{dc\_cha} = U_{dc\_lim} - d_{equ}RI_{max}$; only when the calculated charging threshold is lower than the $V_{REC0} + U_{\Delta}$, the charging threshold should be set as $V_{REC0} + U_{\Delta}$ to avoid the ESSs charging falsely.

5. Simulation and Experimental Results

5.1. Simulation Platform and Parameters

Using Matlab/Simulink, we built a model including a 24-pulse rectifier unit, trains and a supercapacitor ESS, which is shown as Figure 16.

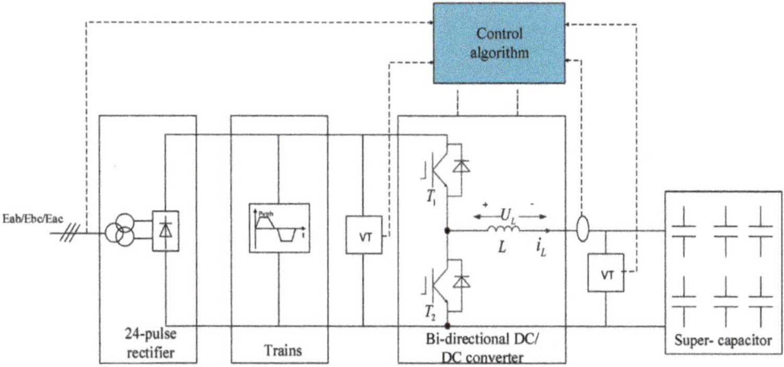

Figure 16. Simulation platform in Matlab/Simulink.

The simulation parameters are shown in Table 1.

Table 1. Simulation parameters.

Substation	U_{1N}	10 kV
	U_{1N}:U_{2N}	10,000:590
Storage system	L	0.83 mH
	L_d	1.5 mH
	C_{sc}	31.5 F
	C_d	5000 uF
	U_{dc_cha}	900 V

5.2. Simulation Results and Analysis

The line voltages of the transformer primary in the simulation were, respectively, given as 10, 9.3 and 10.7 kV. The simulation results are shown in Figure 17; when the medium-voltage grid voltage is 10 kV, the discharge threshold is adjusted to 830 V automatically. Another case, when grid voltage is 9.3 kV, the discharge threshold is adjusted to 770 V. While the grid voltage is 10.7 kV, the discharge threshold value is adjusted to 880 V. The supercapacitor ESS in the process of the above changes achieves the release of energy in a reliable way, and also verifies the feasibility of the policy based on dynamic threshold adjustment.

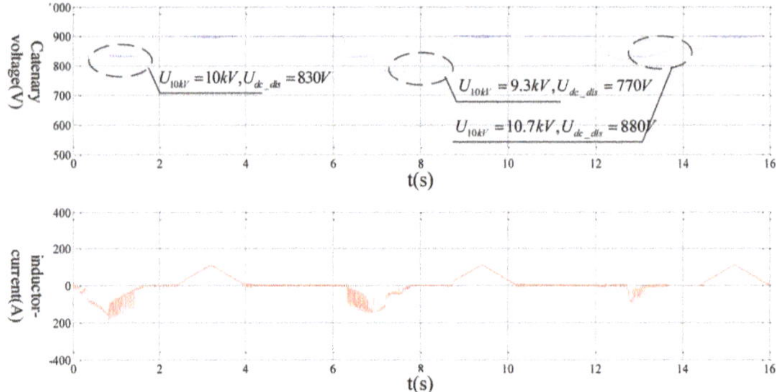

Figure 17. Simulation result of dynamic voltage threshold.

5.3. Experimental Conditions

In order to verify the feasibility of the control algorithm, a 200 kW wayside supercapacitor ESS was developed. The system was installed in the Tongzhou Beiyuan substation of the Batong Line of the Beijing subway system, and tests were conducted during the day when the Batong Line is under normal operation. Based on these experiments, the control method's correctness has been verified. The ESS is shown in Figure 18.

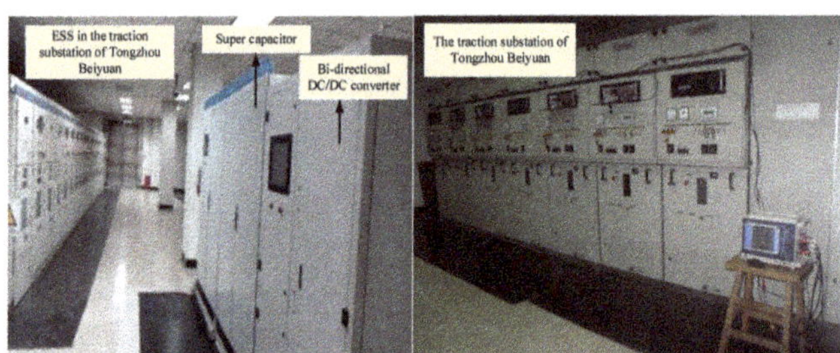

Figure 18. Supercapacitor ESS installed in substation.

The system design parameters are shown in Table 2, and the bidirectional DC/DC converter rated power is 200 kW, wherein the high side input voltage is 500–1000 V, rated working voltage is 750 V and the peak voltage of 1000 V. In addition, the low side output current is 0–400 A. Peaking power of supercapacitor is 200 kW, operating voltage range is from 0 V to 500 V and the operating current is 0–400 A. The maximum storage energy of supercapacitor is 0.944 kWh in total.

Table 2. Parameters of 200 kW wayside supercapacitor storage system.

Parameters		Value
Bi-directional DC/DC converter	Input voltage (V)	500–1000
	Input current (A)	0–267
	Output voltage (V)	0–500
	Output current (A)	0–400
Ultra capacitor	Working voltage (V)	0–500
	Working current(A)	0–400
	Maximum storage energy (kWh)	0.944

The main wiring schematic of the Tongzhou Beiyuan traction substation in the Beijing Subway system is shown in Figure 19.

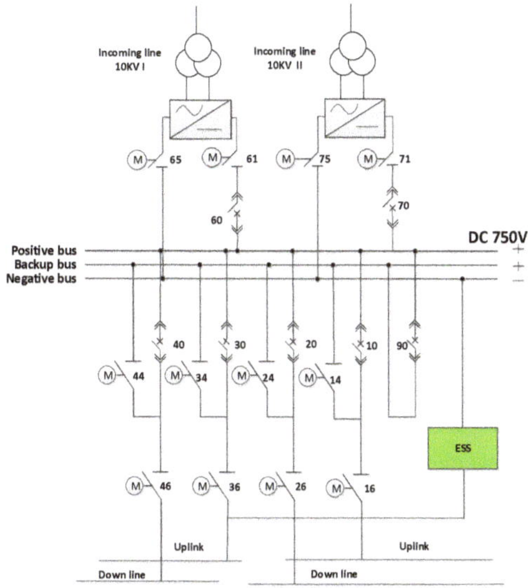

Figure 19. Primary connection of traction substation with supercapacitor storage system.

5.4. Experimental Results and Analysis

When the train is on the state of traction, the DC voltage drops, while along with the storage system's work, the DC voltage dropping degree can be suppressed. On the contrary, the DC voltage boosts when the train brakes. The supercapacitor storage system recycles the regenerative braking energy so as to suppress the DC voltage's increase. These are shown in Figure 20.

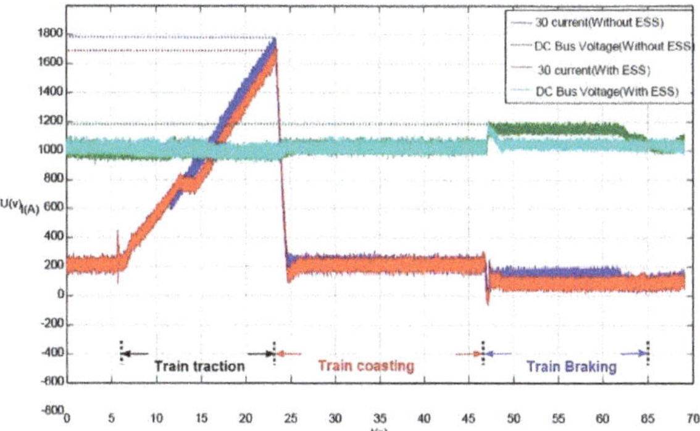

Figure 20. One Traction and braking cycle with and without ESS.

Next, tests are performed under circulation conditions, as shown in Figure 21. Through the above trials, the supercapacitor system can charge and discharge normally. In addition, it can also stabilize the DC voltage and save braking energy, which can be analyzed through the experiment waveform as follows.

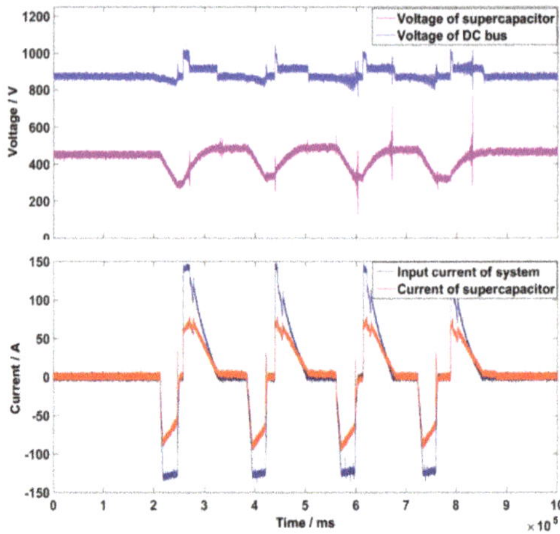

Figure 21. Waveforms in multiple cycles.

In Section 4.1, the influence of the discharge threshold is analyzed, and the analysis and conclusions are verified by experiments during night tests. In the night tests, the train runs four successive cycles (traction-coasting-braking), and the voltage and current of the supercapacitor energy storage prototype has been recorded by oscilloscope. During the experiment, the DC voltage without load is 830 V. In the process of the tests, the U_{dc_dis} was given as 805, 810 and 815 V, and the voltage of the supercapacitor is shown in Figure 22.

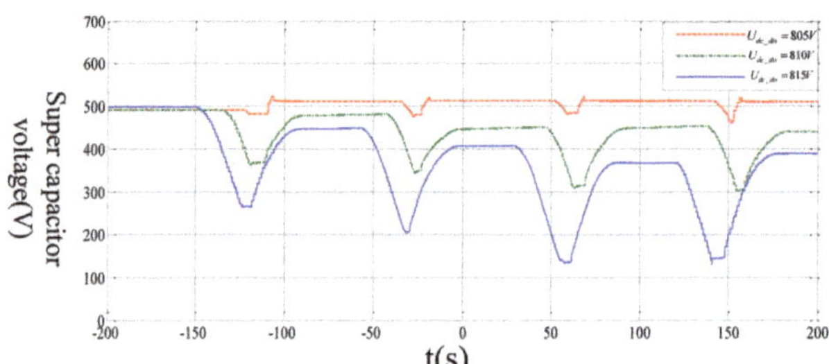

Figure 22. Influence of U_{dc_dis}.

E_{train} is on the behalf of the energy consumed by train at the traction stage, and E_{uc} is the supercapacitor ESS's output energy. When discharge threshold U_{dc_dis} is 805, 810 and 815 V, the respective output coefficients are 2.6%, 13% and 25%. When the discharge threshold is closer to output voltage V_{REC0} without load, the output coefficient of the supercapacitor ESS increases. When its power and traction power train match, the output coefficient can be up to 1, which is consistent with the theoretical analysis in Section 4.1.

The following tests are conducted during the day when there are many trains running with passengers. The charging threshold is constantly set to 890 V during the first test, and due to the presence of the line impedance, the threshold is set too high, thus the ESS basically does not absorb regenerative braking energy, as shown in Figure 23a. When the train is pulled, the bus voltage drops, the energy storage device discharges, the voltage of the supercapacitor changes from 500 V to 200 V. Because of unreasonable charging threshold settings, at the stage of train braking, the DC bus voltage rises, but no more than the charge threshold. Thus in the whole test cycle, energy storage device does not absorb the regenerative braking energy basically. At this time, the regeneration braking energy can only be consumed by the braking resistor, which results in energy waste. The energy saving effect greatly reduces because the inappropriate threshold setting.

Then the dynamic threshold adjustment strategy was used, as shown in Figure 23b. The ESSs charging times are more in number, and absorbing more regeneration braking energy, which means that the energy saving effect of ESS is improving. During the test cycles, due to the appropriate threshold setting and the discharging threshold changes with the DC power supply system, the ESS shows good energy saving effect. At every test cycle, the supercapacitor can be charged to the maximum voltage 500 V when the train is braking and discharged to the minimum voltage 200 V when the train is in the state of traction, which means the ESS has fully absorbed the regenerative braking energy and released the energy to the traction power supply when the DC bus voltage drops.

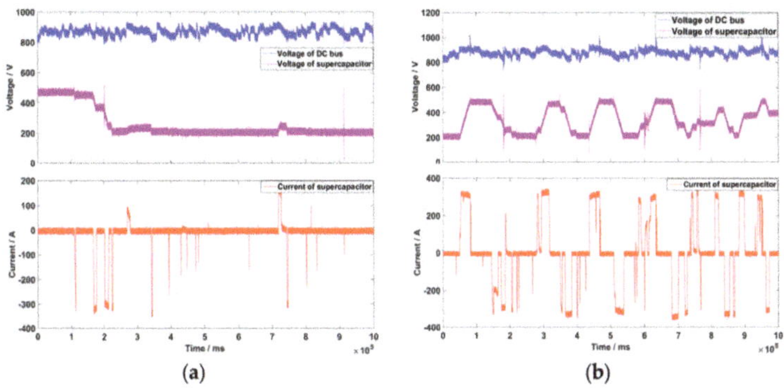

Figure 23. Threshold impacts on energy savings: (**a**) Charging threshold is 890V; (**b**) Dynamic threshold adjustment strategy.

6. Conclusions

In this study, based on the analysis of braking characteristics of urban rail transit train and 24-pulse traction substation characteristics, a control strategy is proposed to apply into the wayside supercapacitor ESS. On the one hand, the energy-saving and voltage stabilization effects of the supercapacitor applied to urban rail transit are verified through simulation and prototype test in Beijing subway, while on the other hand, the feasibility of the dynamic threshold adjustment control strategy is verified. In this paper, the research shows that the threshold of charge and discharge must be adjusted dynamically, and the performance is satisfactory.

Acknowledgments: This work was supported by the National Natural Science Foundation of China under Grant 51577010.

Author Contributions: Fei Lin, Xuyang Li and Yajie Zhao contributed significantly to the analysis and manuscript preparation. Zhongping Yang contributed to the conception of the study.

Conflicts of Interest: The authors declare no conflict of interest.

References

1. Qin, F.; Zhang, X.; Zhou, Q. Evaluating the impact of organizational patterns on the efficiency of urban rail transit systems in China. *J. Transp. Geogr.* **2014**, *40*, 89–99. [CrossRef]
2. Ogasa, M. Energy saving and environmental measures in railway technologies: Example with hybrid electric railway vehicles. *IEEJ Trans. Electr. Electron. Eng.* **2008**, *3*, 15–20. [CrossRef]
3. Ogasa, M. Onboard Storage in Japanese Electrified Lines. In Proceedings of the 14th International Power Electronics and Motion Control Conference, Ohrid, Macedonia, 6–8 September 2010. [CrossRef]

4. González-Gil, A.; Palacin, R.; Batty, P. Sustainable urban rail systems: Strategies and technologies for optimal management of regenerative braking energy. *Energy Convers. Manag.* **2013**, *75*, 374–388. [CrossRef]
5. Richardson, M.B. Flywheel Energy Storage System for Traction Applications. In Proceedings of the 2002 International Conference on Power Electronics, Machines and Drives, Bath, UK, 4–7 June 2002; pp. 275–279.
6. Liu, H.; Jiang, J. Flywheel energy storage—An upswing technology for energy sustainability. *Energy Build.* **2007**, *39*, 599–604. [CrossRef]
7. Thompson, R.C.; Kramer, J.; Hayes, R.J. Response of an urban bus flywheel battery to a rapid loss-of-vacuum event. *J. Adv. Mater.* **2005**, *37*, 42–50.
8. Gabash, A.; Li, P. Active-reactive optimal power flow in distribution networks with embedded generation and battery storage. *IEEE Trans. Power Syst.* **2012**, *27*, 2026–2035. [CrossRef]
9. Gabash, A.; Li, P. Flexible optimal operation of battery storage systems for energy supply networks. *IEEE Trans. Power Syst.* **2013**, *28*, 2788–2797. [CrossRef]
10. Iannuzzi, D. Improvement of the Energy Recovery of Traction Electrical Drives Using Supercapacitors. In Proceedings of the 13th IEEE Power Electronics and Motion Control Conference, Poznan, Poland, 1–3 September 2008; pp. 1469–1474.
11. Barrero, R.; Tackoen, X.; Van Mierlo, J. Improving Energy Efficiency in Public Transport: Stationary Supercapacitor Based Energy Storage Systems for a Metro Network. In Proceedings of the IEEE Vehicle Power and Propulsion Conference, Harbin, China, 3–5 September 2008; pp. 1–8.
12. Ciccarelli, F.; Iannuzzi, D.; Tricoli, P. Control of metro-trains equipped with onboard supercapacitors for energy saving and reduction of power peak demand. *Transp. Res. Part C Emerg. Technol.* **2012**, *24*, 36–49. [CrossRef]
13. Latkovskis, L.; Brazis, V.; Grigans, L. Simulation of on Board Supercapacitor Energy Storage System for Tatra T3A Type Tramcars. In *Modelling Simulation and Optimization*; InTech: Rijeka, Croatia, 2010.
14. Teymourfar, R.; Asaei, B.; Iman-Eini, H. Stationary super-capacitor energy storage system to save regenerative braking energy in a metro line. *Energy Convers. Manag.* **2012**, *56*, 206–214. [CrossRef]
15. Iannuzzi, D.; Tricoli, P. Metro Trains Equipped Onboard with Supercapacitors: A Control Technique for Energy Saving. In Proceedings of the 2010 International Symposium on Power Electronics, Electrical Drives, Automation and Motion (SPEEDAM), Pisa, Italy, 14–16 June 2010; pp. 750–756.
16. Iannuzzi, D.; Lauria, D.; Tricoli, P. Optimal design of stationary supercapacitors storage devices for light electrical transportation systems. *Optim. Eng.* **2012**, *13*, 689–704. [CrossRef]
17. Battistelli, L.; Fantauzzi, M.; Iannuzzi, D.; Lauria, D. Generalized Approach to Design Supercapacitor-Based Storage Devices Integrated into Urban Mass Transit Systems. In Proceedings of the IEEE 2011 International Conference on Clean Electrical Power (ICCEP), Ischia, Italy, 14–16 June 2011; pp. 530–534.
18. Teymourfar, R.; Farivar, G.; Iman-Eini, H.; Asaei, B. Optimal Stationary Super-Capacitor Energy Storage System in a Metro Line. In Proceedings of the 2011 2nd International Conference on Electric Power and Energy Conversion Systems (EPECS), Sharjah, The United Arab Emirates, 15–17 November 2011; pp. 1–5.
19. Grbović, P.J.; Delarue, P.; Le Moigne, P.; Bartholomeus, P. Modeling and control of the ultracapacitor-based regenerative controlled electric drives. *IEEE Trans. Ind. Electron.* **2011**, *58*, 3471–3484. [CrossRef]
20. Grbovic, P.J.; Delarue, P.; Le Moigne, P. Modeling and Control of Ultra-Capacitor Based Energy Storage and Power Conversion System. In Proceedings of the 2014 IEEE 15th Workshop on Control and Modeling for Power Electronics (COMPEL), Santander, Spain, 22–25 June 2014; pp. 1–9.
21. Ciccarelli, F.; Iannuzzi, D.; Lauria, D. Supercapacitors-Based Energy Storage for Urban Mass Transit Systems. In Proceedings of the 2011 14th European Conference on Power Electronics and Applications (EPE 2011), Birmingham, UK, 30 August–1 September 2011; pp. 1–10.
22. Zhang, Y.; Wu, L.; Hu, X.; Liang, H. Model and Control for Supercapacitor-Based Energy Storage System for Metro Vehicles. In Proceedings of the International Conference on Electrical Machines and Systems (ICEMS 2008), Wuhan, China, 17–20 October 2008; pp. 2695–2697.
23. Battistelli, L.; Ciccarelli, F.; Lauria, D.; Proto, D. Optimal Design of DC Electrified Railway Wayside Storage System. In Proceedings of the 2009 International Conference on Clean Electrical Power, Capri, Italy, 9–11 June 2009; pp. 739–745.
24. Barrero, R.; Tackoen, X.; van Mierlo, J. Stationary or onboard energy storage systems for energy consumption reduction in a metro network. *Proc. Inst. Mech. Eng. Part F J. Rail Rapid Transit* **2010**, *224*, 207–225. [CrossRef]
25. Vuchic, V.R. *Urban Transit Systems and Technology*; John Wiley & Sons: Hoboken, NJ, USA, 2007.

26. Wang, X.; Zang, H. Simulation study of DC traction power supply system for urban rail transportation. *Acta Simulata Syst. Sin.* **2002**, *12*, 1692–1697.
27. Han, L. A Study on the operation mode of medium voltage double-ring network and the relationship between interlock and inter-tripping in urban rail transit system. *Urban Rapin Rail Transit* **2004**, *1*, 018.
28. Yang, S.; Goto, K.; Imamura, Y.; Shoyama, M. Dynamic Characteristics Model of Bi-Directional DC-DC Converter Using State-Space Averaging Method. In Proceedings of the 2012 IEEE 34th International Telecommunications Energy Conference (INTELEC), Scottsdale, AZ, USA, 30 September–4 October 2012; pp. 1–5.

Article

Optimal Energy Management, Location and Size for Stationary Energy Storage System in a Metro Line Based on Genetic Algorithm

Huan Xia *, Huaixin Chen, Zhongping Yang, Fei Lin and Bin Wang

School of Electrical Engineering, Beijing Jiaotong University, No.3 Shangyuancun, Beijing 100044, China; 13121389@bjtu.edu.cn (H.C.); zhpyang@bjtu.edu.cn (Z.Y.); flin@bjtu.edu.cn (F.L.); 12121547@bjtu.edu.cn (B.W.)

* Author to whom correspondence should be addressed; huanhuan7000@gmail.com; Tel.: +86-10-5168-4864.

Academic Editor: William Holderbaum

Received: 18 August 2015; Accepted: 14 September 2015; Published: 16 October 2015

Abstract: The installation of stationary super-capacitor energy storage system (ESS) in metro systems can recycle the vehicle braking energy and improve the pantograph voltage profile. This paper aims to optimize the energy management, location, and size of stationary super-capacitor ESSes simultaneously and obtain the best economic efficiency and voltage profile of metro systems. Firstly, the simulation platform of an urban rail power supply system, which includes trains and super-capacitor energy storage systems, is established. Then, two evaluation functions from the perspectives of economic efficiency and voltage drop compensation are put forward. Ultimately, a novel optimization method that combines genetic algorithms and a simulation platform of urban rail power supply system is proposed, which can obtain the best energy management strategy, location, and size for ESSes simultaneously. With actual parameters of a Chinese metro line applied in the simulation comparison, certain optimal scheme of ESSes' energy management strategy, location, and size obtained by a novel optimization method can achieve much better performance of metro systems from the perspectives of two evaluation functions. The simulation result shows that with the increase of weight coefficient, the optimal energy management strategy, locations and size of ESSes appear certain regularities, and the best compromise between economic efficiency and voltage drop compensation can be obtained by a novel optimization method, which can provide a valuable reference to subway company.

Keywords: energy storage system; super-capacitor; energy management; configuration; economic efficiency; voltage drop compensation; genetic algorithm

---

1. Introduction

In recent years, with the rapid development of the Chinese economy, growing environmental pollution, and traffic congestion in major cities are becoming serious social issues. For the purpose of improving the urban environment and energy efficiency, the development of modern urban rail transit, which has the significant advantages of large capacity, punctuality, safety, energy conservation, and environmental protection, becomes a social consensus [1,2]. Low running resistance and the reuse of braking energy are two main factors that make urban rail transit better than other means of transport in energy efficiency. Recent studies have shown that up to 40% of the energy supplied to electrical rail guided vehicles could be recovered through regenerative braking [1]. In a metro network system, the trains are accelerated and braked frequently. Since most of the rectifiers in the metro network are unidirectional, the regenerative braking energy cannot be returned to the supply network, and if there are no adjacent accelerating trains or energy storage system to absorb the regenerative energy, the surplus braking energy has to be wasted on the mechanical braking or on-board resistors. If different trains are close to each other and they start all together, contact lines will become overloaded and the pantograph voltages of trains will drop significantly, which results in high lines loss and the opening of minimum voltage protective action of trains by limiting the current. Hence, the installation of energy storage systems in urban railway transit has become a universal concern, which can recycle the regenerative braking energy, prevent regeneration cancellation, shave the peak power of substations, and compensate the voltage drops of pantograph quickly.

Current research activities have presented the application of batteries, flywheels, super-capacitors, and hybrid energy storages as energy storage devices [3–8]. Among the different storage systems available, super-capacitors seem to be the most appropriate for the application in a metro system for the advantages of rapid charging and discharging frequencies, a long cycle life, and high power density, which highly match the characteristics of metro system, such as short running time between stations, frequent accelerating and braking, booming power within a short time, *etc.* Super-capacitor energy storage systems (ESS) can be either stationary or on-board [8–11]. The allocation on board of the storage system increases the train mass and requires additional space for their accommodation. Thus, stationary ESSes set inside traction substations (TSSs) are preferred for metro systems, and their best energy management, location, and size will be discussed in this paper.

Several papers have dealt in depth with optimization of energy management strategies of stationary ESSes [12–14]. Among them [12] proposes a control strategy based on the maximum kinetic energy recovery throughout braking operations of the running vehicles. The strategy stays on the knowledge of the state of charge of ESS and the actual vehicle speeds. Reference [13] proposes a optimization procedure based on a linearized modeling of the electrical LRV network, the target of the control strategy is the optimal tracking of the storage device voltage subject to the minimization of the substations supplied power. Optimal location and size of ESSes are also investigated in detail in [15–22]. Reference [15] discusses the configuration of ESSes for voltage drop compensation, which takes account of the topology of the line and the movement of the vehicles. Reference [16] proposes an optimization method based on a genetic algorithm, which can obtain certain preferable location and size for ESSes.

However, there are still some drawbacks on the above research. Firstly, some of the references involve only small amounts of substations and vehicles when modeling the urban railway network [12–15] and some of them do not take into account the time-variation (network topology change with vehicle movement) and nonlinearity (nonlinearity of substation and regenerative braking) of the network structure. Secondly, and most importantly, the optimization research of energy management strategy and configuration for ESSes will influence each other, and they both affect the performances of urban railway network, while the configuration optimization research of ESSes in reference [16–22] is on the premise that energy management strategy of ESSes is fixed and invariable.

In this paper improved energy management strategy of ESSes and novel optimization method are proposed. Compared to previous work [16–22], the improved energy management strategy can manage and coordinate the energy flow of multiple ESSes, which can achieve smoother changes of voltages and currents in the system and improve the energy savings of ESSes effectively, and the new proposed optimization algorithm can further improve the performance of ESSes by optimizing energy management parameters, location, and size of ESSes simultaneously, which has rarely been studied in previous work about the optimization of ESSes, and the evaluation functions of proposed optimization algorithm in this paper are more appropriate, which are put forward from the perspectives of economic efficiency and voltage drop compensation.

The organization of this work is as follow: the simulation platform of urban rail power supply system, which includes trains and super-capacitor energy storage systems has been established in Section 2; additionally, particular data of the researched Chinese metro line is given. Then Section 3 sets up two evaluation functions from the perspectives of economic efficiency and voltage drop compensation. In Section 4, a novel optimization method based on genetic algorithm (GA) is put forward, which can optimize energy management strategy, location and size of ESSes simultaneously. Finally in Section 5, the result of the simulation comparison is presented and discussed.

2. Modeling

2.1. Model of Metro Power Supply Network

The model of metro system's DC traction power supply network is shown in Figure 1 [16]. In order to show the behavior of the metro power supply network as correctly as possible, all components of the metro network, which includes irreversible traction substations (TSS), trains, metro lines, and stationary energy storage systems (ESSes), will be modeled appropriately to maintain original characters of the network structure's time-variation and nonlinearity.

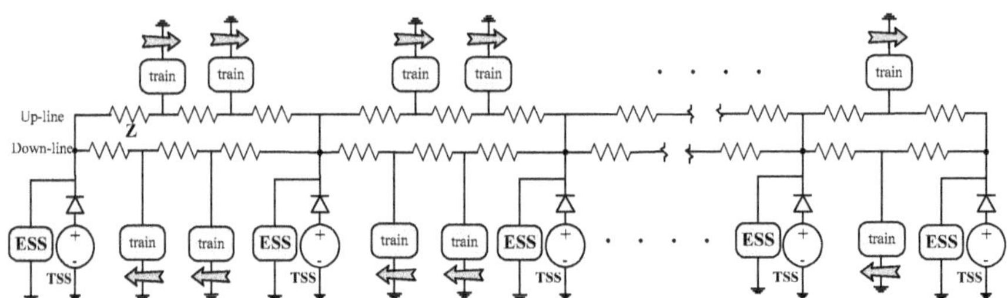

Figure 1. The model of metro system's DC traction power supply network.

2.1.1. Traction Substation (TSS) Model

As shown in Figure 2, the substation is modeled by an ideal DC voltage source connected in series with its equivalent internal resistance R_S and the diode D, which to simulate output characteristics. When the output current of substation is increased, the voltage of substation decreases correspondingly to limit its output power. U_0 is the no-load voltage of substations.

$$i_{\text{uout}} = i_{\text{uin}} + i_{\text{din}} + i \tag{1}$$

$$i = i_{\text{uc}} + i_{\text{sub}} \tag{2}$$

$$i_{\text{sub}} \geq 0 \tag{3}$$

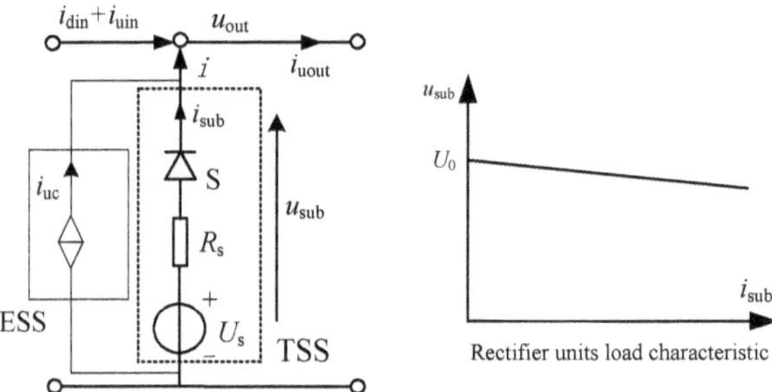

Figure 2. TSS model.

2.1.2. Train Model

As shown in Figure 3, the train model is modeled by a controlled current source which draws electric power at the accelerating time and delivers braking power at the regenerative time. The impedance of the line connected to the trains is expressed as Z, the value of which is time-varying, and it is linear with the line length that is determined by the present position of trains.

When pantograph voltage exceeds U_b, the braking resistor R_b will consume the braking energy. R_f is vehicle filter resistance; L_f is vehicle filter inductance; C_{fc} is the support capacitor of train; P_{aux} is auxiliary power; and P is the electric power of train.

$$u_{\text{out}} = u_{\text{in}} + Ri_{\text{out}} + L\frac{di_{\text{out}}}{dt} \tag{4}$$

$$i_{\text{out}} = i_{\text{in}} + i \tag{5}$$

$$i = -i_{inv} - \frac{p_{aux}}{u_{fc}} - C_{fc}\frac{du_{fc}}{dt} \tag{6}$$

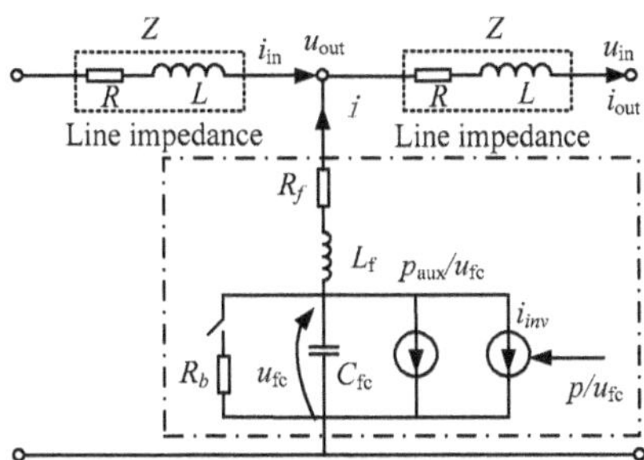

Figure 3. Train model.

2.1.3. Energy Storage System (ESS) Model

The ESS model consists of the super-capacitors, controlled current source, and energy management strategy controller, is shown in Figure 4. The ESS model is connected in parallel with the output of the substation, and it can deliver or draw the electric power from the metro power supply network through the current source which is controlled by the energy management strategy and configuration of super-capacitors in real time.

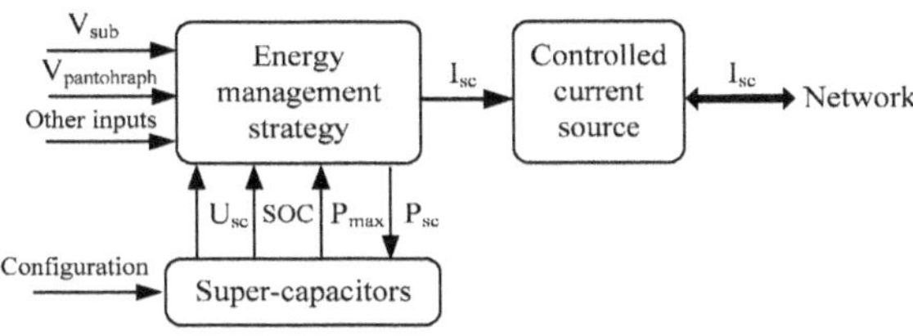

Figure 4. Stationary ESS model.

The SOC (State of Charge) of super-capacitors is defined as follows, it represents the storage energy of ESS, which is proportional to the square of the terminal voltage.

$$0.25 \leq SOC = \frac{E_{\text{sc}}}{E_{\text{scmax}}} = \frac{0.5CU_{\text{sc}}^2}{0.5CU_{\text{scmax}}^2} = \frac{U_{\text{sc}}^2}{U_{\text{scmax}}^2} \leq 1 \tag{7}$$

In a practical application, the function of controlled current source in the model is generally implemented by the unidirectional DC/DC converter. In order to maintain the normal operation of DC/DC converter, the terminal voltage of super-capacitors should be set between 50% and 100% the maximum voltage, so the range of SOC varies from 0.25 to 1.

2.2. Simulation Platform of Metro System for Power Flow Calculation

As above, the model of a DC metro power supply network (DC-PSN) is set up by a novel approach of component segmentation. In order to calculate the power flow of the DC metro power supply network, an integrated simulation platform, which includes DC metro power supply network (DC-PSN), train performance simulator (TPS), and super-capacitor energy storage system (SCESS) is established in the Matlab environment, as shown in Figure 5 [16].

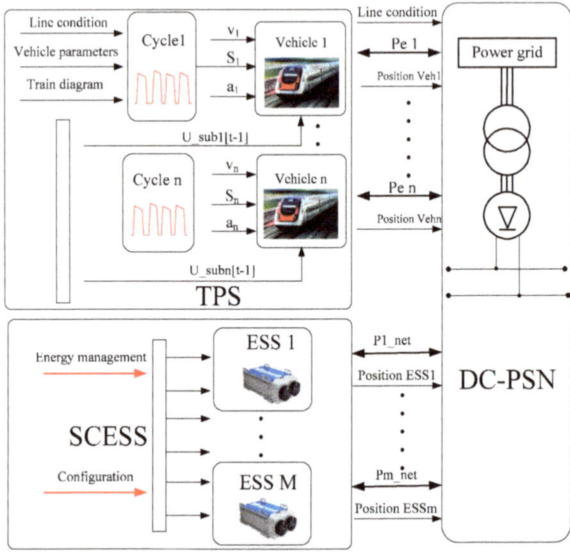

Figure 5. Simulation platform of metro system for power flow calculation.

2.2.1. DC Metro Power Supply Network (DC-PSN)

In the previous section, the paper has presented the structure and model of a DC-PSN. In the power flow calculation of the DC metro power supply network, because of its time-variation (network topology changes with train movement) and nonlinearity (nonlinearity of substation and regenerative braking) of the network structure, a new power flow calculation method by component segmentation is presented. The simulation result shows excellent rapidity and astringency can be obtained by this method. Moreover, the structure and model of the DC-PSN can be extended easily.

2.2.2. Train Performance Simulator (TPS)

As shown in Figure 5, the output of TPS is not only associated with line condition, vehicle data, and timetable, but is also constrained by real-time train pantograph voltage. From the TPS we can get positions of up-line and down-line trains and their corresponding electric power, which offer essential data for subsequent power flow calculation of the DC supply network.

2.2.3. Super-Capacitor Energy Storage System (SCESS)

SCESS set certain energy management strategy, location, and size of ESSes on different substations, which determine the power direction and value of ESSes in real time. The installation of ESSes will change the power flow of the DC metro power supply network and the system performances can be improved significantly by setting the most appropriate energy management strategy, location, and size.

2.3. Case Data

A particular case of Beijing Subway line is studied in this paper. The total length of the line is about 11.3 km along with 12 stations, of which seven are traction substations and their distribution is shown in Table 1. The vehicle data and metro DC network parameters are shown as Table 2. These parameters are provided by the Beijing Subway Company.

Table 1. TSS spacing distances.

Traction substation	1–2	2–3	3–4	4–5	5–6	6–7
Substation spacing (km)	1.1	1.9	2.2	2.3	2.1	2.7

Table 2. Vehicle data.

Parameter	Value	Parameter	Value
Formation	3M3T	Inverter efficiency	0.97
Load condition	312.9t (AW3)	Motor efficiency	0.915
Rated voltage	750 Volt	Gearing efficiency	0.93
AC motor/M	180 kW × 4	Max speed	80 km/h
SIV power	160 kVA × 2	Max acceleration	1 m/s^2
SIV power factor	0.85	Min deceleration	−1 m/s^2
Floating voltage U_s	836 Volt	Equivalent internal resistance R_s	0.07 Ω
Contract line impendence	0.007 Ω/km	Rail impendence	0.009 Ω/km
Pantograph impendence	0.015 Ω	–	–

2.4. Simulation Output

Under simulation conditions, super-capacitor ESSes of 14 kWh are configured in every other substation and controlled with a traditional double-loops control strategy [16]. The simulation output waveforms is shown in Figure 6, which include speed and electric power of an up-line train, voltage and current of the train pantograph, voltage and current of a substation, charging energy, and SOC of the ESS in the substation. SOC of ESS varies between 0.25 and 1.

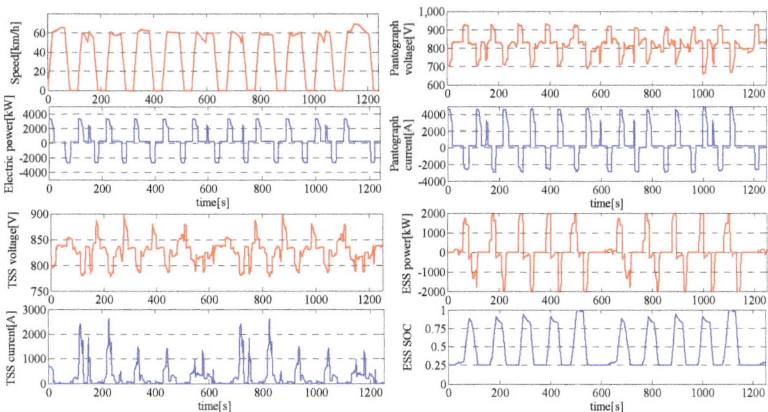

Figure 6. The output waveforms of simulation platform.

3. Objective Function

3.1. Objective Function

In order to evaluate the system performances in terms of energy saving, voltage drop compensation, and installation cost for different energy management strategy and configuration of ESSes, the paper puts forward two evaluation functions and one objective function.

3.1.1. Economic Efficiency, $e\%$

Economic efficiency $e\%$ is put forward from the viewpoint of considering energy savings and installation cost in a unified way to evaluate the economic return rates of ESSes for Subway Company. Economic efficiency $e\%$ is a percentage calculated by dividing the total electricity price of the substations by economic savings (returns minus costs).

As shown in Figure 7, one super-capacitor ESS, which includes the connection unit, DC/DC converter, and super-capacitor strings, is installed on the traction substation. The circuit structure of super-capacitor ESS is shown in Figure 8. The installation cost of ESSes is determined by various factors, which include the capacity, equipment, control circuit, maintenance cost, *etc.* The cost of DC/DC converters and super-capacitor strings are determined mainly by the maximum power of the ESS. In order to ease the DC/DC converter design, maximum voltage of super-capacitor strings should be lower than network voltage (836 V at no load). Hence, six super-capacitor modules (BMOD0063P125) are put in series to form a super-capacitor string, which has terminal voltage of 750 V and maximum continuous power of 180 kW, and the configuration of super-capacitors installed in every substation could be adjusted by changing the number of paralleled super-capacitor strings. The Parameters of super-capacitor modules (BMOD0063P125) are shown as Table 3.

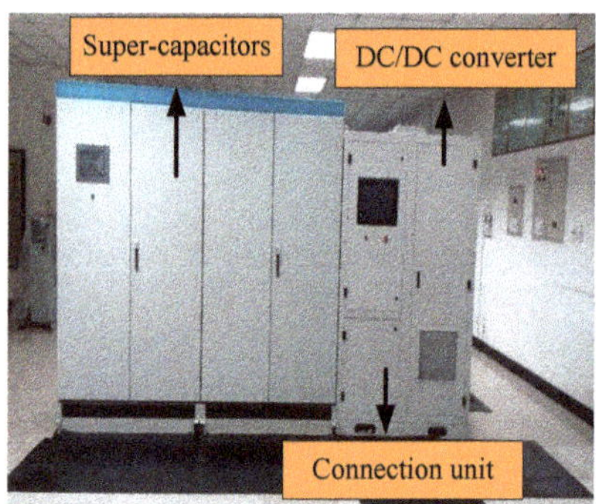

Figure 7. Super-capacitor ESS installed in traction substation.

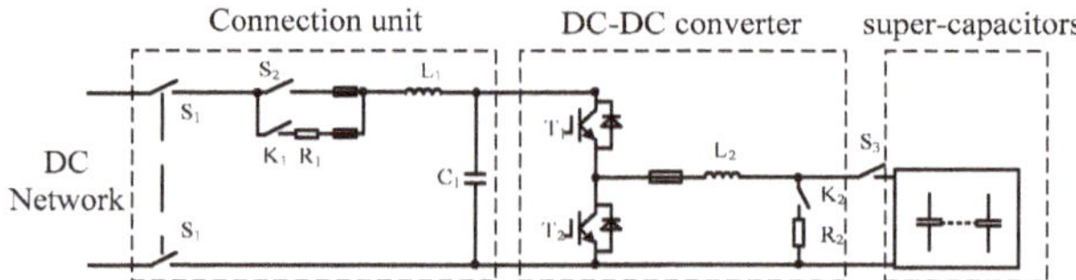

Figure 8. The circuit structure of super-capacitor ESS.

Table 3. Parameters of super-capacitor module (BMOD0063P125).

Parameter	Value	Parameter	Value
Rated voltage	125 V	Capacitance	63 F
Maximum continuous current	240 A	Maximum continuous power	30 kW
Maximum ESR_{DC}, initial	0.018 Ω	Energy	0.137 kWh
Price	5,333 $		

The cost of investment for a super-capacitor ESS on substation k during their life time of l years, can be calculated by:

$$Cost_k = \begin{cases} 0, n_k = 0 \\ (C + n_k \times p \times m) \times (1+r)^l, 0 < n_k \leq 18 \end{cases} \tag{8}$$

where n_k is the number of paralleled super-capacitor strings on substation k; p is the maximum power of one super-capacitor string; and m is dollar per power constant for super-capacitors and DC/DC converter. If n_k equals 0, a super-capacitor ESS would not be installed on substation k. If n_k is more than 0, the cost of investment for super-capacitor ESS $Cost_k$ includes two parts. $n_k \times p \times m$ is the cost of DC/DC converters and super-capacitor strings that are determined by the maximum power of the ESS; C is other part of installation cost from protective device, breaker, maintenance cost, *etc.*, which has a small relationship to power of the ESS. r is the rate of return constant. Considering the limited free space of each metro substation, the number of paralleled super-capacitor strings n_k on each substation is no bigger than 18 in this paper.

By taking the sum of output energy consumption of all TSSes along the metro line, the total energy consumption of the substations in kWh during one year can be calculated from the following formula:

$$E_{\text{sub}} = \sum_{1}^{k} \left[\int_0^T (I_{\text{sub}} \cdot U_{\text{sub}}) dt \right] \times \frac{365}{3600000} \tag{9}$$

where k is the number of traction substations; T is the running time in one day. U_{sub}, I_{sub} are, respectively, the voltage and current of substation.

The application of ESSes in a metro system can reduce the total energy consumption of the substations because of the recycle of trains' regenerative braking energy, but the installation cost should also be considered as well. The total profit obtained by ESSes in l years should be the difference between the saved electricity price and the installation cost of ESSes.

$$P_{\text{sub}}^{\text{nosc}} = E_{\text{sub}}^{\text{nosc}} \times \varepsilon \times \frac{l(2 + (l-2)i)}{2} \tag{10}$$

where $E_{\text{sub}}^{\text{nosc}}$ is one year energy consumption of the substations in absence of ESSes; ε is electricity price in dollar per kWh, $P_{\text{sub}}^{\text{nosc}}$ is l years' electricity price of the substations in absence of ESSes, and i is the yearly inflation of electricity price.

$$P_{\text{sub}}^{\text{sc}} = E_{\text{sub}}^{\text{sc}} \times \varepsilon \times \frac{l(2 + (l-2)i)}{2} + \sum_{1}^{k} Cost_k \tag{11}$$

where $E_{\text{sub}}^{\text{sc}}$ is one year' energy consumption of the substations in presence of ESSes, $Cost_k$ is installation cost of ESS on substation k, and $P_{\text{sub}}^{\text{sc}}$ is l years' expenditure of the substations in presence of ESSes, which includes the electricity price and installation cost of ESSes.

In this paper, economic efficiency $e\%$ is defined as the following formula:

$$e\% = \frac{P_{sub}^{nosc} - P_{sub}^{sc}}{P_{sub}^{nosc}} \times 100\% \quad (12)$$

when economic efficiency $e\%$ equals 0, it means the saved electricity price is the same as the installation cost of ESSes. Necessary parameters for calculating the economic efficiency $e\%$ of ESSes are given in Table 4.

Table 4. Necessary parameters for calculating economic efficiency.

Parameter	Value	Parameter	Value
p	180 kW	ε	0.16 $/kWh
m	0.244 $/W	r	5%
C	0.16 M$	i	5%
l	10 years		

3.1.2. Voltage Drop Compensation, $v\%$

If different trains are close to each other and they start all together, contact lines will become overloaded and the pantograph voltages of trains will drop significantly, which results in high line loss and the opening of minimum voltage protective action of trains by limiting the current. The installation of ESSes in the metro system can shave the peak power of substations, improve the load capacity of the system, and compensate the pantograph drops quickly. Voltage drop compensation $v\%$, in this paper, evaluates in percent the voltage drop compensation at the pantograph, giving the rate about how much the voltage drops improvement is when the ESSes are installed.

$$v\% = \frac{\sum_1^j \int_{U_p<U_r} \left(U_r - U_p^{nosc}\right) dt - \sum_1^j \int_{U_p<U_r} \left(U_r - U_p^{sc}\right) dt}{\sum_1^j \int_{U_p<U_r} \left(U_r - U_p^{nosc}\right) dt} \times 100\% \quad (13)$$

where, *Up* is the pantograph voltage of trains; *Ur* is the rated voltage of trains' pantograph; *j* is the amount of up-line and down-line trains. From the Equation (14), voltage drop compensation $v\%$ is calculated based on the integral of voltage drops improvement when *Up* is less than *Ur*, which is more appropriate and comprehensive than the maximum voltage drop compensation just in a moment in [15].

3.1.3. Objective Function, *ObjV*

Given economic efficiency $e\%$ and voltage drop compensation $v\%$, the objective function for optimal energy management strategy and configuration of ESSes is shown as below:

$$ObjV = \omega \times e\% + (1 - \omega) \times v\% \quad (14)$$

where ω is the weight coefficient of economic efficiency $e\%$, it represents the emphasis degrees of economic efficiency $e\%$. When ω is set to 1, it means that economic efficiency is the only evaluation index considered in the optimization.

4. Novel Optimization Method Based on a Genetic Algorithm

The traditional optimization method based on a genetic algorithm proposed in [16] can optimize the location and size of ESSes significantly, but the adopted energy management strategy is constant. The energy management is also important for the performance improvement of a metro supply network, and the optimization of energy management and configuration for ESSes will influence each other. Thus, the proposed novel optimization method in this paper, which combines a genetic algorithm and a simulation platform of urban rail power supply system, is meant to optimize energy management, location, and size of ESSes simultaneously.

4.1. Improved Energy Management Strategy

In order to improve economic efficiency $e\%$ and voltage drop compensation $v\%$ of ESSes, an improved energy management strategy is put forward, which decides the charging and discharging current of the ESS by detecting the voltage of substation, ESS and train pantographs, as shown in Figure 9.

The improved energy management strategy can be divided into three parts: SOC constraint, current loop, and energy management. Due to the function of the SOC constraint, the working range of the SOC is 0.25–1, and the terminal voltage of ESSes is limited between 375 V and 750 V. Energy management can switch four work states to produce appropriate reference $P_{sc}{}^*$ for the ESS according to the substation voltage and pantograph voltage of the trains. The current loop can control the charging and discharging current of the super-capacitor ESS according to the reference $I_{sc}{}^*$.

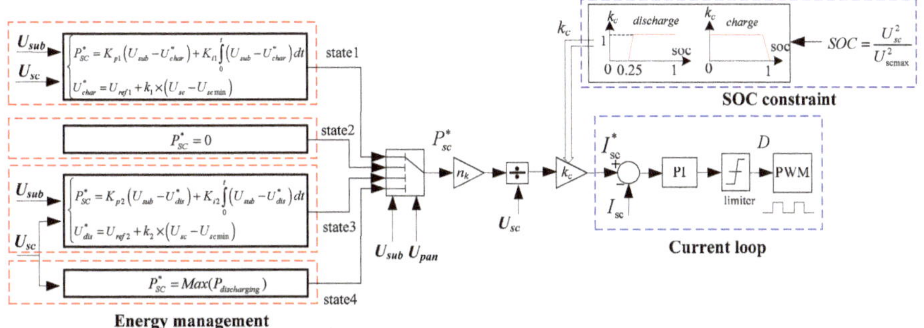

Figure 9. The improved energy management strategy of stationary ESSes.

The charging and discharging current reference $I_{sc}{}^*$ of the super-capacitor ESS can be calculated by Equation (15). And work states of super-capacitor ESS are shown in Figure 10.

$$I_{sc}^* = \frac{n_k \times k_c \times P_{sc}^*}{U_{sc}} \tag{15}$$

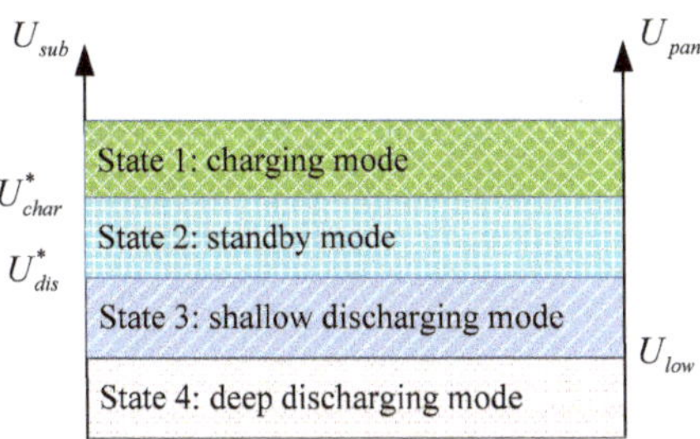

Figure 10. Work states of super-capacitor ESS.

State 1: When the voltage of substation is higher than charging threshold value $U_{char}{}^*$, the magnitude of the charging current reference $P_{sc}{}^*$ is determined by the PI controller according to the difference value between the present substation voltage and the threshold value $U_{char}{}^*$.

From Equation (16), if the electric braking power of train is small, the super-capacitor ESS will absorb all the regenerative braking energy and maintain the substation voltage at $U_{char}{}^{*}$. Then, if the electric braking power of train is excessive, the super-capacitor ESS will absorb the braking energy with maximum charging current. The value of $U_{char}{}^{*}$ will increase with the increase of ESSes' terminal voltage, which enlarges the charging current of ESSes with smaller terminal voltage of ESSes significantly, as shown in Figure 11. The value of U_{scmin} is set to 375 V in this paper.

$$\begin{cases} P_{sc}^{*} = K_{p1} \times (U_{sub} - U_{char}^{*}) + K_{i1} \times \int_{0}^{t} (U_{sub} - U_{char}^{*}) dt \\ U_{char}^{*} = U_{ref1} + k_{1} \times (U_{sc} - U_{scmin}) \\ K_{p1} \geq 0, K_{i1} \geq 0, U_{char}^{*} > U_{no\_load}, k_{1} \geq 0, U_{scmax} \geq U_{sc} \geq U_{scmin} \end{cases} \quad (16)$$

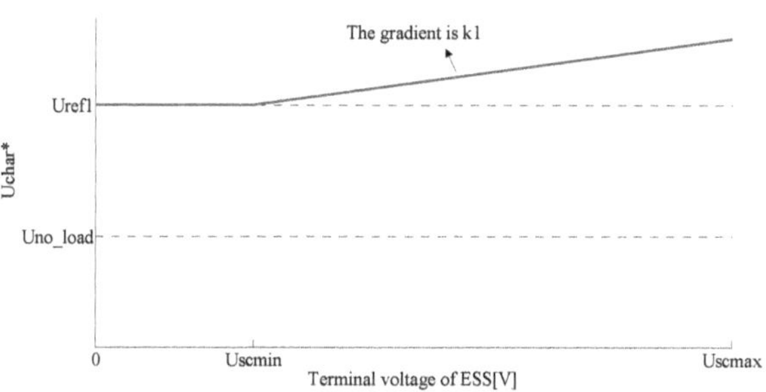

Figure 11. The values of $U_{char}{}^{*}$.

When the train is braking in one substation, by traditional energy management, the ESS installed in the substation will draw high power of regenerative energy and take no account of its terminal voltage and stored energy [16]. When the ESS is charged up to 100% with the regenerative energy, its terminal voltage would be 750 V and its charging current will be interrupted instantaneously, which leads to the drastic changes of substation current and line current, then all the regenerative energy of trains flows to the ESSes in the near substations as shown in Figure 12. ESSes are charged in turn and both with large current. On the contrary, by the improved energy management strategy, the ESS can adjust the threshold value $U_{char}{}^{*}$ according to its terminal voltage and achieve smoother changes of terminal voltage and charging current. Consequently, the regenerative energy is distributed to ESSes more evenly. It is worth mentioning that the improved energy management strategy reduces the line loss greatly and it also contributes to balance the terminal voltage for all different strings of ESS on a substation.

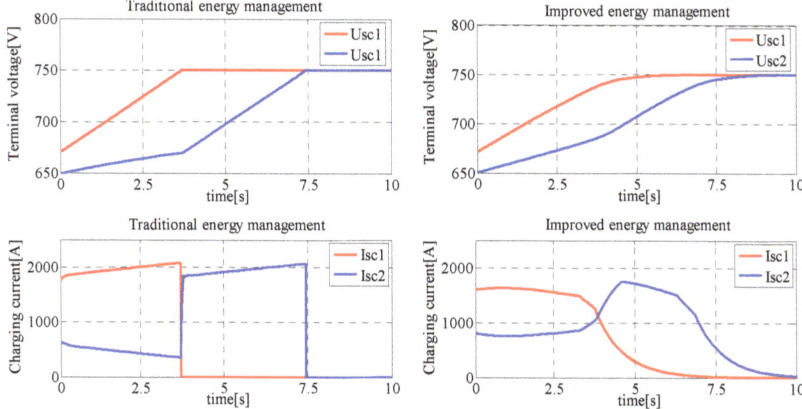

Figure 12. Terminal voltage and charging current of ESSes.

State 2: When the voltage of substation fluctuates between the charging threshold value $U_{char}{}^*$ and the discharging threshold value $U_{dis}{}^*$, super-capacitor ESS maintain the standby state.

$$P_{SC}^* = 0 \tag{17}$$

State 3: When the voltage of substation is less than the discharging threshold value $U_{dis}{}^*$ and pantograph voltage of trains within one substation spacing range of ESS is higher than the low voltage threshold U_{low}, discharging power reference $P_{sc}{}^*$ of ESS is determined by the substation voltage U_{sub} and ESS terminal voltage U_{sc} simultaneously as follow:

$$\begin{cases} P_{SC}^* = K_{p2} \times (U_{sub} - U_{dis}^*) + K_{i2} \times \int\limits_0^t (U_{sub} - U_{dis}^*)dt \\ U_{dis}^* = U_{ref2} + k_2 \times (U_{sc} - U_{scmin}) \\ K_{p2} \geq 0, K_{i2} \geq 0, U_{dis}^* < U_{no\_load}, k_2 \geq 0, U_{scmax} \geq U_{sc} \geq U_{scmin} \end{cases} \tag{18}$$

The value of $U_{dis}{}^*$ will increase with the increase of ESSes' terminal voltage as shown in Figure 13, which enlarges the discharging current of ESSes with larger terminal voltage and balances SOC of ESSes significantly. When the accelerated train draws the energy in one substation, all ESSes nearby can deliver energy to shave the power of substations and compensate the voltage drops of the pantograph. As shown in Figure 14, by traditional energy management, the ESS installed in the substation will deliver highest power of energy and take no account of its terminal voltage and stored energy. When terminal voltage of one ESS decreases to U_{scmin}, its discharging current will be interrupted instantaneously, which also leads to drastic changes of substation current and line current. On the contrary, the improved energy management strategy can achieve smoother changes of voltages and currents in the system, and ESSes with higher SOC tend to deliver more energy to the supply network. Thus, by the improved energy management strategy, the flow of energy can be managed more steadily and effectively, and the line loss can be reduced greatly.

State 4: When the voltage of substation is less than discharging threshold value $U_{dis}{}^*$ and the pantograph voltages of trains within one substation spacing range of ESS are less than the low voltage threshold U_{low}, super-capacitor ESS will deliver the energy with maximum discharging power. According to appropriate setting of K_{p2}, K_{i2}, k_2, U_{ref2}, U_{low}, ESS will retain proper energy when the pantograph voltage of a nearby train is acceptable, and when the pantograph voltage of a nearby train is very low, ESS delivers maximum discharging power to shave the peak power of the substation and compensate the pantograph voltage drop.

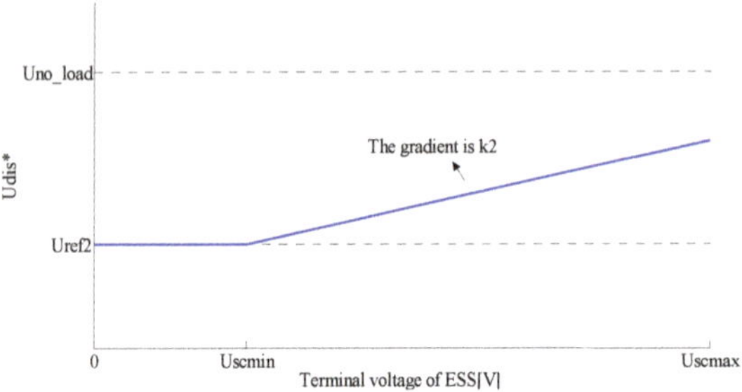

Figure 13. The values of U_{dis}^{*}.

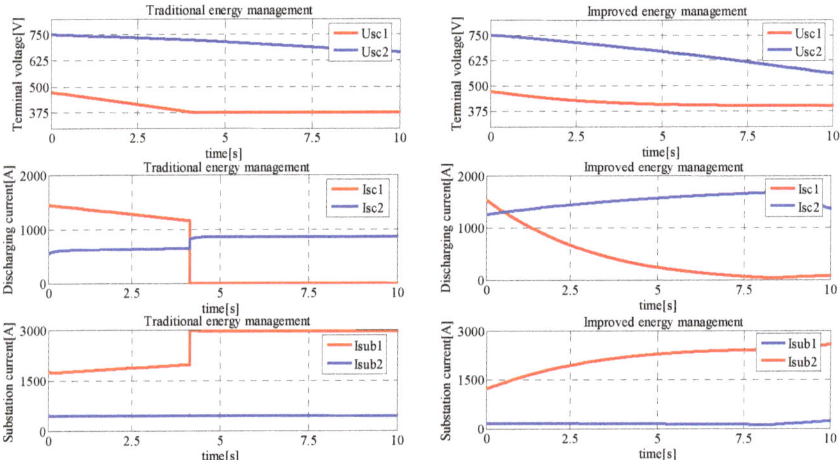

Figure 14. Terminal voltage and discharging current of ESSes.

In improved energy management strategy, K_{p1}, K_{i1}, k_1, K_{p2}, K_{i2}, k_2, U_{low}, U_{ref1}, U_{ref2} are nine undetermined parameters. In order to obtain best performance of system based on economic efficiency $e\%$ and voltage drop compensation $v\%$, the most appropriate parameters of improved energy management strategy and ESS configuration on each substation will be obtained simultaneously by the optimization method based on a genetic algorithm.

4.2. Novel Optimization Method

4.2.1. Genetic Algorithm

The genetic algorithm (GA) is a global optimal searching algorithm based on Darwin's nature evolution theory and Mendel's genetics and mutation theory. It consists of three parts: encoding, fitness evaluation, and genetic manipulation [23–25]. Combined with paper demands, the basic procedures of the genetic algorithm are shown as follows.

Encoding

The energy management strategy and configuration of ESSes installed in seven TSSs can be encoded by 16 numbers as shown in Figure 15, where each X chromosomere presents a population individual. The first nine numbers represent nine pending parameters of improved energy management strategy, u_1, u_2, u_3 represent U_{low}, U_{ref1}, U_{ref2}; the last seven numbers represent seven pending numbers of super-capacitor strings installed in seven different traction substations.

$$X = [\underbrace{k_{p1}k_{i1}k_1k_{p2}k_{i2}k_2u_1u_2u_3}_{\text{energy management strategy}}\overbrace{\underbrace{x_1x_2x_3x_4x_5x_6x_7}_{\text{configuration}}}^{TSS\times 7}]$$

Figure 15. Set of X chromosomere.

Objective Function *ObjV*

In this paper, the optimization of energy management strategy and configuration of ESSes is to obtain the maximum objective function *ObjV*, the reciprocal of *ObjV* is the value of fitness. They are calculated as follows:

$$\begin{cases} ObjV[X] = \omega \cdot e\%[X] + (1-\omega) \cdot v\%[X] \\ Fitness[X] = \frac{1}{ObjV[X]} \end{cases} \tag{19}$$

where ω is the weight coefficient of economic efficiency $e\%$. $ObjV[X]$ is the objective function when the energy management strategy, allocation, and size of ESSes are set by X.

Genetic Manipulation

Genetic manipulation includes three basic steps—selection, crossover, and mutation. From the view of operators, the genetic algorithm is well-suited to solve combination optimization problems. Compared with other intelligence algorithms, a genetic algorithm has a higher rate of convergence, more efficient calculation, and higher robustness for combination optimization and discrete optimization.

4.2.2. Process of Novel Optimization Method

The schematic diagram of the novel optimization method, which combines a genetic algorithm and simulation platform of urban rail power supply system, is shown as Figure 16. A genetic algorithm can constantly optimize the chromosomere of the population individuals, which means the energy management strategy and configuration of ESSes are optimized constantly. The newfound energy management strategy and configuration of ESSes would be entered into the simulation platform, and obtain their *ObjV*, $e\%$, $v\%$ through the simulation. According to the *ObjV*, $e\%$, $v\%$, the genetic algorithm can continue the further and cyclic optimization. According to the optimization results by a large number of simulation calculations, the genetic algorithm converges to the global optimum with the increase of evolution generation.

For every different objective function, the genetic algorithm will take 5.5 days to obtain the corresponding optimal solution. Of course, if several workstations work simultaneously, the total simulation time can be effectively decreased. The simulation platform of urban rail power supply system is established by software Matlab 7.10.0(2010a). The hardware performance of our workstations that implement the simulation platform is shown as Table 5. It is worth mentioning that increasing the population size or improving the genetic algorithm by means of a hybrid algorithm can improve the convergence speed and decrease the evolution generation.

The relevant parameters of the genetic algorithm are given by Table 6; *NIND* is population size, *PRECI* is the length of individual, *MAXGEN* is maximum evolution generation, P_c is the crossover rate, P_m is the mutation rate, and *GGAP* is generational gap.

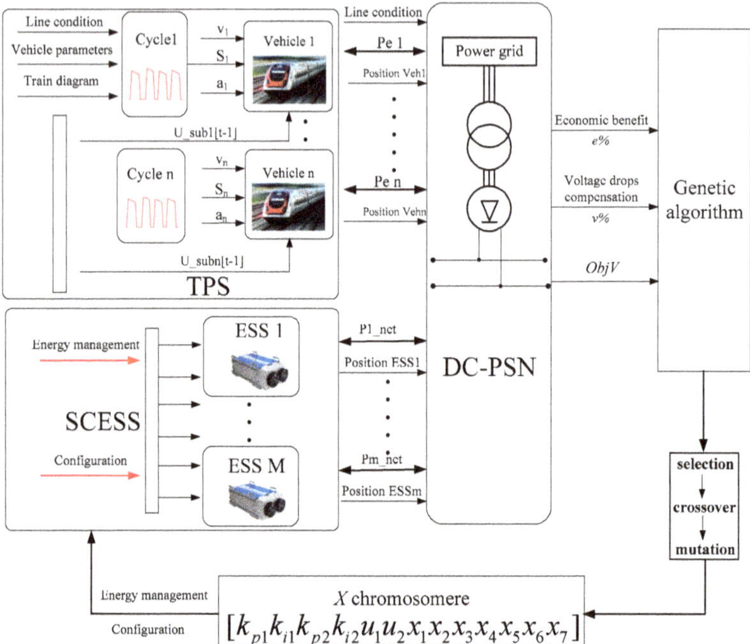

Figure 16. The schematic diagram of novel optimization method.

Table 5. The parameters of hardware platform.

Hardware	Parameter
CPU	Intel(R) Xeon(R) CPU E5649 @ 2.53GHz × 2
RAM	64 GB
GPU	NVIDIA Quadro 4000

Table 6. The parameters of improved genetic algorithm.

NIND	*PRECI*	*MAXGEN*	P_c	P_m	*GGAP*
40	20	100	0.7	0.015	0.95

4.3. Optimization Result Analysis

As shown in Figure 17, the simulation comparison result between two different optimization methods with corresponding optimum *ObjV* are obtained separately under different values of weight coefficient ω. Based on a genetic algorithm, both optimization methods can obtain optimal *ObjV* with an increase of evolution generation, but the novel optimization method can obtain much higher *ObjV*. The values of maximum *ObjV* as well as corresponding economic efficiency $e\%$ and voltage drop compensation, $v\%$ based on two optimization methods and different values of weight coefficient ω are shown in Table 7. The values of parameters to determine energy management strategy and configuration of ESSes on every substation can be obtained separately based on two optimization methods and different values of weight coefficient ω, as shown in Tables 8 and 9.

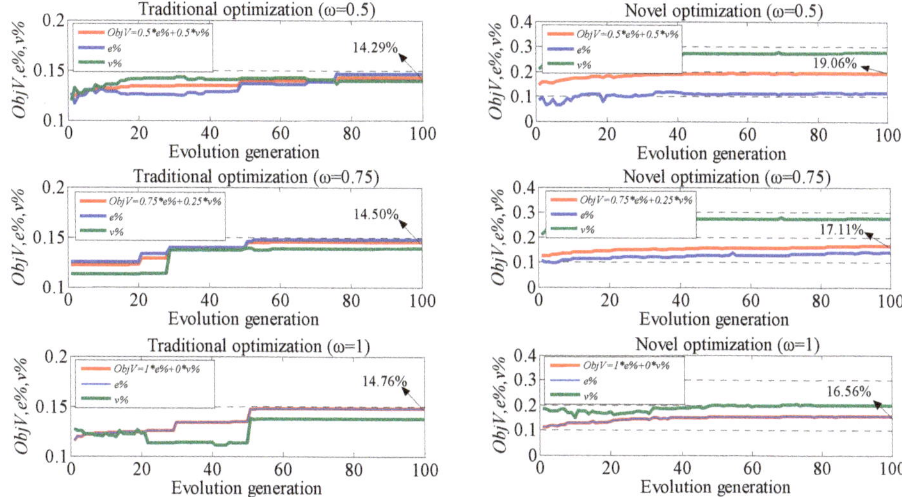

Figure 17. Simulation comparisons of two optimization methods.

Table 7. Maximum *ObjV* obtained by different optimization method.

Optimization Method	ω	Maximum ObjV	Economic Efficiency *e*%	Voltage Compensation Rate *v*%
Traditional optimization	0.5	14.29%	14.65%	13.93%
Traditional optimization	0.75	14.50%	14.72%	13.84%
Traditional optimization	1	14.76%	14.76%	13.70%
Novel optimization	0.5	19.06%	15.06%	23.05%
Novel optimization	0.75	17.11%	15.79%	21.06%
Novel optimization	1	16.56%	16.56%	17.20%

Table 8. The parameters of optimal energy management strategies.

Optimization Method	ω	Energy Management Strategy of ESSes								
		kp1	*ki1*	*k1*	*kp2*	*ki2*	*k2*	U_{low}	U_{ref1}	U_{ref2}
Traditional optimization	-	50	50	-	50	50	-	-	850.0	800.0
Novel optimization	0.5	298	90	0.011	0.158	44.39	0.075	771.0	836.1	802.0
Novel optimization	0.75	193	83	0.006	1.20	40.34	0.037	772.0	836.2	806.4
Novel optimization	1	18	76	0.002	19.33	39.85	0.008	779.1	836.5	811.8

Table 9. Optimized location and size of ESSes.

Optimization Method	ω	TSS No. and Set Numbers of ESSes						
		1	2	3	4	5	6	7
Traditional optimization	0.5	0	16	15	0	10	0	13
Traditional optimization	0.75	0	14	15	0	10	0	14
Traditional optimization	1	0	14	15	0	10	0	13
Novel optimization	0.5	0	18	11	0	17	0	17
Novel optimization	0.75	0	18	10	0	10	0	17
Novel optimization	1	0	14	16	0	8	0	7

From Figure 17 and Table 7, whatever the value of ω, novel optimization method can obtain much higher $ObjV$, $e\%$ and $v\%$ compared to traditional optimization method. With the increase of ω from 0.5 to 0.75 to 1, the maximum $ObjV$ of ESSes obtained by traditional optimization is 14.29%, 14.50%, and 14.76%, which can be increased to 19.06%, 17.11%, and 16.56%, respectively, by the novel optimization method. And both economic efficiency $e\%$ and voltage drop compensation $v\%$ can be improved effectively by the novel optimization compared to traditional optimization. By the novel optimization method, economic efficiency $e\%$ can be improved because of more appropriate energy management, less line loss, and voltage drop compensation $v\%$ can be improved effectively because of the function of U_{low}.

From Table 8, the adopted energy management strategy of the traditional optimization method is constant, and it is only determined by six parameters. By contrast, the energy management strategy obtained by the novel optimization method is determined by nine parameters. The novel optimization method can optimize the energy management, location, and size of ESSes simultaneously. Under different values of weight coefficient ω, the best energy management strategy is different and among the nine relevant parameters appear some regularities. k_{p1}, k_1, k_{p2}, k_2 are more important factors that affect the performance of the metro system, and k_{i1}, k_{i2}, U_{low}, U_{ref1}, U_{ref2} have smaller changes. Without regard to the integral term, the best energy management strategy of ESSes for different value of weight coefficient ω is shown in Figure 18. For charging energy management strategy, the value of k_{p1} (the slope of charging current *vs.* U_{sub}) and k_1 (the slope of charging current *vs.* U_{sc}) decrease with the increase of weight coefficient ω. For discharging energy management strategy, the value of k_{p2} (the slope of discharging current *vs.* U_{sub}) increases and k_2 (the slope of discharging current *vs.* U_{sc}) decrease with the increase of weight coefficient ω.

Table 9 shows the optimal location and size of ESSes obtained by two different optimization methods. By contrast, two optimization methods ultimately configure super-capacitor ESSes in same location of substations, and the size of ESSes tend to be smaller with the increase of weight coefficient ω. Configuring ESSes in fewer substations with one or two substation spacing and decreasing the size of ESS installed in one substation can reduce the installation cost, but the distance between the train and ESS will also increase, which causes higher line loss and less energy recovered and voltage drop compensation $v\%$ will also decrease. The best compromise between economic efficiency $e\%$ and voltage drop compensation $v\%$ under different value of weight coefficient ω can be obtained by two optimization methods. Compared to the traditional optimization method, the best configuration of ESSes obtained by the novel optimization method changes are more intense and it can achieve much higher $ObjV$ under different values of weight coefficient ω.

The maximum $ObjV$ with corresponding economic efficiency $e\%$ and voltage drop compensation $v\%$ for different value of weight coefficient ω can be obtained by novel optimization method as shown in Table 10 and Figure 19. From Figure 19, when ω increases from 0.3 to 1, Economic efficiency $e\%$ increases from 0.1465 to 0.1656, and voltage drop compensation $v\%$ decreases from 0.2335 to 0.1720. According to its own optimization requirement and the concrete result obtained by the novel optimization method in Figure 19, Subway Company could choose the best value of weight coefficient ω for itself.

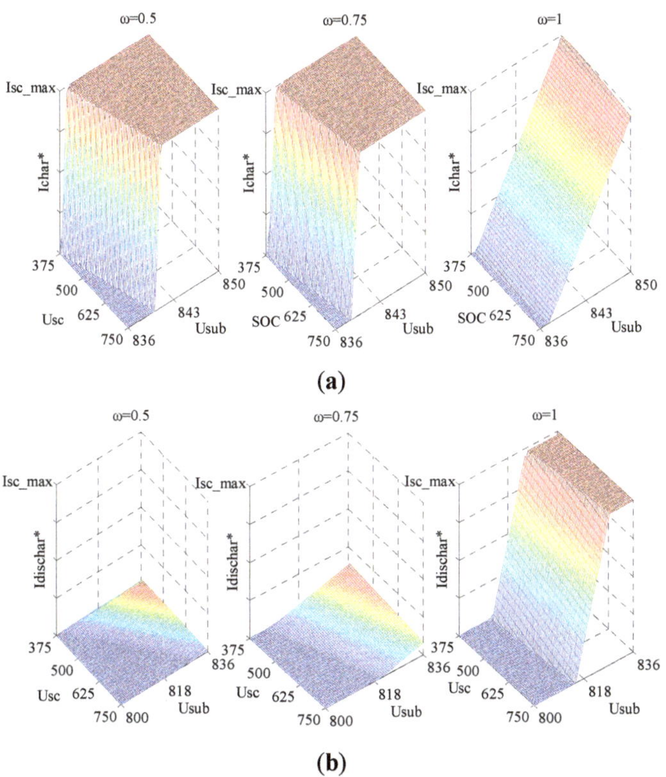

Figure 18. Best energy management strategy of ESSes. (**a**) Charging energy management strategy; and (**b**) discharging energy management strategy.

Table 10. Optimal *ObjV*, $e\%$ and $v\%$ obtained by novel optimization method.

ω	0.3	0.4	0.5	0.6	0.7	0.8	0.9	1
ObjV	0.2074	0.1989	0.1906	0.1826	0.1747	0.1685	0.1663	0.1656
$e\%$	0.1465	0.1485	0.1506	0.1509	0.1525	0.1631	0.1654	0.1656
$v\%$	0.2335	0.2325	0.2305	0.2301	0.2265	0.1902	0.1744	0.1720

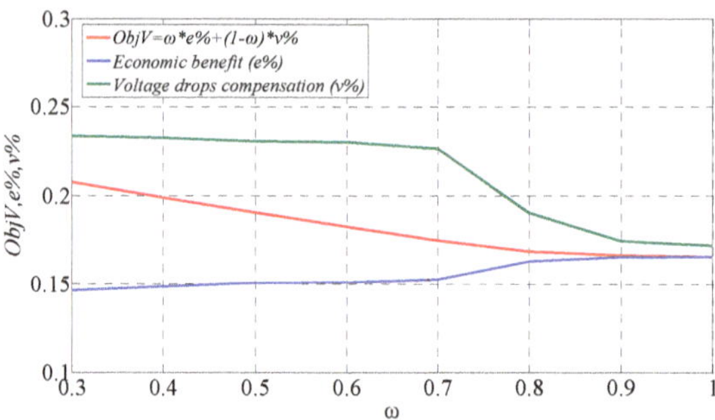

Figure 19. Optimal *ObjV*, $e\%$ and $v\%$ obtained by the novel optimization method.

5. Conclusions

Firstly, this paper establishes the proper simulation platform of a metro system that contains seven substations to simulate the electrical power flow by Matlab/Simulink. Then, two evaluation functions are set up from the perspectives of economic efficiency and voltage drop compensation. Ultimately, a novel optimization method is put forward, which can optimize the energy management strategy, location, and size of ESSes simultaneously by the combination of a genetic algorithm and simulation platform of a metro system. With actual parameters of a Chinese metro line applied in the simulation comparison, the proposed novel optimization method can achieve much better performance of a metro system from the perspectives of *ObjV* and two evaluation functions. The simulation result obtained by the novel optimization method shows that with the increase of weight coefficient ω, the optimal energy management strategy is different and the nine relevant parameters appear with some regularities, among them k_{p1}, k_1, k_{p2}, and k_2 are more important factors that affect the performance of the metro system. Additionally, novel optimization methods can also optimize the configuration of ESSes, which can achieve the best compromise between economic efficiency $e\%$ and voltage drop compensation $v\%$. The novel optimization method and its optimized result can provide valuable reference to Subway Company.

Acknowledgments: This research was supported by the Major State Basic Research Development Program of China (973 Program: 2011CB711100) and I13L00100 from Beijing Laboratory of Urban Rail Transit.

Author Contributions: Zhongping Yang contributed to the conception of the study, Huan Xia and Huaixin Chen contributed significantly to analysis and manuscript preparation, Fei Lin helped perform the analysis with constructive discussions and Bin Wang provided the line and vehicle data.

Conflicts of Interest: The authors declare no conflict of interest.

References

1. Barrero, R.; van Mierlo, J.; Tackoen, X. Energy savings in public transport. *IEEE Veh. Technol. Mag.* **2008**, *3*, 26–36. [CrossRef]
2. Barrero, R.; Tackoen, X.; Van Mierlo, J. Improving energy efficiency in public transport: Stationary supercapacitor based energy storage systems for a metro network. In Proceedings of the IEEE Vehicle Power and Propulsion Conference (VPPC'08), Harbin, China, 3–5 September 2008; pp. 1–8.
3. Ogasa, M. Energy saving and environmental measures in railway technologies: Example with hybrid electric railway vehicles. *IEEJ Trans. Electr. Electron. Eng.* **2008**, *3*, 15–20. [CrossRef]
4. Yamanoi, T.; Kawahara, K.; Kani, Y.; Kodama, Y. Measurement of utilized regenerative power in DC feeding system. In Proceedings of the 2013 IEEJ Industry Applications Society Conference, Yamaguchi, Japan, 28–30 August 2013.

5. Avanzo, S.D.; Iannuzzi, D.; Murolo, F.; Rizzo, R.; Tricoli, P. A sample application of supercapacitor storage systems for suburban transit. In Proceedings of the Electrical Systems for Aircraft, Railway and Ship Propulsion (ESARS), Bologna, Italy, 19–21 October 2010; pp. 1–7.
6. Ibrahim, H.; Ilinca, A.; Perron, J. Energy storage systems—Characteristics and comparisons. *Renew. Sustain. Energy Rev.* **2008**, *12*, 1221–1250. [CrossRef]
7. Gabash, A.; Li, P. Active-reactive optimal power flow in distribution networks with embedded generation and battery storage. *IEEE Trans. Power Syst.* **2012**, *27*, 2026–2035. [CrossRef]
8. Gabash, A.; Li, P. Flexible optimal operation of battery storage systems for energy supply networks. *IEEE Trans. Power Syst.* **2013**, *28*, 2788–2797. [CrossRef]
9. Grbovi, P.J.; Delarue, P.; Le Moigne, P.; Bartholomeus, P. The ultracapacitor-based controlled electric drives with braking and ride-through capability: Overview and analysis. *IEEE Trans. Ind. Electron.* **2011**, *58*, 925–936. [CrossRef]
10. Iannuzzi, D. Improvement of the energy recovery of traction electrical drives using supercapacitors. In Proceeding of 13th Power Electronics and Motion Control Conference (EPE-PEMC 2008), Poznan, Poland, 1–3 September 2008; pp. 1469–1474.
11. Ciccarelli, F.; Iannuzzi, D.; Tricoli, P. Control of metro-trains equipped with onboard supercapacitors for energy saving and reduction of power peak demand. *Transp. Res. C Emerg. Technol.* **2012**, *24*, 36–49. [CrossRef]
12. Ciccarelli, F.; Del Pizzo, A.; Iannuzzi, D. Improvement of energy efficiency in light railway vehicles based on power management control of wayside lithium-ion capacitor storage. *IEEE Trans. Power Electron.* **2014**, *29*, 275–286. [CrossRef]
13. Ciccarelli, F.; Iannuzzi, D.; Spina, I. Comparison of energy management control strategy based on wayside ESS for LRV application. In Proceedings of the 39th IEEE Annual Conference on Industrial Electronics Society (IECON 2013), Vienna, Austria, 10–13 November 2013; pp. 1548–1554.
14. Battistelli, L.; Fantauzzi, M.; Iannuzzi, D.; Lauria, D. Energy management of electrified mass transit systems with energy storage devices. In Proceedings of the 2012 International Symposium on Power Electronics, Electrical Drives, Automation and Motion (SPEEDAM), Sorrento, Italy, 20–22 June 2012; pp. 1172–1177.
15. Iannuzzi, D.; Pighetti, P.; Tricoli, P. A study on stationary supercapacitor sets for voltage droops compensation of streetcar feeder lines. In Proceedings of the 2010 Electrical Systems for Aircraft, Railway and Ship Propulsion (ESARS), Bologna, Italy, 19–21 October 2010; pp. 1–8.
16. Wang, B.; Yang, Z.; Lin, F.; Zhao, W. An improved genetic algorithm for optimal stationary energy storage system locating and sizing. *Energies* **2014**, *7*, 6434–6458. [CrossRef]
17. Iannuzzi, D.; Ciccarelli, F.; Lauria, D. Stationary ultracapacitors storage device for improving energy saving and voltage profile of light transportation networks. *Transp. Res. C Emerg. Technol.* **2012**, *21*, 321–337. [CrossRef]
18. Radcliffe, P.; Wallace, J.S.; Shu, L.H. Stationary applications of energy storage technologies for transit systems. In Proceedings of the 2010 IEEE Electric Power and Energy Conference (EPEC), Halifax, NS, Canada, 25–27 August 2010; pp. 1–7.
19. Barrero, R.; Tackoen, X.; Van Mierlo, J. Stationary or onboard energy storage systems for energy consumption reduction in a metro network. *Proc. Inst. Mech. Eng. F J. Rail Rapid Transit* **2010**, *224*, 207–225. [CrossRef]
20. Teymourfar, R.; Asaei, B.; Iman-Eini, H. Stationary super-capacitor energy storage system to save regenerative braking energy in a metro line. *Energy Convers. Manag.* **2012**, *56*, 206–214. [CrossRef]
21. Iannuzzi, D.; Pagano, E.; Tricoli, P. The use of energy storage systems for supporting the voltage needs of urban and suburban railway contact lines. *Energies* **2013**, *6*, 1802–1820. [CrossRef]
22. Battistelli, L.; Ciccarelli, F.; Lauria, D.; Proto, D. Optimal design of DC electrified railway stationary storage system. In Proceedings of the 2009 International Conference on Clean Electrical Power, Capri, Italy, 9–11 June 2009; pp. 739–745.
23. Guo, P.; Wang, X.; Han, Y. The enhanced genetic algorithms for the optimization design. In Proceedings of the 3rd International Conference on Biomedical Engineering and Informatics (BMEI 2010), Yantai, China, 16–18 October 2010; Volume 7, pp. 2990–2994.

24. Tegani, I.; Aboubou, A.; Becherif, M.; Ayad, M.Y.; Kraa, O.; Bahri, M.; Akhrif, O. Optimal sizing study of hybrid wind/PV/diesel power generation unit using genetic algorithm. In Proceedings of the 4th International Conference on Power Engineering, Energy and Electrical Drives (POWERENG 2013), Istanbul, Turkey, 13–17 May 2013; pp. 134–140.
25. Jiao, L.Y.; Lei, H.Z. The Application of genetic algorithm in fitting the spatial variogram. In Proceedings of the 2011 International Conference on Computer Science and Network Technology (ICCSNT), Harbin, China, 24–26 December 2011.

Article

Analysis of a Battery Management System (BMS) Control Strategy for Vibration Aged Nickel Manganese Cobalt Oxide (NMC) Lithium-Ion 18650 Battery Cells

Thomas Bruen [1,*], James Michael Hooper [1], James Marco [1], Miguel Gama [2] and Gael Henri Chouchelamane [2]

1. Warwick Manufacturing Group (WMG), University of Warwick, Coventry CV4 7AL, UK; j.m.hooper@warwick.ac.uk (J.M.H.); james.marco@warwick.ac.uk (J.M.)
2. Jaguar Land Rover, Banbury Road, Warwick CV35 0XJ, UK; mgamaval@jaguarlandrover.com (M.G.); gchouch1@jaguarlandrover.com (G.H.C.)

* Correspondence: thomas.bruen@warwick.ac.uk; Tel.: +44-02476-573-061

Academic Editor: William Holderbaum
Received: 31 January 2016; Accepted: 21 March 2016; Published: 1 April 2016

Abstract: Electric vehicle (EV) manufacturers are using cylindrical format cells as part of the vehicle's rechargeable energy storage system (RESS). In a recent study focused at determining the ageing behavior of 2.2 Ah Nickel Manganese Cobalt Oxide (NMC) Lithium-Ion 18650 battery cells, significant increases in the ohmic resistance (R_O) were observed post vibration testing. Typically a reduction in capacity was also noted. The vibration was representative of an automotive service life of 100,000 miles of European and North American customer operation. This paper presents a study which defines the effect that the change in electrical properties of vibration aged 18650 NMC cells can have on the control strategy employed by the battery management system (BMS) of a hybrid electric vehicle (HEV). It also proposes various cell balancing strategies to manage these changes in electrical properties. Subsequently this study recommends that EV manufacturers conduct vibration testing as part of their cell selection and development activities so that electrical ageing characteristics associated with road induced vibration phenomena are incorporated to ensure effective BMS and RESS performance throughout the life of the vehicle.

Keywords: vehicle vibration; electric vehicle (EV); hybrid electric vehicle (HEV); Li-ion battery ageing; battery management system (BMS)

---

1. Introduction

Within the road transport sector, a main driver for technological innovation is the need to reduce fuel consumption and vehicle exhaust emissions. Legislative requirements are motivating original equipment manufacturers (OEMs) to develop and integrate new and innovative technologies into their fleet. Consequently, over the last few years, different types of electric vehicles (EVs) have been built alongside conventional internal combustion engine (ICE) cars. Within the field of EVs, a key enabling technology is the design and integration of rechargeable energy storage systems (RESS) [1,2]. Multi-cell RESSs require a battery management system (BMS) to ensure safe and consistent operation over the life of the vehicle and to report the status of the RESS to the wider vehicle control systems. One of the key challenges is to monitor the variations in capacity and impedance between cells. Often, the RESS is limited by characteristics of the weakest cell [3,4], and accounting for these differences minimizes the impact of cell variation on RESS performance.

Many OEMs are employing cylindrical format cells (e.g., 18650) for the design and construction of the RESS [5–8]. Cylindrical cells are often chosen in EV applications over their prismatic and pouch cell counterparts because of a combination of factors. For example, 18650 format cells are produced in very large quantity which makes them cost effective [8–10]. Similarly, they may have built-in safety systems such as a positive temperature coefficient (PTC) resistor that prevents high current surge and the use of a current interrupt device (CID) to protect the cell in the event of excessive internal pressure [8–10].

To ensure in-market reliability and customer satisfaction, OEMs perform a variety of life representative durability tests during the design and prototype stages of the development process. Firstly, these tests ensure that new vehicle sub-assemblies and components are fit-for-purpose. Secondly, it allows OEMs to obtain characterization data for models that are used for the development of core functionality within the BMS. Thirdly, it ensures that the product meets strict requirements for vehicle homologation.

Vibration durability is one of these tests, and plays an important role in the selection of components. As discussed within [11–15] poorly integrated components, assemblies or structures subjected to vibration can result in a significantly reduced service life or the occurrence of catastrophic structural failure through fatigue cracking or work hardening of materials [13,16,17].

With respect to 18650 format battery cells, vibration-induced degradation in cell electrical performance has been previously reported. In [18], 18650 format Li-ion cells, of unknown chemistry, were subjected to a vibration profile along the Z-axis of the cells. The cells were clamped to an electromagnetic shaker (EMS) table and excited for 186 h with a swept-sine wave from 4 to 20 Hz and back to 4 Hz in 30 s. The authors reported that most of the cells exhibited an increase in resistance along with a reduction in their 1C discharge capacity. Additionally, they described the occurrence of complete cell level failures, such as an internal short circuit. The latter failure mode was attributed within the research to the central mandrel becoming loose during the vibration test, which in turn damaged the upper and lower cell components, including the current collector and tabs.

In [19], commercially available nickel manganese cobalt oxide (NMC) Samsung 2.2 Ah 18650 cylindrical cells (model number ICR18650-22F) were evaluated for electrical degradation when subjected to vibration profiles representative of 100,000 miles of European and North American customer use for chassis mounted RESS. The two batches of cells were evaluated to two different random vibration cycles:

- Society of Automotive Engineers (SAE) J2380 [20];
- WMG/Millbrook Proving Ground (WMG/MBK) profiles [21,22].

Both batches of cells displayed a significant increase in ohmic resistance (R_O) regardless of vibration profile utilized for their assessment. Increases in R_O were within the range of 17.4% to 128.1%. A reduction in capacity was also observed in some samples evaluated, with 12.22% being the greatest reduction observed. It was acknowledged within [19] that the impact of vibration-induced ageing may require greater levels of cell balancing by the battery management system (BMS), when the cells are aggregated to form a rechargeable energy storage system (RESS). As a result, the authors are proposing to utilize the electrical data obtained from the vibration aged Samsung 18650 cells defined in [19] to determine, via simulation techniques, the specific impact on the BMS strategy of a hybrid electric vehicle (HEV) and plug-in hybrid electric vehicle (PHEV).

This paper is structured as follows: Section 2 of this paper provides a detailed overview of the experimental method employed to assess vibration-induced ageing. Section 3 summarizes the experimental results. The cell modelling techniques are introduced in Section 4 and the models are parameterized in Section 5. Two case studies, considering the impact of vibration-aged cells in series and parallel connections, are presented in Section 6, using the models parameterized from the experimental data. Further work and conclusions are presented in Sections 7 and 8 respectively.

2. Experimental Method—Vibration Ageing of Cells

The following section is a summary of the experimental method employed for the vibration ageing of the cells, whose electrical data is used to parameterize the cell models. Additional information with respect to the test fixtures, facility, instrumentation and sample arrangement are defined in [19]. Note that only samples evaluated to the SAE J2380 standard, and assessed at 50% SOC in [19] are employed within this study. The experimental procedure followed during this test programme is summarized in Figure 1.

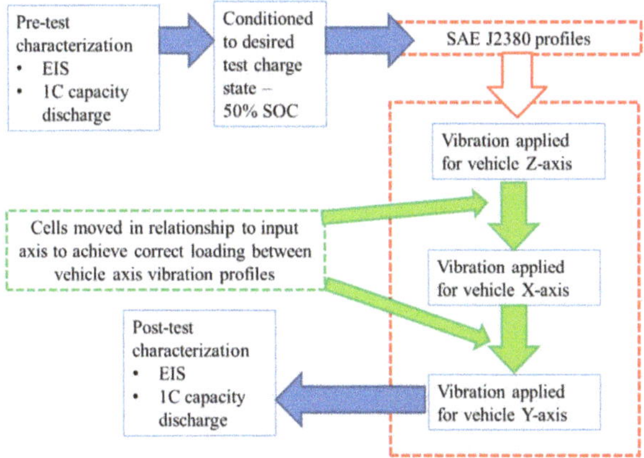

Figure 1. Schematic of test process.

2.1. Test Samples

Table 1 defines the details of sample preparation, cell SOC and cell orientation of five Samsung 2.2 Ah 18650 cells (NMC). Three samples were subject to vibration in accordance to SAE J2380, whilst the remaining two were defined as control cells. The control samples were not subjected to any vibration loading [19]. During testing, the control samples were either co-located within the same environmental conditions as the test cells or kept in permanent storage.

Table 1. Test sample information.

Sample No in [1] *	Test Profile	SOC (%)	Cell Orientation (Vehicle Axis: Cell Axis)
4	Control sample—In permanent storage	50%	Control
5	Control sample—Followed J2380 test samples	50%	Control
13	J2380	50%	Z:Z
14	J2380	50%	Z:X
15	J2380	50%	Z:Y

* Samples 1 to 3, 6 to 12 and 16 to 18 are omitted as they were not evaluated in accordance with SAE J2380 or not conditioned to 50% SOC in [1].

2.2. Pre-test Characterization

The five cells presented in Table 1 were electrically characterized as described in the remainder of this section. Additional information on the cell electrical characterization methods are discussed in [19]. It must be noted that other electrical characterization tests were conducted on these cells, however only the characterization tests relating to the data required for cell modelling are described within this Section (and illustrated by Figure 1).

2.2.1. Electrochemical Impedance Spectroscopy (EIS)

EIS data was recorded 4 h after the last pulse of the pulse power tests, as suggested by Barai *et al.* [23] and was performed at 50% SOC. The EIS measurement was carried out in a galvanostatic mode using a ModuLab® (Solartron, Leicester, UK) electrochemical system model 2100 A fitted with a 2 A booster and driven by Modulab® ECS software. The EIS spectra were collected within the frequency range of 10 mHz to 10 kHz using 10 frequency points per decade. The amplitude of the applied current was 200 mA (RMS). No DC current was superimposed on the RMS value.

2.2.2. 1 C Capacity Discharge

The cells were fully charged using a constant current phase of 1.1 A (C/2) to 4.2 V followed by a constant voltage phase at 4.2 V until the current reduced to 0.05 A (C/44). The cells were allowed to rest for 4 h prior to being fully discharged at 1 C to 2.75 V, which is the lower voltage threshold as defined by the manufacturer. The charge extracted from the cells during the discharge was recorded as a measure of the 1 C capacity.

2.3. Conditioning to Desired Test Charge State

Following electrical characterization, each cell was pre-conditioned to a defined SOC (either 25% 50% or 75% SOC) prior to durability testing and allocated a test orientation with respect to the vehicle Z-axis (discussed further in Section 2.4). The cell SOC was adjusted by fully charging the cells with a constant current of 1.1 A (C/2) to 4.2 V followed by a constant voltage phase at 4.2 V until the current fell to 0.05 A (C/44). At the end of charge, the cells were allowed to rest for 4 hours prior to being discharged at 1 C for 30 min, to achieve a cell SOC of 50%. The cells were allowed to relax for 4 h before the application of vibration energy.

2.4. Application of SAE J2380 Vibration Profiles and Cell Orientation

SAE J2380 was selected as the desired vibration durability method. The SAE J2380 profiles are presented in Figure 2.

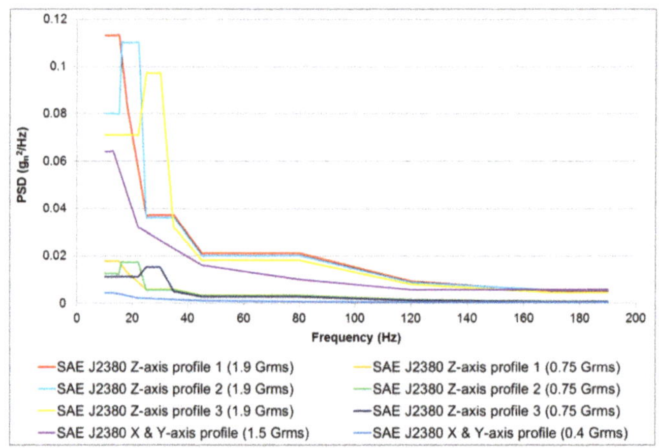

Figure 2. Society of Automotive Engineers (SAE) J2380 vibration power spectral density (PSD) profiles for testing samples 13 to 15 [20].

SAE J2380 is currently the only internationally recognized vibration test standard that has been correlated to 100,000 miles of road vehicle durability (North American vehicle usage by the 90 percentile customer) [20–22], and has been devised to assess multiple types of chassis mounted RESS (and their associated sub components). Unlike traditional vibration standards (such as UN38.3 Test 3) which have been devised to assess the fail-safe performance of EV batteries when subjected to extreme vibration conditions (such as during shipment via air or during a vehicle to vehicle crash) SAE J2380 has been synthesized from "road induced vibration excitation" associated with normal customer usage. This specification applies vibration loading through a "random" excitation which is more representative of road-induced structural vibration than contemporary vibration specifications [13]. Also unlike other EV standards (such as UN38.3 Test 3 or ECE R100) SAE J2380 applies vibration to each cell in the X, Y and Z axis, as opposed to applying vibration in a single axis for the duration of the test [1,21,22]. It is beyond the scope of this paper to discuss in detail the derivation of the vibration profiles. However, this information is discussed within [13,21,22,24,25].

As part of the experimental procedure, each profile is sequentially applied to the cells to achieve the desired 100,000 miles of representative EV life. For a complete execution of SAE J2380 the three different combinations of vibration loads with respect to each cell orientation are defined below:

- Z:Z to X:X to Y:Y
- Z:X to X:Y to Y:Z
- Z:Y to X:Z to Y:X

Using the above notation, for each pair of letters, the first letter refers to the vehicle axis, whilst the second refers to cell orientation. For simplicity this paper identifies the cell orientation in relationship to the vertical (Z axis) of the vehicle. For example a cell that was subject to the vibration sequence of Z:X to X:Y to Y:Z, is referred to as being evaluated in the Z:X orientation. Figure 3a illustrates the axis convention for the vehicle axis, whilst Figure 3b illustrates the axis convention for the 18650 cell.

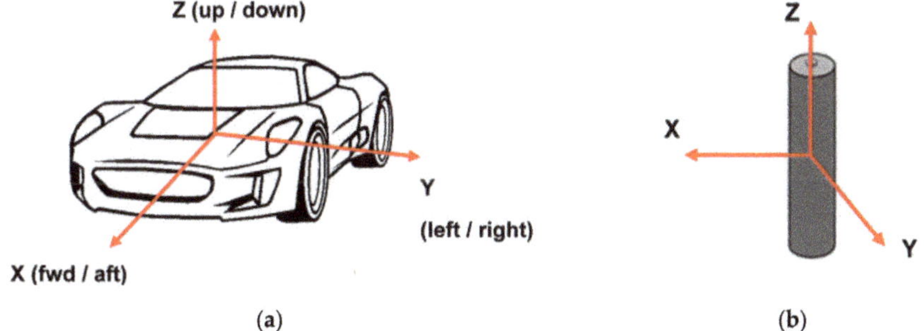

Figure 3. (**a**) Axis convention of vehicle vibration durability profiles; (**b**) Axis convention of cells.

Due to limited equipment availability, a single axis Derriton (Hastings, UK; model number: VP85, serial number: 74) shaker was employed for the durability testing. Because the orientation of the EMS could not be changed, the cells had to be rotated on the durability fixture between X, Y and Z axis profile changes to achieve the correct loading. This test methodology is termed as not testing "with respect to gravity" and does not allow for changes in sample mass during the re-orientation of cells with respect to the input axis of vibration. While the authors believe that this limitation did not significantly impact the results, this factor is discussed further in [19], where alternative experimental methods are assessed.

Vibration testing was conducted within an air-conditioned room at a temperature of 21 ± 5 °C. The closed loop application of the vibration profile was achieved using an averaging control strategy, as defined within [26] which included ±3 dB alarm limits and ±6 dB experimental abort limits. Once the cells were installed to the durability fixture and mounted onto the EMS table, the Z-axis vibration profile of SAE J2380 was applied first. The calculation of Grms levels (defined in Table 2) is discussed further in [25].

Table 2. Society of Automotive Engineers (SAE) J2380 vibration profiles schedule.

Profile Description and GRMS Level	Duration (HH:MM)	Test Cumulative Duration (HH:MM)
Z Axis Schedule		
Subject cells to 9 min of Z-axis profile 1 at 1.9 Grms in the Z axis orientation of the cells under assessment.	00:09	00:09
Subject cells to 5 h and 15 min of Z-axis profile 1 at 0.75 Grms in the Z axis orientation of the cells under assessment.	05:15	05:24
Subject cells to 9 min of Z-axis profile 2 at 1.9 Grms in the Z axis orientation of the cells under assessment.	00:09	05:33
Subject cells to 5 h and 15 min of Z-axis profile 2 at 0.75 Grms in the Z axis orientation of the cells under assessment.	05:15	10:48
Subject cells to 9 min of Z-axis profile 3 at 1.9 Grms in the Z axis orientation of the cells under assessment.	00:09	10:57
Subject cells to 5 h and 15 min of Z-axis profile 3 at 0.75 Grms in the Z axis orientation of the cells under assessment.	05:15	16:12
X Axis Schedule		
Subject cells to 5 min of X & Y-axis profile at 1.5 Grms in the X axis orientation of the cells under assessment.	00:05	16:17
Subject cells to 19 h of X & Y-axis profile at 0.4 Grms in the X axis orientation of the cells under assessment.	19:00	35:17
Subject cells to 5 min of X & Y-axis profile at 1.5 Grms in the X axis orientation of the cells under assessment.	00:05	35:22
Subject cells to 19 h of X & Y-axis profile at 0.4 Grms in the X axis orientation of the cells under assessment.	19:00	54:22
Y Axis Schedule		
Subject cells to 5 min of X & Y-axis profile at 1.5 Grms in the Y axis orientation of the cells under assessment.	00:05	54:27
Subject cells to 19 h of X & Y-axis profile at 0.4 Grms in the Y axis orientation of the cells under assessment.	19:00	73:27
Subject cells to 5 min of X & Y-axis profile at 1.5 Grms in the Y axis orientation of the cells under assessment.	00:05	73:32
Subject cells to 19 h of X & Y-axis profile at 0.4 Grms in the Y axis orientation of the cells under assessment.	19:00	92:32
Total	-	92:32

On completion of the Z-axis schedule, the cells were left to stabilize for 4 h. The cells were then moved on the durability fixture to the corresponding vehicle X-axis and subjected to the X-axis vibration profile (Table 2). Finally, the cells were repositioned on the durability fixture to facilitate the application of the vehicle Y-axis vibration profile (Table 2). At the end of the vibration profile, the cells were left to stabilize for a further 4 h prior to visual inspection.

2.5. Post-test Characterization

Post-testing, the cell characterization measurements defined in Section 2.2 were repeated, with experimental values recorded in Tables 3 and 4.

3. Vibration Ageing Results

The following Section defines the results relating to changes in the EIS and capacity performance of the cells tested at 50% SOC and highlights the specific changes with these performance characteristics post vibration testing to SAE J2380.

3.1. EIS Results for Post Vibration Aged Cells

Figure 4 shows the ohmic resistance (R_O) of the cells at Start of Test (SOT) and End of Test (EOT) as measured by EIS. A complete explanation of EIS results is beyond the scope of this study and is already well documented in a number of academic and educational texts [27,28]. Figure 4 presents typical Nyquist plots of the cells pre and post vibration test for the cells condition to 50% SOC. Table 3 quantifies the increase in ohmic resistance obtained from the respective EIS data.

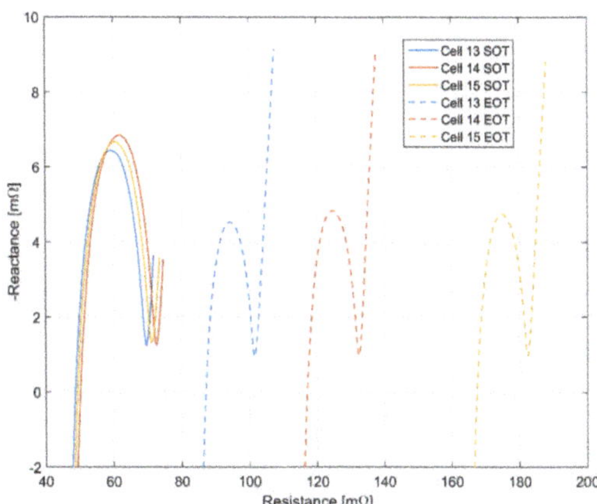

Figure 4. Electrochemical impedance spectroscopy (EIS) start of test (SOT) and end of test (EOT) results.

Typically, all samples displayed a significant increase ohmic resistance. Samples oriented in Z:Y axis and pre-conditioned to 50% SOC exhibited the highest increase out of the 50% SOC samples evaluated to SAE J2380. This is presented below in Table 3.

Table 3. Electrochemical impedance spectroscopy (EIS) ohmic resistance (R_O) results for all tested samples.

Sample No	SOC	Orientation	SOT (mΩ)	EOT (mΩ)	Percentage Change (%)
15	50 %	Z:Y	46.4	164.5	254.53
14	50 %	Z:X	47.3	114.2	141.44
13	50 %	Z:Z	46.0	84.0	82.61
5	50 %	Control	49.6	60.8	22.58
			SOT	**EOT**	
Standard deviation for tested 50% SOC samples (mΩ)			0.67	40.67	
Mean for tested 50% SOC samples (mΩ)			46.57	120.90	

3.2. 1 C discharge capacity

Table 4. presents the 1C discharge capacity for each cell at SOT and EOT. From Table 4, it can be seen that the results show a tendency for samples orientated in the Z:Y axis and pre-conditioned to 50% SOC to exhibit a higher capacity fade than other 50% SOC samples.

Table 4. Summary of change in 1C discharge capacity performance of all test cells.

Sample No.	SOC (%)	Orientation	Cell Capacity at SOT (Ah)	Cell Capacity at EOT (Ah)	Percentage Change in Ah (%)
15	50%	Z:Y	2.18	2.14	−1.83
13	50%	Z:Z	2.23	2.19	−1.79
14	50%	Z:X	2.15	2.17	0.93
5	50%	Control	2.18	2.19	0.46
			SOT	**EOT**	
Standard deviation for tested 50% SOC samples (Ah)			0.040	0.025	
Mean for tested 50% SOC samples (Ah)			2.19	2.17	

4. Cell Modelling

Equivalent circuit models (ECMs) are commonly used to model cells due to their simplicity, ease of parameterization and real-time suitability compared to physics based models [29]. As well as being used to analyze cell and battery pack performance as part of a model-based design process [30], they can also be used for model-based state estimation of SOC and state of health (SOH) [3,31,32].

ECMs generally consist of a resistor connected in series with a number of resistor-capacitor (RC) pairs. A greater number of RC pairs increases the model bandwidth and accuracy, at the expense of computational complexity. Figure 5a shows a single cell ECM with N_{RC} RC pairs. Several ECMs can be combined in parallel as in Figure 5b and/or in series as in Figure 5c. Compared to other types of cell, 18650 format cells have a low comparatively charge capacity. For larger applications, such as electric vehicles, the battery pack must contain many cells in parallel and series in order to meet power and energy requirements. For example, the Tesla Model S 85 kWh battery pack contains 7104 18650 format cells arranged into 96 series units, with each series unit containing 74 cells connected in parallel [33]. Understanding and managing the variations in the properties of each of these cells is important for optimizing the performance of the battery pack.

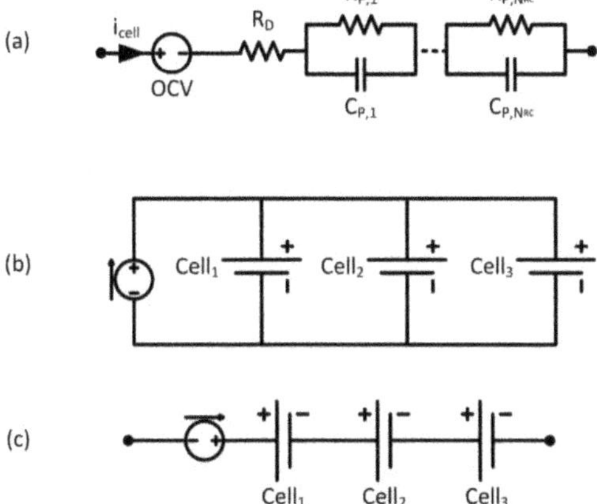

Figure 5. (**a**) A single cell equivalent circuit model (ECM) with N_{RC} pairs; (**b**) Three cells in parallel with a current source; (**c**) Three cells in series with a current source.

Dubarry *et al.* [34] showed that a series string of cells can be equated to a single cell model by accounting for the variations between cells. However, the authors note that this will become complex for larger numbers of cells, and the resulting model lacks the individual cell voltages which are important for assessing when the series string of cells has reach end of charge/end of discharge. In [35] the authors find that string SOC estimation accuracy suffers when an averaged single cell model is used rather than individual cells. In [36], it is suggested that prior to assembling a series-parallel model of cells, the cells should be screened to ensure that only cells with very similar resistance and capacity are connected together. From this, it is assumed that cells in parallel can be equated to a single cell, as can cells in series. Combining both together, an arbitrary series-parallel module of cells can be represented by a single averaged ECM. However, As discussed in [37] there is no guarantee that the cells will remain similar over their life, and a single effective cell will not account for variations in current and SOC which can occur. Individual ECMs can also be used to implement model-based balancing strategies for cells in series [38].

The EIS data collected during the experimental phase were all obtained at 50% SOC. While it is known that cell impedance is a function of SOC, it is generally considered that there is little variation in the central SOC region at moderate temperatures [39,40] For the studies considered in this paper, the SOC of the cells is maintained between 30 and 70%. While the impedance remains constant, the open circuit voltage (OCV) is a nonlinear function of SOC. This can be incorporated into the cell model by using a lookup table.

In this case, OCV is implemented as a state and a parameter F is found using Equation (1), where Q is the 1C discharge capacity in Ah. This is then used to capture the relationship of OCV and current given by Equation (2). Each RC pair voltage is governed by Equation (3), and the terminal voltage of the cell given by Equation (4). Note that the DC resistance in the ECM is referred to as R_D to avoid confusion with R_O obtained from the EIS plots. Model errors and limits in bandwidth mean that the R_D resistance may not equate exactly to the R_O value for a given cell:

$$F(OCV) = \frac{1}{36Q}\frac{dOCV}{dSOC} \tag{1}$$

$$\dot{OCV} = F(OCV)\, i_{cell} \tag{2}$$

$$\dot{v}_p = -\frac{v_p}{R_p C_p} + \frac{i_{cell}}{C_p} \tag{3}$$

$$v_t = OCV + \sum_{n=1}^{N_{RC}} v_{p,n} + R_D i_{cell} \tag{4}$$

These equations can be put into state-space form, which is well suited to modelling arbitrary time-domain signals such as the current applied under real-world driving conditions. For a single cell model with N_{RC} pairs, the state equation are given by Equation (5), and the output equation by Equation (6).

$$\begin{bmatrix} OCV \\ v_{p,1} \\ v_{p,2} \\ \vdots \\ v_{p,N_{RC}} \end{bmatrix} = \begin{bmatrix} 0 & & & & \\ & -\frac{1}{R_{p,1}C_{p,1}} & & & \\ & & -\frac{1}{R_{p,2}C_{p,2}} & & \\ & & & \ddots & \\ * & & & & -\frac{1}{R_{p,N_{RC}}C_{p,N_{RC}}} \end{bmatrix} \begin{bmatrix} OCV \\ v_{p,1} \\ v_{p,2} \\ \vdots \\ v_{p,N_{RC}} \end{bmatrix} + \begin{bmatrix} F(OCV) \\ \frac{1}{C_{p,1}} \\ \frac{1}{C_{p,2}} \\ \vdots \\ \frac{1}{C_{p,N_{RC}}} \end{bmatrix} i_{cell} \tag{5}$$

$$v_t = \begin{bmatrix} 1 & 1 & \cdots & 1 \end{bmatrix} \begin{bmatrix} OCV \\ v_{p,1} \\ v_{p,2} \\ \vdots \\ v_{p,N_{RC}} \end{bmatrix} + R_D i_{cell} \tag{6}$$

The ECM impedance can be considered in the frequency domain, as a function of complex angular frequency s. The impedance for an ECM with N_{RC} RC pairs is given in Equation (7):

$$Z(s) = R_D + \sum_{n=1}^{N_{RC}} \frac{R_{p,n}}{R_{p,n}C_{p,n}s + 1} \tag{7}$$

An averaged parallel model (APM)—a single-cell effective model of the parallel-connected unit of cells— can also be created by combining each cell as per Equation (8). This averages out any variations in cell properties and as such cannot be used to explore the wider impact of combining cells in parallel. Similarly, an average series model (ASM) for cell impedance from a series string of N cells is given by Equation (9):

$$Z_{eff} = \frac{1}{\sum_{k=1}^{N} \frac{1}{Z_k}} \tag{8}$$

$$Z_{eff} = \frac{1}{N} \sum_{k=1}^{N} Z_k \tag{9}$$

The output voltage can then be multiplied by N to obtain the string voltage. Using the method described and validated in [37], the individual ECMs can be combined into a parallel cell model, with the applied current as the input, and the terminal voltage as the output. Unlike the single effective cell model, this solution does not average out the variations in cell properties, and it means that the current through each cell can be calculated. In the paper, individual ECMs were parameterized from cells with different impedances resulting from charge-discharging cycling-induced ageing. A dynamic current load was applied to the cells while connected in parallel, and individual cell currents were also measured. The same applied load was used as an input to the parallel cell model so that cell currents and voltage could be calculated. The simulated cell currents were found to be accurate to 2% of the measured currents.

A summary of the parallel cell model is provided below, with the full derivation in [37]. The individual cells are combined into a single system by creating block diagonals of the state matrices in Equation (5) and selecting one set of output matrices from Equation (6) (since all of the cells are at the same voltage, only one output voltage is required and in this case the first cell is chosen to calculate voltage), which takes the form of Equation (10). The input matrix has to be updated at each time step by solving Equation (1). This system requires an input vector of cell currents, which is typically not available for a system of parallel cells. However, by applying Kirchoff's laws, the cell currents can be calculated from knowledge of the cells' states, model parameters and the known current applied to the parallel stack. This results in a linear system of equations which solves for a vector of cell currents based on each cell being at the same terminal voltage as its neighbor. These cell currents are given by Equation (11), which can be substituted into Equation (10) to give Equation (12):

$$\begin{bmatrix} \dot{x}_1 \\ \dot{x}_2 \\ \vdots \\ \dot{x}_N \end{bmatrix} = \begin{bmatrix} A_1 & & & \\ & A_2 & & \\ & & \ddots & \\ & & & A_N \end{bmatrix} \begin{bmatrix} x_1 \\ x_2 \\ \vdots \\ x_N \end{bmatrix} + \begin{bmatrix} B_1 & & & \\ & B_2 & & \\ & & \ddots & \\ & & & B_N \end{bmatrix} \begin{bmatrix} i_{cell,\,1} \\ i_{cell,\,2} \\ \vdots \\ i_{cell,\,N} \end{bmatrix}$$

$$[v_t] = \begin{bmatrix} C_1 & 0 & 0 & \cdots & 0 \end{bmatrix} \begin{bmatrix} x_1 \\ x_2 \\ \vdots \\ x_N \end{bmatrix} + \begin{bmatrix} D_1 & 0 & 0 & \cdots & 0 \end{bmatrix} \begin{bmatrix} i_{cell,\,1} \\ i_{cell,\,2} \\ \vdots \\ i_{cell,\,N} \end{bmatrix} \quad (10)$$

$$\begin{bmatrix} i_{cell\ 1} \\ i_{cell\ 2} \\ \vdots \\ i_{cell\ N} \end{bmatrix} = GR^{-1}\left(Ex + Fi_1\right)$$

$$G = \begin{bmatrix} 1 & -1 & 0 & 0 \\ 0 & 1 & -1 & 0 \\ 0 & \ddots & 1 & -1 \\ 0 & \cdots & 0 & 1 \end{bmatrix}$$

$$R = \begin{bmatrix} 1 & 0 & 0 & 0 \\ 0 & -(R_{D,1}+R_{D,2}) & R_{D,2} & 0 \\ 0 & R_{D,2} & \ddots & R_{D,N-1} \\ 0 & 0 & R_{D,N-1} & -(R_{D,N-1}+R_{D,N}) \end{bmatrix} \quad (11)$$

$$E = \begin{bmatrix} 0 & 0 & \cdots & & & & \cdots & 0 \\ -1 & -1 & 1 & 1 & 0 & & \cdots & 0 \\ 0 & \cdots & -1 & -1 & 1 & 1 & 0 & 0 \\ 0 & \cdots & & 0 & -1 & -1 & 1 & 1 \end{bmatrix}$$

$$F = \begin{bmatrix} 1 \\ -(R_{D,1}+2R_c) \\ 0 \\ \vdots \\ 0 \end{bmatrix}$$

$$\begin{aligned} \dot{\mathrm{x}} &= \mathrm{A}'\mathrm{x} + \mathrm{B}'^{\mathrm{i}_1} \\ \mathrm{y} &= \mathrm{C}'\mathrm{x} + \mathrm{D}'\mathrm{i}_1 \\ \mathrm{A}' &= \mathrm{A} + \mathrm{BGR}^{-1}\mathrm{E} \end{aligned} \quad (12)$$

$$B' = BGR^{-1}F$$

$$C' = C + DGR^{-1}E$$

$$D' = DGR^{-1}F$$

This augmented solution demonstrates that the entire system of parallel states can be solved for while maintaining the same input (applied current) and output (cell voltage) and state-space structure as for a single cell model. This allows for the same simulation and analysis methods to be used as for single cells or cells in series.

5. Model Parameterization

ECMs are not based on physical elements, and as such must be parameterized using system identification techniques rather than taking physical measurements of the cells. Each set of EIS data described in Section 3.1 was used to derive ECM parameters. To obtain the parameters, a nonlinear least squares optimization routine was used to apply the cost function in Equation (13). This adjusts the parameter vector θ in order to minimize the sum-of-squares difference between the experimentally measured impedance Z_E, and the model impedance Z_M, over N_f frequency points. The model impedance is calculated using Equation (7). The nonlinear least squares algorithm finds a local minimum to the cost function. There may be several local minima and depending on the starting point of the optimization, the global minimum may not be found. To reduce the sensitivity to initial conditions a multi-start algorithm was employed, which runs the local solver several times from different starting points, and then chooses the best solution from all of the runs:

$$\arg\min_{\theta} \sum_{n=1}^{N_f} \left(Z_n^E - Z_n^M(\theta) \right)^2$$

$$\theta = \left[R_D, R_{p,1}, C_{P,1}, R_{p,2}, \ldots, C_{P,N_{RC}} \right] \tag{13}$$

The solution is also sensitive to the number of RC pairs chosen. The decision was made to cover the full frequency range of the EIS data up to the inductive region. For this type of cell, the crossing point of the impedance into the inductive region occurs at around 200 Hz. This is faster than a BMS will typically sample at [41] but previous research has found that there can be high frequency dynamics within parallel stacks [37] that the BMS may need to take account of in order to properly manage the individual cells.

An example of the Nyquist fits for cell 5 is shown in Figure 6a, and the associated errors in Figure 6b. It was found that a 4RC model provides an acceptable fit to the EIS data, with further RC pairs not significantly reducing the error. Several of the RC pairs are required to approximate a resistor-constant phase element pair which would better represent the curve shape, but cannot be easily translated to the time domain [42].

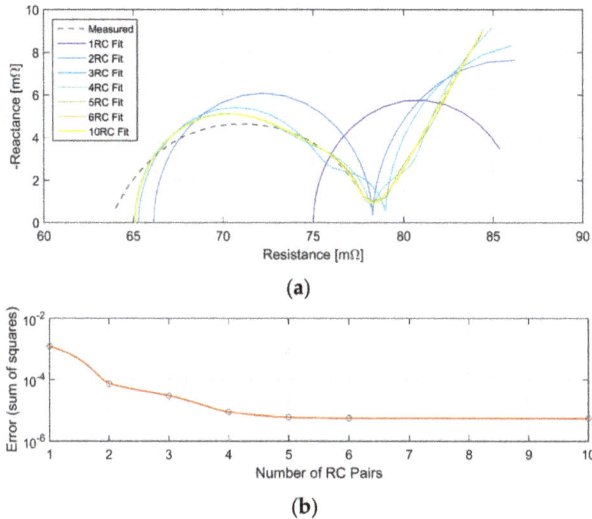

Figure 6. (**a**) Comparison of impedance accuracy of cell 5 for various model orders; (**b**) Sum of squares error between model and measured impedance for various model orders.

The model parameter results for the cells before and after ageing, are given in Table 5. As mentioned in Section 4, the limitations of the model mean that the R_D values do not exactly match the R_O values from Table 3, but the differences are all under 5 mΩ.

Table 5. Parameterization results for cells 5, 13, 14 and 15.

Cell Age	SOT				EOT			
Cell Number	**5**	**13**	**14**	**15**	**5**	**13**	**14**	**15**
R_D	0.05416	0.05034	0.05183	0.05097	0.06506	0.08858	0.11846	0.16881
R_{p1}	0.01084	0.01067	0.01086	0.01074	0.00915	0.00921	0.00933	0.00927
C_{p1}	0.38418	0.39693	0.38167	0.39326	0.11962	0.13251	0.11753	0.1235
R_{p2}	0.00847	0.00774	0.0087	0.00834	0.00401	0.00369	0.00449	0.00426
C_{p2}	3.0674	3.4627	2.9801	3.1816	2.1007	2.6789	1.8483	2.0417
R_{p3}	0.00174	0.00171	0.00177	0.0018	0.00253	0.00268	0.00243	0.00246
C_{p3}	355.19	465.36	331.11	339.43	481.78	508.56	570.11	540.8
R_{p4}	0.00869	0.01002	0.00881	0.00826	0.02331	0.02562	0.02754	0.02574
C_{p4}	3150.5	3191.9	3155.6	3041.6	1459.9	1454.3	1498	1513.7

6. Simulation Case Studies

The ECM representations outlined in Section 4 have been implemented using the ode15 s solver in MATLAB (MathWorks, Natick, MA, USA), which was chosen because of its suitability to finding the solution to numerically stiff systems with a wide range of poles.

6.1. Parallel Cells Subjected to a HEV Profile

When cells with different impedances are connected in parallel, the individual cell currents can be significantly different. These currents are typically not monitored by a BMS and as such cannot be actively regulated during usage. To consider the impact that vibration may have on battery packs containing cells connected electrically in parallel, the ECMS for cells 13–15 are combined in parallel as in Figure 5b. This is compared with the APM defined by Equation (8).

A charge-sustaining drive cycle, typical of a HEV, is used for this simulation. Cells 13–15 are combined in parallel, and maintained at approximately 50% SOC. This drive cycle was derived from real-world driving data and is scaled such that the maximum current is 1C, and the cells started at 51% SOC, and ended at around 49% SOC. Figure 7a shows the applied current and individual cell currents (both as C-rates) for the aged cells, along with a zoomed-in Section. Figure 7b shows the corresponding cell currents. The terminal voltage simulation results in Figure 8 also show that even for aged cells, for a given current input, the calculated voltage out is almost identical to the parallel cell model, with a root mean squared (RMS) error of 0.15 mV.

In order to compare the relative loading of each cell, the nominal charge throughput is calculated. This is given for cell n by Equation (14), and can be considered an indication of the average current undergone by each cell, relative to the case of all cells being equal:

$$q_{t,n} = \frac{100}{N_p}\sqrt{\frac{\sum i_{cell,n}^2}{\sum i_{app}^2}} \tag{14}$$

The heat power and energy generated by the cell can be calculated using Equations (15) and (16) respectively. The heat energy per cell can be scaled relative to the heat energy calculated from the APM:

$$P_{heat} = (v_t - OCV)\, i_{cell} \tag{15}$$

$$E_{heat} = \int P_{heat} dt \tag{16}$$

The results, given in Table 6 show that the average loading of cell 13 is 27.5% higher than the nominal case, and cell 15 is 26.8% lower. This means that cell 13 undergoes 175% the loading of cell 15. The loading of the cells is relatively consistent over the course of the drive cycle. Current distribution within a parallel stack is driven by differences in OCV and impedance. The small SOC change means that the OCV of each cell remains similar, with less than 5 mV difference between cells. Therefore the difference in impedance (resulting from cell vibration) is what is causing cell currents to differ in this case. Previous work has shown that this can cause significant dynamic variation in cell currents [37]. However, that study used cells aged by different mechanism. Those cells had a very similar DC resistance, but differences in solid-electrolyte interphase (SEI) build up resulted in a notably different charge transfer resistance. This meant that the impedance variation between the cells was a function of frequency. However, for these cells aged primarily through vibration, the DC resistance has shifted significantly, with little SEI build-up. This means that there is always a large difference in impedance between cells regardless of the frequency, and the difference in cell currents is primarily driven by the ohmic resistance of the model. This, along with the similar OCVs of each cell, means that the proportion of heat energy generated by each cell is very similar to the proportion of current loading.

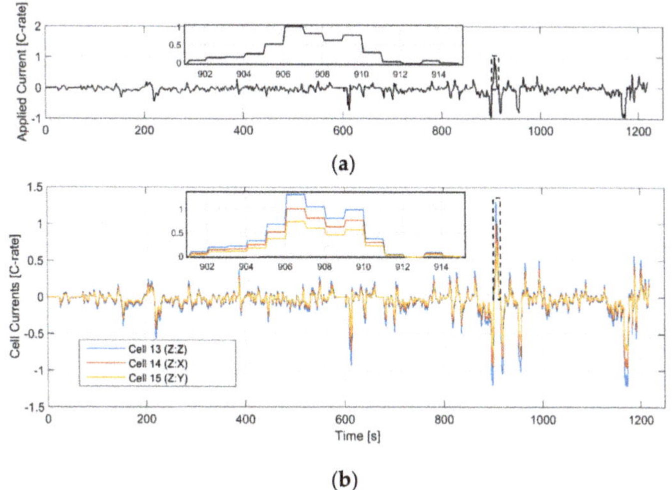

Figure 7. (**a**) Current applied to the parallel unit of cells; (**b**) Individual cell currents.

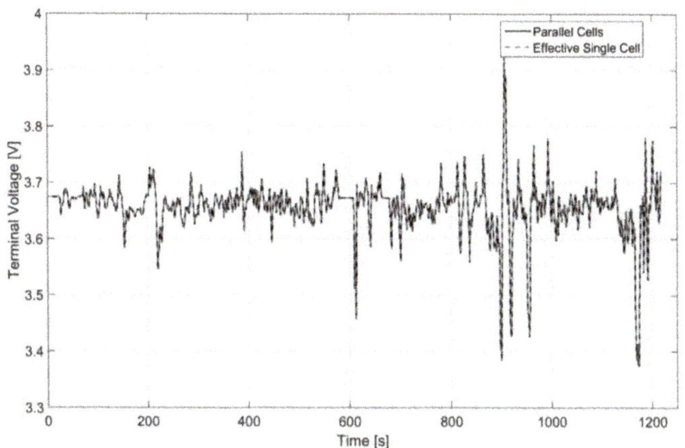

Figure 8. Terminal voltage over the drive cycle for the parallel cell model and effective single cell model.

Table 6. Summary of key results from parallel cell simulation.

Cell Sample Number		13	14	15
Relative current loading (%)	New	101.9	98.2	99.9
	Aged	127.5	99.4	73.2
Relative heat energy (%)	New	101.9	98.2	99.9
	Aged	126.8	99.5	73.8

In Figure 9, the nominal charge throughput is plotted as a function of ohmic resistance. A power-law trend line has also been calculated using a least squares approach. A power law was chosen because, unlike a polynomial or exponential fit, it behaves logically when extrapolating:

as resistance tends to infinity, the charge throughput tends to zero, and as resistance tends to zero, charge throughput tends to infinity.

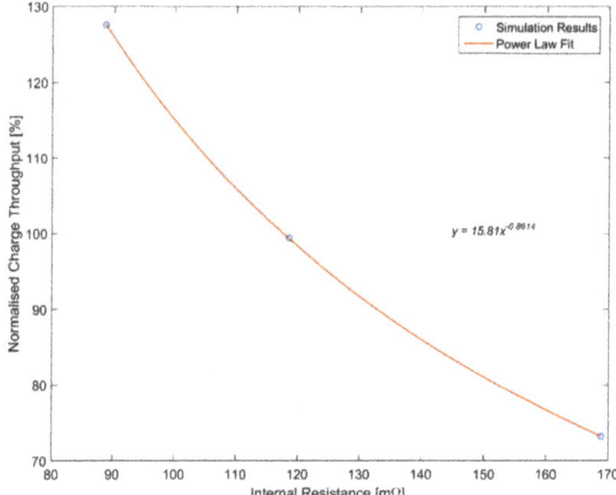

Figure 9. Normalized charge throughput as a function of R_O resistance.

Since these currents are typically not measured by a BMS, it is important to design the pack such that these variations are accounted for and individual cells are not taken outside of their intended operating window during the entirety of the battery pack's life. The differences in current may cause the cells to age by different amounts, which could be unpredictable and must be carefully analyzed and managed. The thermal management system of the battery pack must also be designed, from a hardware and BMS perspective, to be able to reduce any temperature differences which may arise because of impedance variation. Impedance generally decreases with temperature [39], which could further increase the variation in cell currents if not properly managed.

6.2. Charging Cells in Series

When cells are connected in series, the cell current is the same for each cell, but each cell can be at a different voltage. Differences in voltage can occur due to cells being at unequal SOCs and impedance variation between cells. Variations can occur for a variety of reasons, including differences in capacity, self-discharge rate and temperature [43,44].

Constant current-constant voltage charging is commonly used to charge cells. However, this cannot be directly applied with a module of cells in series. The charging strategy considered here is to charge at a constant current (C/2 charging rate) until any one cell reaches the maximum cell voltage. The applied current is then reduced by 10%, and constant current charging continues until one cell again reaches maximum voltage. The current is reduced by another 10%, and this process is repeated until the charging current is below a certain threshold. For this study, ECMs for cells 13–15 are connected in series as in Figure 5c, and this is compared against three ASMs in series. The ASM is found by applying Equation (9) to the measured impedance of cells 13–15, and then identifying the parameters as described in Section 5.

As can be seen in Figure 10a, there is almost a 90 mV difference between cells 14 and 15, despite there being very little difference in SOC. This means that the higher impedance cell reaches its maximum voltage before the other, and so the lower impedance cell is not charged as much as it could be. Figure 10b shows that the ASM SOC is 1.5% higher than the lowest SOC of the string of individual ECMs despite all cells starting from the same SOC. For a series string of cells, the lowest cell SOC is a strong indicator of the overall pack performance [3]. This difference would increase if charging occurred over a wider SOC window or at a higher current.

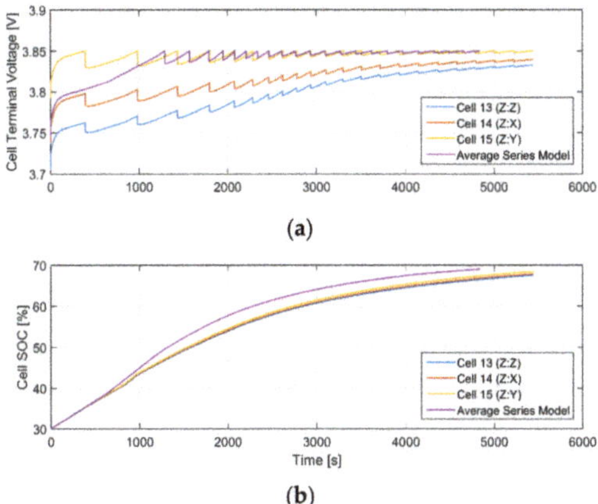

Figure 10. (**a**) Terminal voltage for the three individual cells and average series cell; (**b**) SOC for the three individual cells and average series cell.

Cell balancing systems are used to remove variations in energy between cells in series [45,46]. Cell voltage is commonly used as a metric of imbalance because it is a direct measurement rather than an estimate, as is the case for SOC or charge level. In this case the SOCs remain similar throughout charging, so there is no significant SOC imbalance. However, there is a significant voltage imbalance due to the applied load and differences in impedance. As such if balancing was performed over the charging period and activated based on the voltage difference, the result would be that once the pack had finished charging and the applied load was zero, the SOCs and terminal voltages of the cells would actually be imbalanced.

In Figure 11, the three cell string is compared with the same charging simulation for three of the ASMs in series. Figure 11a shows that charging current is generally higher for the ASM string, and as such reaches end of charge 549 s quicker. Figure 11b shows that the string voltage is also higher for the ASM case, increasing the available power of the string.

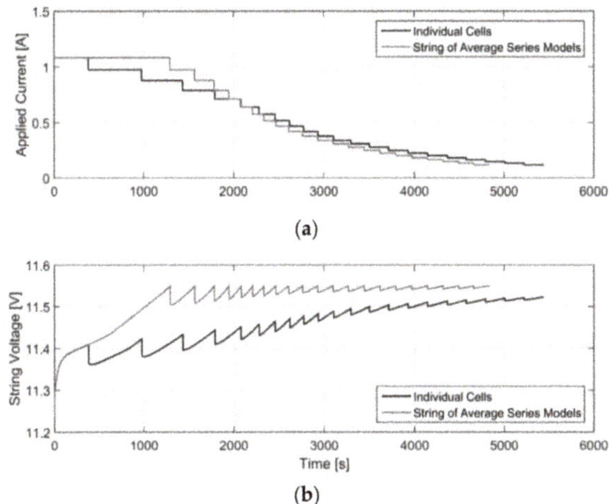

Figure 11. (**a**) Charging current for a string of individual cells, and a string of mean series model cells; (**b**) Total voltage for a string of individual cells, and a string of mean series model cells.

7. Further Work

One of the limitations of the methodology employed within this study is that electrical characterization data was only measured at SOT and EOT from the vibration aged cells. As a result, no discussion or conclusions can be made about the rate of degradation throughout the vehicle's life and therefore the gradual change on cell impedance on the robustness of the BMS algorithms for managing change imbalance within the complete battery pack. It is recommended that a future study should characterize the cells at intermediate points during the test program, e.g., intervals representative of 10,000 miles of vehicle use. This would facilitate further investigation into both the absolute value of degradation, but also the expected in-service rate of capacity and power fade over the life of the vehicle. The variance in impedance between the aged cells means that a balancing system is necessary to manage the available capacity over the life of a battery pack. Model-based design using these ECMs could be used to design a balancing control system to maximize the available battery pack energy and account for the variations in impedance and capacity. Other BMS functionality, such as detecting a cell end of life condition or failure through vibration can also be explored using model-based techniques. Vibration causes a notable shift in ohmic resistance, which is distinct from other major types of ageing [39], and it may be possible for a BMS to classify the type of ageing based on this information.

8. Conclusions

Using experimental data and previously validated mathematical cell models, it has been shown that the ageing caused by vibration can have a significant impact on various BMS functions and the battery pack itself. For cells connected in parallel, the thermal management system must be designed to be reduce variations in temperature which can occur through different current loading between cells. Significant differences in cell current are also shown to occur owing to the impedance differences between cells aged through vibration. Since the individual cell currents are not measured, it is important that the battery pack is designed so that each cell is not taken outside its intended operating window. When aged cells are connected in series, variations in impedance limit the ability to fully charge the series string, reducing the available power and increasing the charging time for the complete battery system.

Acknowledgments: The research presented within this paper is supported by the Engineering and Physical Science Research Council (EPSRC-EP/I01585X/1) through the Engineering Doctoral Centre in High Value, Low Environmental Impact Manufacturing. The research was undertaken in collaboration with the WMG Centre High Value Manufacturing Catapult (funded by Innovate UK) and Jaguar Land Rover. The authors would also like to express their gratitude to Millbrook Proving Ground Ltd (Component Test Laboratory) for their support and advice throughout the durability test program.

Author Contributions: Thomas Bruen—Primary researcher and lead author. James Michael Hooper—Experimental researcher (durability characterization) and co-author. James Marco—Academic research supervision and co-author. Miguel Gama—Industrial research support and peer-review. Gael Henri Chouchelamane—Experimental researcher (electrical characterization) and co-author.

Conflicts of Interest: The authors declare no conflict of interest.

References

1. Jackson, N. Technology Road Map, R+D Agenda and UK Capabilities. In Proceedings of the Cenex Low Carbon VehicleEvent, Millbrook Proving Ground, Bedfordshire, UK, 15–16 September 2010; pp. 1–16.
2. Parry-Jones, R.; Cable, V. *Driving Success—A Strategy for Growth and Sustainability in the UK Automotive Sector*; Automotive Council UK: London, UK, 2013.
3. Lu, L.; Han, X.; Li, J.; Hua, J.; Ouyang, M. A review on the key issues for lithium-ion battery management in electric vehicles. *J. Power Sources* **2013**, *226*, 272–288. [CrossRef]
4. Xing, Y.; Ma, E.W.M.; Tsui, K.L.; Pecht, M. Battery Management Systems in Electric and Hybrid Vehicles. *Energies* **2011**, *4*, 1840–1857. [CrossRef]

5. Day, J. Johnson Controls' Lithium-Ion Batteries Power Jaguar Land Rover's 2014 Hybrid Range Rover. Available online: http://johndayautomotivelectronics.com/johnson-controls-lithium-ion-batteries-power-2014-hybrid-range-rover/ (accessed on 28 November 2015).
6. Rawlinson, P.D. Integration System for a Vehicle Battery Pack. US8833499 B2, 16 September 2014.
7. Berdichevsky, G.; Kelty, K.; Straubel, J.B.; Toomre, E.; Motors, T. *The Tesla Roadster Battery System*, 2nd ed.; Tesla Motors: Palo Alto, CA, USA, 2007.
8. Kelty, K. The battery technology behind the wheel. Available online: http://asia.stanford.edu/us-atmc/wordpress/wp-content/uploads/2010/12/ee402s-04022009-tesla.pdf (accessed on 23 March 2016).
9. Paterson, A. Our Guide to Batteries. Available online: http://www.jmbatterysystems.com/JMBS/media/JMBS/Technology/Axeon-Guide-to-Batteries-2nd-edition.pdf (accessed on 23 March 2016).
10. Anderman, M. *Tesla Motors: Battery Technology, Analysis of the Gigafactory, and the Automakers' Perspectives*; Advanced Automotive Batteries: 9204 Citron Way, Oregon House, CA, USA, 2014; pp. 1–39.
11. Karbassian, A.; Bonathan, D.P. Accelerated Vibration Durability Testing of a Pickup Truck Rear Bed (2009-01-1406). *SAE Int.* **2009**, 1–5. [CrossRef]
12. Risam, G.S.; Balakrishnan, S.; Patil, M.G.; Kharul, R.; Antonio, S. Methodology for Accelerated Vibration Durability Test on Electrodynamic Shaker. *SAE Int.* **2006**, *1*, 1–9.
13. Harrison, T. *An Introduction to Vibration Testing*; Bruel and Kjaer Sound and Vibration Measurement: Naerum, Denmark, 2014.
14. Hooper, J.M.; Marco, J. Experimental Modal Analysis of Lithium-Ion Pouch Cells. *J. Power Sources* **2015**, *285*, 247–259. [CrossRef]
15. Moon, S.-I.; Cho, I.-J.; Yoon, D. Fatigue life evaluation of mechanical components using vibration fatigue analysis technique. *J. Mech. Sci. Technol.* **2011**, *25*, 611–637. [CrossRef]
16. Halfpenny, A.; Hayes, P. Fatigue Analysis of Seam Welded Structures using nCode DesignLife. In Proceedings of 2010 European HyperWorks Technology Conference, Versailles, France, 27–29 October 2010; pp. 1–21.
17. Halfpenny, A. Methods for Accelerating Dynamic Durability Tests. In Proceedings of the 9th International Conference on Recent Advances in Structural Dynamics, Southampton, UK, 17–19 July 2006; pp. 1–19.
18. Brand, M.J.; Schuster, S.F.; Bach, T.; Fleder, E.; Stelz, M.; Glaser, S.; Muller, J.; Sextl, G.; Jossen, A.A.; Gläser, S.; *et al.* Effects of vibrations and shocks on lithium-ion cells. *J. Power Sources* **2015**, *288*, 62–69. [CrossRef]
19. Hooper, J.; Marco, J.; Chouchelamane, G.; Lyness, C. Vibration Durability Testing of Nickel Manganese Cobalt Oxide (NMC) Lithium-Ion 18650 Battery Cells. *Energies* **2016**, *9*, 27. [CrossRef]
20. *Vibration Testing of Electric Vehicle Batteries*; Society of Automotive Engineers (SAE): Warrendale, PA, USA, 2013; pp. 1–7.
21. Hooper, J. Study into the Vibration Inputs of Electric Vehicle Batteries. Master's Thesis, Cranfield University, Cranfield, Bedfordshire, UK, December 2012.
22. Hooper, J.M.; Marco, J. Characterising the in-vehicle vibration inputs to the high voltage battery of an electric vehicle. *J. Power Sources* **2014**, *245*, 510–519. [CrossRef]
23. Barai, A.; Chouchelamane, G.H.; Guo, Y.; McGordon, A.; Jennings, P. A study on the impact of lithium-ion cell relaxation on electrochemical impedance spectroscopy. *J. Power Sources* **2015**, *280*, 74–80. [CrossRef]
24. Hooper, J.; Marco, J. Understanding Vibration Frequencies Experienced by Electric Vehicle Batteries. In Proceedings of the 4th Hybrid and Electric Vehicles Conference (HEVC), London, UK, 6–7 Novermber 2013; pp. 1–6.
25. Harrison, T. *Random Vibration Theory*; Bruel and Kjaer Sound and Vibration Measurement: Naerum, Denmark, 2014.
26. Harrison, T. *The Vibration System*; Bruel and Kjaer Sound and Vibration Measurement: Naerum, Denmark, 2014.
27. Chouchelamane, G. *Electrochemical Impedance Spectroscopy*; Warwick Manufacturing Group: Warwick, UK, 2013.
28. Birkl, C.R.; Howey, D.A. Model Identification and Parameter Estimation for LiFePO4 Batteries. In Proceedings of the 4th Hybrid and Electric Vehicles Conference (HEVC), London, UK, 6–7 Novermber 2013; pp. 1–6.
29. Hu, X.; Li, S.; Peng, H. A comparative study of equivalent circuit models for Li-ion batteries. *J. Power Sources* **2012**, *198*, 359–367. [CrossRef]

30. He, H.; Xiong, R.; Fan, J. Evaluation of Lithium-Ion Battery Equivalent Circuit Models for State of Charge Estimation by an Experimental Approach. *Energies* **2011**, *4*, 582–598. [CrossRef]
31. Seaman, A.; Dao, T.-S.; McPhee, J. A survey of mathematics-based equivalent-circuit and electrochemical battery models for hybrid and electric vehicle simulation. *J. Power Sources* **2014**, *256*, 410–423. [CrossRef]
32. Fleischer, C.; Waag, W.; Heyn, H.-M.; Sauer, D.U. On-line adaptive battery impedance parameter and state estimation considering physical principles in reduced order equivalent circuit battery models. Part 1. Requirements, critical review of methods and modeling. *J. Power Sources* **2014**, *260*, 276–291. [CrossRef]
33. The Tesla Battery Report. Available online: http://advancedautobat.com/industry-reports/2014-Tesla-report/Extract-from-the-Tesla-battery-report.pdf (accessed on 22 October 2014).
34. Dubarry, M.; Vuillaume, N.; Liaw, B.Y. From single cell model to battery pack simulation for Li-ion batteries. *J. Power Sources* **2009**, *186*, 500–507. [CrossRef]
35. Truchot, C.; Dubarry, M.; Liaw, B.Y. State-of-charge estimation and uncertainty for lithium-ion battery strings. *Appl. Energy* **2014**, *119*, 218–227. [CrossRef]
36. Kim, J.; Cho, B.H. Screening process-based modeling of the multi-cell battery string in series and parallel connections for high accuracy state-of-charge estimation. *Energy* **2013**, *57*, 581–599.
37. Bruen, T.; Marco, J. Modelling and experimental evaluation of parallel connected lithium ion cells for an electric vehicle battery system. *J. Power Sources* **2016**, *310*, 91–101. [CrossRef]
38. Bruen, T.; Marco, J.; Gama, M. Model Based Design of Balancing Systems for Electric Vehicle Battery Packs. *IFAC-PapersOnLine* **2015**, *48*, 395–402. [CrossRef]
39. Waag, W.; Käbitz, S.; Sauer, D.U. Experimental investigation of the lithium-ion battery impedance characteristic at various conditions and aging states and its influence on the application. *Appl. Energy* **2013**, *102*, 885–897. [CrossRef]
40. Andre, D.; Meiler, M.; Steiner, K.; Wimmer, C.; Soczka-Guth, T.; Sauer, D.U. Characterization of high-power lithium-ion batteries by electrochemical impedance spectroscopy. I. Experimental investigation. *J. Power Sources* **2011**, *196*, 5334–5341. [CrossRef]
41. Mueller, K.; Tittel, D.; Graube, L.; Sun, Z.; Luo, F. Optimizing BMS Operating Strategy Based on Precise SOH Determination of Lithium Ion Battery Cells. In Proceedings of the International Federation of Automotive Engineering Societies 2012 World Automotive Congress, Beijing, China, 27–30 November 2012; pp. 807–819.
42. Waag, W.; Käbitz, S.; Sauer, D.U. Application-specific parameterization of reduced order equivalent circuit battery models for improved accuracy at dynamic load. *Measurement* **2013**, *46*, 4085–4093. [CrossRef]
43. Xu, J.; Li, S.; Mi, C.; Chen, Z.; Cao, B. SOC Based Battery Cell Balancing with a Novel Topology and Reduced Component Count. *Energies* **2013**, *6*, 2726–2740. [CrossRef]
44. Zhang, Z.; Sisk, B. Model-Based Analysis of Cell Balancing of Lithium-ion Batteries for Electric Vehicles. *SAE Int. J. Alt. Power* **2013**, *2*, 379–388. [CrossRef]
45. Moore, S.W.; Schneider, P.J. A Review of Cell Equalization Methods for Lithium Ion and Lithium Polymer Battery Systems. *SAE Publ.* **2001**, *2001010959*, 1–7.
46. Gallardo-Lozano, J.; Romero-Cadaval, E.; Milanes-Montero, M.I.; Guerrero-Martinez, M.A. Battery equalization active methods. *J. Power Sources* **2014**, *246*, 934–949. [CrossRef]

Article

Application of Model Predictive Control to BESS for Microgrid Control

Thai-Thanh Nguyen, Hyeong-Jun Yoo and Hak-Man Kim *

Department of Electrical Engineering, Incheon National University, 12-1 Songdo-dong, Yeonsu-gu, Incheon 406-840, Korea; ntthanh@inu.ac.kr (T.-T.N.); yoohj@inu.ac.kr (H.-J.Y.)

* Author to whom correspondence should be addressed; hmkim@inu.ac.kr; Tel.: +82-32-835-8769; Fax: +82-32-835-0773.

Academic Editor: William Holderbaum
Received: 27 June 2015; Accepted: 5 August 2015; Published: 19 August 2015

Abstract: Battery energy storage systems (BESSs) have been widely used for microgrid control. Generally, BESS control systems are based on proportional-integral (PI) control techniques with the outer and inner control loops based on PI regulators. Recently, model predictive control (MPC) has attracted attention for application to future energy processing and control systems because it can easily deal with multivariable cases, system constraints, and nonlinearities. This study considers the application of MPC-based BESSs to microgrid control. Two types of MPC are presented in this study: MPC based on predictive power control (PPC) and MPC based on PI control in the outer and predictive current control (PCC) in the inner control loops. In particular, the effective application of MPC for microgrids with multiple BESSs should be considered because of the differences in their control performance. In this study, microgrids with two BESSs based on two MPC techniques are considered as an example. The control performance of the MPC used for the control microgrid is compared to that of the PI control. The proposed control strategy is investigated through simulations using MATLAB/Simulink software. The simulation results show that the response time, power and voltage ripples, and frequency spectrum could be improved significantly by using MPC.

Keywords: microgrid; model predictive control; predictive power control; battery energy storage system (BESS); frequency control

---

1. Introduction

Microgrids are becoming popular in distribution systems because they can improve the power quality and reliability of power supplies and reduce the environmental impact. Microgrid operation can be classified into two modes: grid-connected and islanded modes. In general, microgrids are comprised of distributed energy resources (DERs) including renewable energy sources, distributed energy storage systems (ESSs), and local loads [1–3]. However, the use of renewable energy sources such as wind and solar power in microgrids causes power flow variations owing to uncertainties in their power outputs. These variations should be reduced to meet power-quality requirements [4,5]. This study focuses on handling the problems that are introduced by wind power.

To compensate for fluctuations in wind power, various ESSs have been implemented in microgrids. Short-term ESSs such as superconducting magnetic energy storage (SMES) systems [6], electrical double-layer capacitors (EDLCs) [7], and flywheel energy storage systems (FESSs) [8–10] as well as long-term ESSs such as battery energy storage systems (BESSs) [11,12] are applied to microgrid control. ESSs can also be used to control the power flow at point of common coupling in the grid-connected mode as well as to regulate the frequency and voltage of a microgrid in the islanded mode. Among these ESSs, BESSs have been implemented widely owing to their versatility, high energy density, and efficiency. Moreover, their cost has decreased whereas their performance and lifetime has increased [13].

In practice, BESSs with high performance such as smooth and fast dynamic response during charging and discharging are required for microgrid control. This performance depends on the control performance of the power electronic converter. Proportional-integral (PI) control is a practical and popular control technique for BESS control systems. However, PI control might show unsatisfactory results for nonlinear and discontinuous systems [10]. Meanwhile, model predictive control (MPC) is considered an attractive alternative to promote the performance of future energy processing and control systems [14]. Predictive strategies are based on the inherent discrete nature of a power converter. Owing to the finite number of switching states of a power converter, all possible states are considered for predicting the system behavior. Then, each prediction is used to evaluate a cost function. Consequently, the switching state with the minimum cost function is selected and applied to the converter [15]. One of the advantages of an MPC is the easy inclusion of constraints and nonlinearities. Therefore, MPC has been widely applied to drive applications [15–18] and power converters such as active front-end rectifiers [19], matrix converters [20], and multilevel converters [21]. Recently, it has been applied to a bidirectional AC-DC converter for use in BESSs [22–24].

Only a few literatures were found on the application of MPC to microgrid control. Most existing studies focused on MPC for a distributed generator in a microgrid with voltage and/or power control [25–27]. A modified MPC method for voltage control of a BESS in the islanded mode operation of a microgrid was presented in [27]. However, this study did not deal with frequency control in the islanded mode operation of a microgrid. MPC based on PI control in the outer control loop and predictive current control (PCC) in the inner control loop for BESS was presented in [28]. Coordinated predictive control of a wind/battery microgrid system was proposed to maintain the system voltage and frequency by adjusting the output power of BESS. PCC was used to control the current in the inner control loop, whereas PI regulators were used to regulate the voltage and power in the outer control loop. Owing to the use of PI regulators in the outer control loop, the dynamic response time under such MPC techniques was similar to that under PI control techniques with outer and inner control loops using PI regulators.

Another MPC technique is based on predictive power control (PPC), in which the power is predicted and controlled directly. This MPC technique could be applied to microgrid control because it affords advantages such as fast dynamic response for power control; however, studies have not yet explored the application of the PPC-based MPC technique to microgrid control. Furthermore, this MPC technique can only be used for power control. To overcome this problem, PI regulators can be used in an additional control loop to control the frequency and voltage. Therefore, this MPC technique uses PI regulators in the outer control loop and PPC in the inner control loop. It is similar to previous MPC techniques in which PI control is used in the outer control loop and PCC is used in the inner control loop. However, an MPC technique based on PI and PPC requires more computation time than does one based on PI and PCC, owing to the predicting powers in the inner PPC control loop. Therefore, in a microgrid with a single BESS, MPC based on PI and PCC is a suitable alternative for microgrid control. Another approach to overcome this limitation of the MPC control technique is to use a droop control scheme. Thus, a PPC-based MPC technique can be applied to microgrids consisting of multiple BESSs with different functionalities. This study deals with the effective application of an MPC technique to a microgrid with two BESSs as an example of multiple BESSs in a microgrid.

This study discusses the effective application of two MPC techniques to BESSs for microgrid control based on the characteristics of the MPC techniques as well as the functionalities of BESSs. One BESS is based on PI control in the outer and PCC in the inner control loops (PI (outer) + PCC (inner)); it is used for smoothing wind power fluctuations both in the grid-connected and the islanded modes. The other BESS is based on PPC (one loop); it controls the tie-line powers at the point of common coupling in the grid-connected mode and the frequency in the islanded mode. Additionally, to reduce the power losses of converters, the reduction of the switching frequency of the converter is considered an additional control variable in the MPC algorithm. The control performances of the two types of MPC techniques are compared to the PI control technique using PI regulators in the outer and inner control loops (PI (outer) + PI (inner)). The tuning of PI regulator parameters must be taken into account to effectively compare the control performance of MPC techniques to the PI control technique. Several tuning techniques have been used to select the PI regulator parameters. In this study, the tuning technique provided by MATLAB/Simulink software is used. The efficacy of the proposed control system is verified via simulations in the MATLAB/Simulink environment.

The remainder of this paper is organized as follows. Section 2 introduces the discrete-time model of the converter for prediction and MPC algorithms. Two types of MPC techniques are introduced in this section. Section 3 describes the microgrid system used to test the performance of the proposed control strategies. Section 4 presents a comparison of the MPC and PI control techniques and the considerations for the effective application of MPC-based BESSs to microgrid control. Section 5 presents the simulation results for microgrid control in the grid-connected and islanded modes. The performances of the MPC techniques are compared to those of the PI control technique. Finally, Section 6 summarizes the main conclusions of this study.

2. MPC for BESS

2.1. Discrete-Time Model of Converter

The predicted variables of BESS are determined based on the discrete-time model of the converter. In this study, the BESS uses a two-level voltage source converter (VSC) converter, shown in Figure 1, connected to the three-phase AC power supply voltage v_g through filter inductance L and resistance R. The equations for each phase are given by Equations (1)–(3):

$$v_{aN} = L\frac{di_a}{dt} + Ri_a + v_{ga} \tag{1}$$

$$v_{bN} = L\frac{di_b}{dt} + Ri_b + v_{gb} \tag{2}$$

$$v_{cN} = L\frac{di_c}{dt} + Ri_c + v_{gc} \tag{3}$$

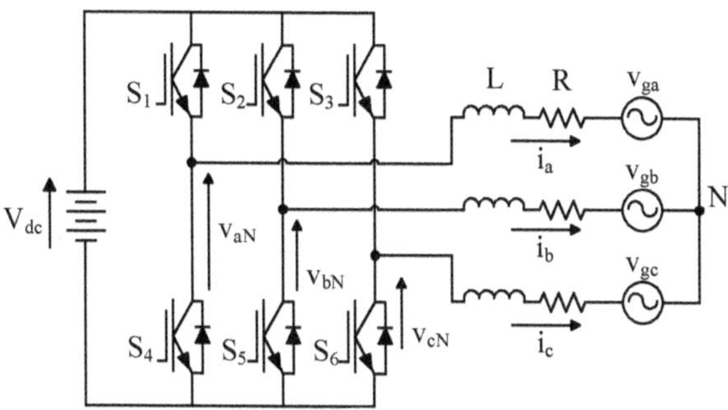

Figure 1. Configuration of BESS.

These equations can be represented by the space-vector equations given in Equation (4).

$$\frac{2}{3}\left(v_{aN} + \boldsymbol{a}v_{bN} + \boldsymbol{a}^2v_{cN}\right) = L\frac{d}{dt}\left(\frac{2}{3}\left(i_a + \boldsymbol{a}i_b + \boldsymbol{a}^2i_c\right)\right) + R\left(\frac{2}{3}\left(i_a + \boldsymbol{a}i_b + \boldsymbol{a}^2i_c\right)\right) + \frac{2}{3}\left(v_{ga} + \boldsymbol{a}v_{gb} + \boldsymbol{a}^2v_{gc}\right) \tag{4}$$

where $\boldsymbol{a} = e^{j2\pi/3}$.

Equation (4) can be simplified by considering the following definitions.

$$v = \frac{2}{3}\left(v_{aN} + \boldsymbol{a}v_{bN} + \boldsymbol{a}^2 v_{cN}\right) \tag{5}$$

$$i = \frac{2}{3}\left(i_a + \boldsymbol{a}i_b + \boldsymbol{a}^2 i_c\right) \tag{6}$$

$$v_g = \frac{2}{3}\left(v_{ga} + \boldsymbol{a}v_{gb} + \boldsymbol{a}^2 v_{gc}\right) \tag{7}$$

The voltage v in Equation (5) is determined by the switching states of the converter and the DC link voltage (V_{DC}), as given in Equation (8).

$$v = \frac{2}{3}V_{DC}\left(S_a + \boldsymbol{a}S_b + \boldsymbol{a}^2 S_c\right) \tag{8}$$

Where the switching signals S_a, S_b, and S_c are defined as follows:

$$S_a = \begin{cases} 1 \text{ if } S_1 \text{ on and } S_4 \text{ off} \\ 0 \text{ if } S_1 \text{ off and } S_4 \text{ on} \end{cases} \tag{9}$$

$$S_b = \begin{cases} 1 \text{ if } S_2 \text{ on and } S_5 \text{ off} \\ 0 \text{ if } S_2 \text{ off and } S_5 \text{ on} \end{cases} \tag{10}$$

$$S_c = \begin{cases} 1 \text{ if } S_3 \text{ on and } S_6 \text{ off} \\ 0 \text{ if } S_3 \text{ off and } S_6 \text{ on} \end{cases} \tag{11}$$

The combination of S_a, S_b, and S_c creates eight switching states and eight voltage vectors, as shown in Table 1.

Table 1. Switching states and voltage vectors [29].

x	S_a	S_b	S_c	**Voltage vectors**
1	0	0	0	$v_0 = 0$
2	1	0	0	$v_1 = \frac{2}{3}V_{dc}$
3	1	1	0	$v_2 = \frac{1}{3}V_{dc} + j\frac{\sqrt{3}}{3}V_{dc}$
4	0	1	0	$v_3 = -\frac{1}{3}V_{dc} + j\frac{\sqrt{3}}{3}V_{dc}$
5	0	1	1	$v_4 = -\frac{2}{3}V_{dc}$
6	0	0	1	$v_5 = -\frac{1}{3}V_{dc} - j\frac{\sqrt{3}}{3}V_{dc}$
7	0	1	1	$v_6 = \frac{1}{3}V_{dc} - j\frac{\sqrt{3}}{3}V_{dc}$
8	1	1	1	$v_7 = 0$

Substituting Equations (5)–(7) in Equation (4), we get

$$v = L\frac{di}{dt} + Ri + v_g \tag{12}$$

From Equation (12), the discrete-time model of the converter is determined by approximating the derivative load current di/dt in terms of a forward Euler approximation, as shown in Equation (13).

$$\frac{di}{dt} \approx \frac{i(k+1) - i(k)}{T_s} \tag{13}$$

By substituting Equation (13) in Equation (12), the future current at the sampling instant $k + 1$ is represented as

$$i^p(k+1) = \left(1 - \frac{RT_s}{L}\right)i(k) + \frac{T_s}{L}\left(v(k) - v_g(k)\right) \tag{14}$$

where $i(k)$ and $v_g(k)$ are the three-phase current and voltage of the BESS measured at sampling instant k, respectively; $v(k)$ is the voltage vector according to the eight switching states of the converter; and T_s is the sampling time.

Based on the measured voltage and current of BESS at sampling instant k, the variables at sampling instant $k + 1$ are predicted as given in Equation (14). For a small sampling time (T_s), the predicted grid voltage at sampling instant $k + 1$ can be assumed equal to the measured grid voltage at the kth sampling instant ($v_g(k + 1) = v_g(k)$) owing to the fundamental grid frequency [29]. As a result, the predicted instantaneous real and reactive powers can be expressed as follows:

$$P^p(k+1) = 1.5Re\left\{\bar{i}^p(k+1)v_g^m(k)\right\} \tag{15}$$

$$Q^p(k+1) = 1.5Im\left\{\bar{i}^p(k+1)v_g^m(k)\right\} \tag{16}$$

where $\overset{-p}{i}(k+1)$ is the complex conjugate of the predicted current vector $i^p(k+1)$.

Equations (14)–(16) show that the predictive current and power highly rely on system model, converter, and filter parameters. Any change in the model parameters can provide inaccuracy in the predictive variables. Reference [29] shown that the current or power ripple could be affected by the parameter variations, whereas the dynamic response was almost unchanged. In case of extreme variations in the model parameters, an online parameter estimation algorithm should be included in the MPC strategy [30,31]. However, MPC can effectively handle the small change in inductive filter parameters. The comparison between MPC with and without the online filter estimation was presented in [29,32]. The major errors were observed at low values of the filter parameters. In addition, only a small difference was observed at high values of the filter parameters. In this study, a high value of the filter parameter is chosen to avoid the major errors by filter parameter variations. Thus, the model parameters is assumed unchanged during simulation for the sake of simplicity.

2.2. Principle of MPC

MPC is based on the inherent discrete nature of a power converter, which has a finite number of switching states. All possibilities of variables (current or real/reactive powers) of the converter according to switching states can be predicted. The predicted variables are compared to the reference control signal, and the predicted variable that is closest to the reference control signal is chosen as shown in Figure 2. Then, the switching state related to this predicted variable is applied to control the converter.

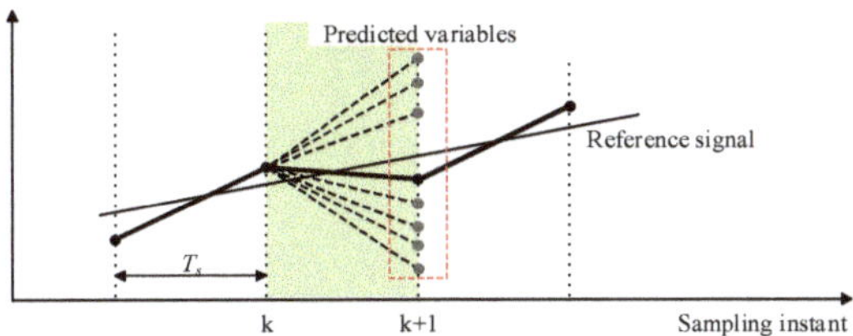

Figure 2. Principle of MPC.

Figure 3 shows two types of MPC techniques applied for BESSs: MPC based on PI control in the outer and PCC in the inner control loops (Figure 3a) and MPC based on PPC (Figure 3b). As shown in Figure 3a, PI control in the outer control loop is used to regulate the real/reactive powers as well as voltage of the microgrid. The reference current obtained by the outer control loop is used for the inner PCC control loop based on Equation (14). As shown in Figure 3b, in comparison, PPC based on Equations (15) and (16) can control real/reactive powers directly. To control the frequency of the microgrid, the frequency droop control scheme is suitable for a BESSs control system. However, conventional droop control can cause a steady-state error [33]. Thus, this study proposes an improved droop control scheme in which the steady-state error is removed by a new feedback signal through the PI regulator [9].

The objective of the MPC scheme is to minimize the error between the reference values and the measured values. This can be achieved by introducing a cost function g_C for PCC and g_S for PPC, as shown in the following equations.

$$g_C = \left| i_\alpha^*(k+1) - i_\alpha^p(k+1) \right|^2 + \left| i_\beta^*(k+1) - i_\beta^p(k+1) \right|^2 + \lambda_C \cdot n \tag{17}$$

$$g_S = |P^*(k+1) - P^p(k+1)|^2 + |Q^*(k+1) - Q^p(k+1)|^2 + \lambda_S \cdot n \tag{18}$$

where $i_\alpha^*(k+1)$ and $i_\beta^*(k+1)$ are the real and imaginary parts of the reference current, $i_\alpha^p(k+1)$ and $i_\beta^p(k+1)$ are the real and imaginary parts of the predicted current vectors $i^p(k+1)$ according to Equation (14), $P^*(k+1)$ and $Q^*(k+1)$ are the real and reactive reference powers, $P^p(k+1)$ and $Q^p(k+1)$ are the predicted real and reactive powers according to Equations (15) and (16), $\lambda_C \cdot n$ and $\lambda_S \cdot n$ represent the reduction of switching frequency of the converter where n is the number of switches that change when the switching states $S = (S_a, S_b, S_c)$ are applied, and λ_C and λ_S are the weighting factor for PCC and PPC, respectively.

The cost functions g_C and g_S have two terms with different goals. The primary goal is the current control in case of g_C or power control in case of g_S, which must be achieved to provide a proper system behavior. The secondary goal is the reduction of switching frequency ($\lambda_C \cdot n$ and $\lambda_S \cdot n$) in both cost functions. The importance of second term corresponds to the weighting factors λ_C and λ_S that can impose a trade-off with the primary control objective. The algorithm to adjust the weighting factors proposed in [29] is used in this study. Total harmonic distortion (THD) is used to estimate the trade-off between the primary and secondary goals.

The switching frequency of the converter depends on the change in the switching state, which can be only one or zero. Therefore, the number of switches that change from $S(k-1)$ to $S(k)$ is defined as given in Equation (19):

$$n = |S_a(k) - S_a(k-1)| + |S_b(k) - S_b(k-1)| + |S_c(k) - S_c(k-1)| \tag{19}$$

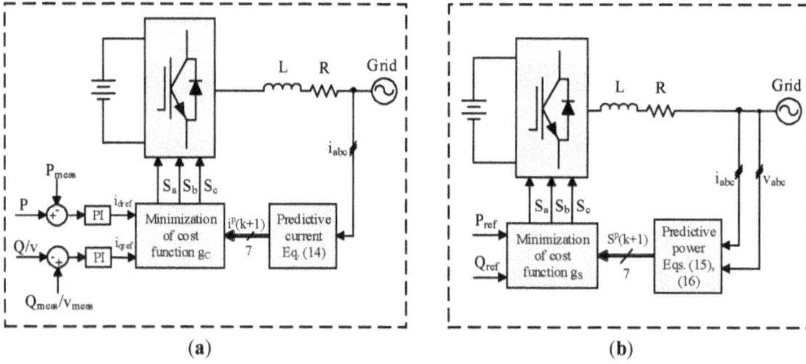

Figure 3. MPC block diagrams: (**a**) MPC based on PI control in the outer and PCC in the inner control loops; (**b**) MPC based on PPC.

The control strategy of MPC techniques involves the following four steps:

(1) The three-phase current and voltage of the BESS are measured, and the values of reference signals are obtained from the outer control loop.
(2) The discrete-time model of the converter is used to predict the values of current or real/reactive powers in the next sampling interval (k + 1) for each voltage vector according to Equations (14)–(16).
(3) The cost function g_C or g_S based on Equations (17) and (18) is used to compute the errors between the reference and the predicted current or real/reactive powers for each voltage vector.
(4) The minimum value of the cost function gives the minimum error between the reference and the measured signals. The voltage vector with respect to the minimum cost function is selected, and the corresponding switching state signals are generated to apply to the converter.

3. Test Microgrid

The test microgrid system (Figure 4) used in this study includes several components: A diesel generator, a consumer load, a wind generator, and two BESSs. Table 2 shows the parameters of the test microgrid system. In this study, the fixed-speed wind energy conversion system (WECS), a type of WECS [34], is used for simplicity. Two BESSs with different control strategies according to the operation mode of the microgrid, as shown in Table 3, are used. In the grid-connected mode, the voltage and frequency of the microgrid is set by the utility grid. Therefore, the main function of the BESS is to control the real and reactive powers. On the other hand, in the islanded mode, the microgrid is disconnected from the utility grid and controls its own frequency and voltage.

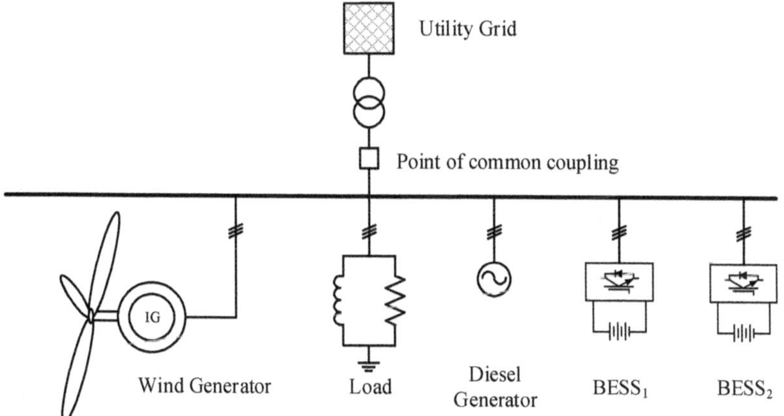

Figure 4. Configuration of microgrid.

Table 2. Parameters of test microgrid.

Components	Rating
Wind generator	150 kVA
$BESS_1$	450 kWh
$BESS_2$	200 kWh
Load	500 kW; 100 kVAR
Diesel generator	500 kVA
Mean wind speed	9 m/s
System frequency	60 Hz
Transformer	700 kVA; 6.6 kV/380 V

Table 3. Control strategies of BESSs.

Operation modes	$BESS_1$	$BESS_2$
Grid-connected	Tie-line powers at point of common coupling	Smoothing wind power
Islanded	Frequency control	Smoothing wind power
	Reactive power at point of common coupling	Voltage control

4. Control Performance of MPC Techniques

4.1. Comparison of Control Performance of MPC and PI Control Techniques

The control performances of two MPC techniques according to the change in real power are compared to that of the PI control technique proposed in [35]. Tuning the PI parameters is an important factor for comparison. Several functions as well as linear analysis tools provided by MATLAB/Simulink are used for tuning. First, the function "getlinio" is used to obtain the linearized input/output of the plant. The linear approximation of the plant is estimated based on the linearized input/output by using the "linearize" function. Then, the linear analysis tool in Simulink is used to estimate the frequency response of a plant based on the linear approximation of the plant. Finally, the PID tuner in Simulink is used to automatically tune the PI parameters based on the frequency response estimation.

Figure 5 shows the simulation results of three types of control techniques. The real power changes from 0 to 50 kW at 1.0 s. The response of the PPC technique is clearly much quicker than that of other techniques. In the case of MPC based on PI in the outer and PCC in the inner control loops and PI control technique using PI regulators in the outer and inner control loops, the dynamic response is similar owing to the action of the PI controller in the outer control loop. Both MPC technique based on PI and PCC and PI technique show good reference tracking under the steady-state condition. However, the power ripple obtained by MPC technique is smaller than that obtained by PI control technique owing to PCC in the inner control loop in MPC technique. Figure 5 shows that MPC techniques can significantly improve the performance of a control system for BESSs in terms of the response time and power ripple.

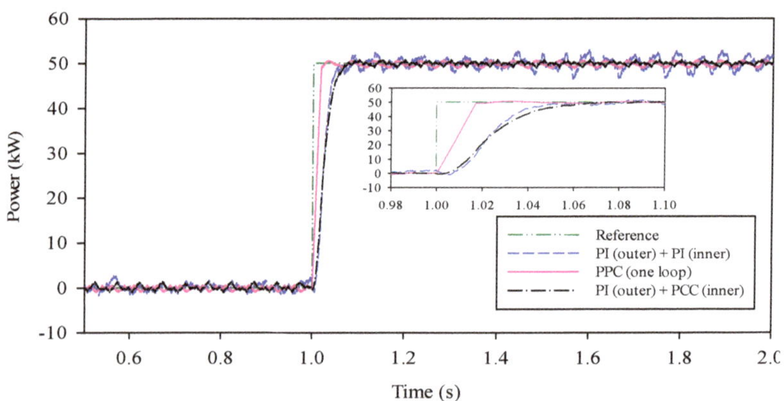

Figure 5. Response of different control techniques for change in reference power.

4.2. Effective Application of MPC Techniques to Microgrid Control

Table 4 shows the characteristics of the MPC and PI control techniques. Among these two MPC techniques, the PPC technique shows the best control performance; however, it can only be used for controlling the power. On the other hand, the MPC technique based on PI in outer and PCC in inner control loops is more flexible owing to the use of a PI regulator in the outer control loop; this technique can be used to control the power, frequency, and voltage. The ripple in case of both MPC techniques is smaller than that in case of the PI control technique.

Table 4. Characteristics of MPC and PI control techniques.

Characteristics	PI (outer) + PI (inner)	PI (outer) + PCC (inner)	PPC (one loop)
Ability to control	P/Q, f/v	P/Q, f/v	P/Q
Response time	Long	Long	Short
Ripple	Large	Small	Small

In this study, two BESSs with different functionalities are proposed to control the microgrid, as shown in Table 3. $BESS_1$ is used to control the power at the point of common coupling and the frequency in the islanded mode, in which case fast dynamic response under disturbances is required for the control system. Therefore, PPC-based MPC is suitable for application to $BESS_1$ because its control performance shows the shortest response time compared to other cases. Furthermore, $BESS_2$ is used for handling fluctuations in wind power in both grid-connected and islanded modes. Thus, the control performance of the MPC technique based on PI control in the outer control loop and PCC control in the inner control loop is suitable for $BESS_2$ owing to gradual fluctuations in wind power. The microgrid voltage is controlled by $BESS_2$ and the frequency, by $BEES_1$ and $BESS_2$ through the improved frequency droop control scheme.

5. Simulation Results

5.1. Control Microgrid in Grid-Connected Mode

BESSs can operate in the charging or discharging mode. Therefore, they can reduce the fluctuations in wind power through effective compensation. Figure 6 shows the action of $BESS_2$ in terms of smoothing the wind power. In the case of $BESS_2$, the MPC technique based on PI control in the outer control loop and PCC in the inner current control loop is applied as the control system. This figure shows that the wind power fluctuations can be reduced significantly by effectively charging or discharging $BESS_2$. Both the MPC and the PI control techniques show good results from the viewpoint of smoothing the wind power. However, the power ripple in case of the MPC technique is much smaller than that in case of the PI control technique.

On the other hand, $BESS_1$ based on the PPC technique controls the power at the point of common coupling. In this study, it is assumed that the real power at the point of common coupling is maintained at zero. Figure 7 shows the simulation result. At 10 s, an additional load of 100 kW is connected to the microgrid. Therefore, $BESS_1$ increases its real power to maintain the power at zero. The subfigure of Figure 7 shows that the response of the MPC technique is slightly quicker than that of the PI control technique. Additionally, the ripples of the BESS power when using the MPC technique is smaller than that of PI control technique. Both the MPC and the PI control techniques show good performance for controlling the power at the point of common coupling.

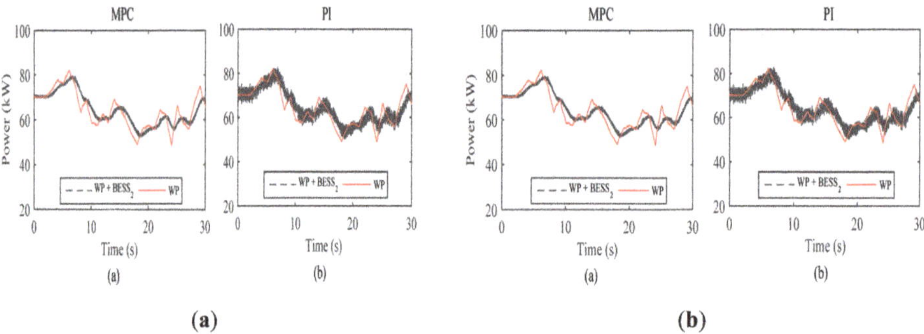

Figure 6. Smoothened wind power: (a) MPC technique; (b) PI technique.

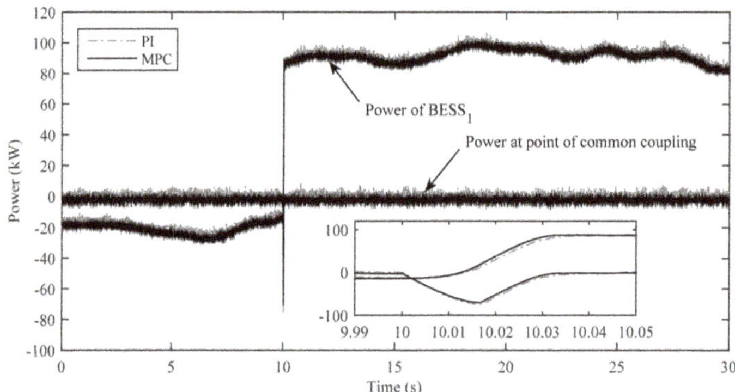

Figure 7. Real power at point of common coupling and real power of $BESS_1$.

5.2. Control Microgrid in Islanded Mode

In the islanded mode, the microgrid frequency is controlled by $BESS_1$, and the microgrid voltage is controlled by $BESS_2$. Figures 8 and 9 respectively show the frequency and voltage of the microgrid. Both the MPC and the PI control techniques can stably control the frequency and voltage of the microgrid. However, as shown in Figure 8, the frequency response under the MPC technique is quicker than that under the PI control technique. Moreover, Figure 9 shows the microgrid voltage. Obviously, the performance of the MPC techniques is much better than that of the PI control technique. The voltage ripple in the case of the MPC technique is much smaller than that in the case of the PI control technique.

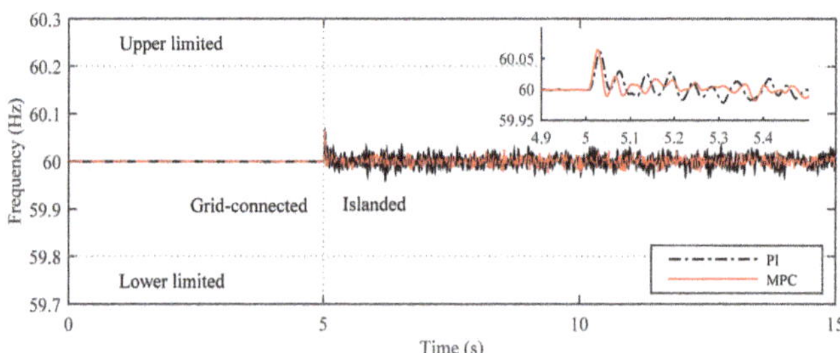

Figure 8. Frequency of microgrid system.

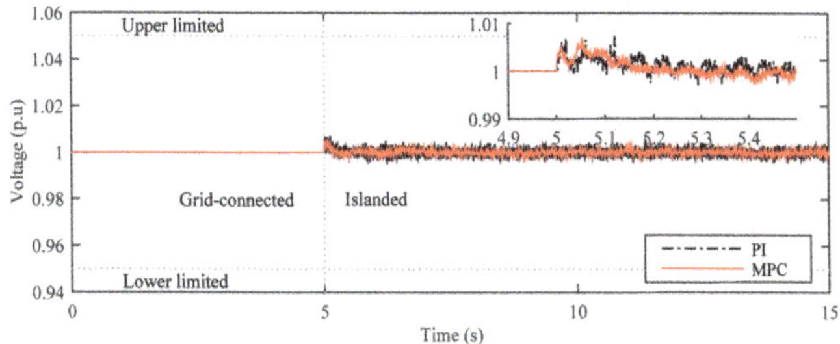

Figure 9. Voltage of microgrid system.

Additionally, the output voltage spectra generated by the converter is one of the important factors. Figure 10 shows a comparison of the voltage spectra of the MPC and PI control techniques. As shown in Figure 10b, the frequency spectrum generated using the PI control technique is concentrated around the carrier frequency owing to PWM. For comparison, Figure 10a shows the frequency spectrum obtained by MPC. The reduction of the switching frequency of the converter is implemented in the cost function of MPC as a secondary control objective to reduce the power losses of converters. Figure 10 shows that the average switching frequency (f_s) obtained by MPC is slightly lower than that obtained by the PI control technique. Moreover, the MPC technique shows significantly lower THD than the PI control technique.

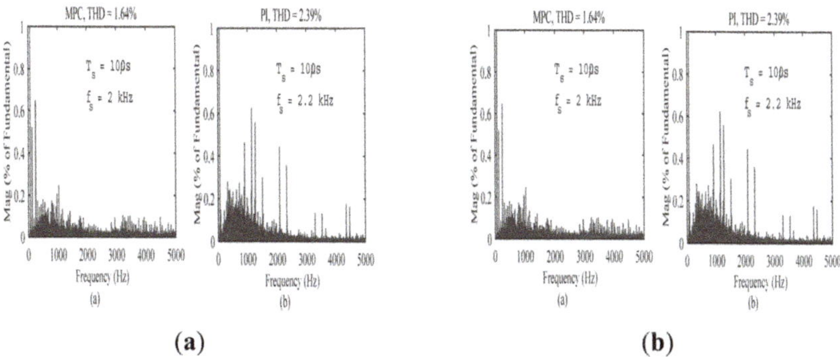

Figure 10. Load voltage spectrum and THD: (**a**) MPC technique; (**b**) PI technique.

6. Conclusions

This study discusses the effective application of two types of MPC techniques to BESSs for microgrid control: MPC based on PPC and MPC based on PI control in the outer control loop and PCC in the inner current control loop. In addition, PI control using a PI regulator in the outer and inner control loops for BESS was compared to these two types of MPC techniques. A reduction switching frequency is implemented in the cost function to reduce the power losses of converters. The simulation results show that the response time, power ripples, and frequency spectrum could be improved significantly by using MPC techniques. Both the average switching frequency and the THD obtained by using MPC techniques were lower than those obtained by using PI control. Using MPC based on PI control in the outer and PCC in the inner control loops did not improve the response time under power changing compared to PI control; however, it could significantly improve the power and voltage ripples under the steady-state condition. Moreover, using PPC-based MPC could reduce the response time under power changing compared to other control techniques. Therefore, in microgrids with multiple BESSs, the PPC-based

MPC technique should be applied for BESSs that control the power at the point of common coupling and the frequency of the microgrid, and an MPC technique based on PI in the outer control loop and PCC in the inner control loop should be applied for BESSs that play the role of smoothing wind power fluctuations. Besides, in case of microgrids with a BESS, PCC-based MPC technique should be a suitable alternative for the BESS owing to its flexible characteristic. MPC technique is easy to implement and it can eliminate the tuning controller parameters effort that has to be done in the PI technique. Furthermore, various control objectives can be included in the MPC strategies.

In the future, we plan to include additional control variables such as considering the state of charge of the battery and coordination control of multiple ESSs in the MPC algorithm.

Acknowledgments: This work was supported by the Power Generation & Electricity Delivery Core Technology Program of the Korea Institute of Energy Technology Evaluation and Planning (KETEP), granted financial resource from the Ministry of Trade, Industry & Energy, Republic of Korea. (No. 20141020402350).

Author Contributions: The paper was a collaborative effort between the authors. The authors contributed collectively to the theoretical analysis, modeling, simulation, and manuscript preparation.

Conflicts of Interest: The authors declare no conflict of interest.

References

1. Mahmoud, M.S.; Hussain, S.A.; Abido, M.A. Modeling and control of microgrid: An overview. *J. Frankl. Inst.* **2014**, *351*, 2822–2859. [CrossRef]
2. Hatziargyriou, N.D. Microgrids. *IEEE Power Energy* **2008**, *6*, 26–29. [CrossRef]
3. Olivares, D.E.; Mehrizi-Sani, A.; Etemadi, A.H.; Canizares, C.A.; Iravani, R.; Kazerani, M.; Hajimiragha, A.H.; Gomis-Bellmunt, O.; Saeedifard, M.; Palma-Behnke, R.; *et al.* Trends in microgrid control. *IEEE Trans. Smart Grid* **2014**, *5*, 1905–1919. [CrossRef]
4. Kim, H.-M.; Lim, Y.; Kinoshita, T. An intelligent multiagent system for autonomous microgrid operation. *Energies* **2012**, *5*, 3347–3362. [CrossRef]
5. Kim, H.-M.; Kinoshita, T. A multiagent system for microgrid operation in the grid-interconnected mode. *J. Electr. Eng. Technol.* **2010**, *2*, 246–254. [CrossRef]
6. Molina, M.G.; Mercado, P.E. Power flow stabilization and control of microgrid with wind generation by superconducting magnetic energy storage. *IEEE Trans. Power Electron.* **2011**, *26*, 910–922. [CrossRef]
7. Inthamoussou, F.A.; Pegueroles-Queralt, J.; Bianchi, F.D. Control of a supercapacitor energy storage system for microgrid applications. *IEEE Trans. Energy Conver.* **2013**, *28*, 690–697. [CrossRef]
8. Islam, F.; Al-Durra, A.; Muyeen, S.M. Smoothing of wind farm output by prediction and supervisory-control-unit-based FESS. *IEEE Trans. Sustain. Energy* **2013**, *4*, 925–933. [CrossRef]
9. Nguyen, T.-T.; Yoo, H.-J.; Kim, H.-M. A flywheel energy storage system based on a doubly fed induction machine and battery for microgrid control. *Energies* **2015**, *8*, 5074–5089. [CrossRef]
10. Chang, X.; Li, Y.; Zhang, W.; Wang, N.; Xue, W. Active disturbance rejection control for a flywheel energy storage system. *IEEE Trans. Ind. Electron.* **2015**, *62*, 991–1001. [CrossRef]
11. Li, X. Fuzzy adaptive kalman filter for wind power output smoothing with battery energy storage system. *IET Renew. Power Gener.* **2012**, *6*, 340–347. [CrossRef]
12. Li, X.; Hui, D.; Lai, X. Battery energy storage station (BESS)-based smoothing control of photovoltaic (PV) and wind power generation fluctuations. *IEEE Trans. Sustain. Energy* **2013**, *4*, 464–476. [CrossRef]
13. Lawder, M.T.; Suthar, B.; Northrop, P.W.C.; De, S.; Hoff, C.M.; Leitermann, O.; Crow, M.L.; Santhanagopalan, S.; Subramanian, V.R. Battery energy storage system (BESS) and battery management system (BMS) for grid-scale applications. *Proc. IEEE* **2014**, *102*, 1014–1030. [CrossRef]
14. Duran, M.J.; Prieto, J.; Barrero, F.; Toral, S. Predictive current control of dual three-phase drives using restraned search techiniques. *IEEE Trans. Ind. Electron.* **2011**, *58*, 3253–3263. [CrossRef]
15. Kouro, S.; Cortés, P.; Vargas, R.; Ammann, U.; Rodríguez, J. Model predictive control—A Simple and powerful method to control power converter. *IEEE Trans. Ind. Electron.* **2009**, *56*, 1826–1838. [CrossRef]
16. Miranda, H.; Cortés, P.; Yuz, J.I.; Rodríguez, J. Predictive Torque control of induction machines based on state-space models. *IEEE Trans. Ind. Electron.* **2009**, *56*, 1916–1924. [CrossRef]

17. Morel, F.; Xuefang, L.S.; Retif, J.M.; Allard, B.; Buttay, C. A comparative study of predictive current control schemes for a permanent-magnet synchronous machine drive. *IEEE Trans. Ind. Electron.* **2009**, *56*, 2715–2728. [CrossRef]
18. Bolognani, S.; Peretti, L.; Zigliotto, M. Design and implementation of model predictive control for electrical motor drives. *IEEE Trans. Ind. Electron.* **2009**, *56*, 1925–1936. [CrossRef]
19. Cortés, P.; Rodríguez, J.; Antoniewicz, P.; Kazmierkowski, M. Direct power control of an AFE using predictive control. *IEEE Trans. Power Electron.* **2008**, *23*, 2516–2523. [CrossRef]
20. Vargas, R.; Rodríguez, J.; Ammann, U.; Wheeler, P.W. Predictive current control of an induction machine fed by a matrix converter with reactive power control. *IEEE Trans. Ind. Electron.* **2008**, *55*, 4362–4371. [CrossRef]
21. Abad, G.; Rodriguez, M.A.; Poza, J. Three-level NPC converter based predictive direct power control of the doubly fed induction machine at low constant switching frequency. *IEEE Trans. Ind. Electron.* **2008**, *55*, 4417–4429. [CrossRef]
22. Torreglosa, J.P.; Garcia, P.; Femadez, L.M.; Jurado, F. Predictive control for the energy management of a fuel-cell-battery-supercapacitor tramway. *IEEE Trans. Ind. Electron.* **2014**, *10*, 276–285. [CrossRef]
23. Hredzak, B.; Agelidis, V.G.; Jang, M. A model predictive control system for a hybrid battery-ultracapacitor power source. *IEEE Trans. Power Electron.* **2014**, *29*, 1469–1479. [CrossRef]
24. Akter, M.P.; Mekhilef, S.; Tan, N.M.L.; Akagi, H. Model predictive control of bidirectional AC-DC converter for energy storage system. *J. Electr. Eng. Technol.* **2015**, *10*, 165–175. [CrossRef]
25. John, T.; Wang, Y.; Tan, K.T.; So, P.L. Model predictive control of distributed generation inverter in a microgrid. In Proceedings of the 2014 IEEE Innovative Smart Grid Technologies (ISGT Asia), Kuala Lumpur, Malaysia, 20–23 May 2014; pp. 657–662.
26. Jafari, H.; Mahmodi, M.; Rastegar, H. Frequency control of micro-grid in autonomous mode using model predictive control. In Proceedings of the 2012 IEEE Iranian Conference on Smart Grids (ICSG), Tehran, Iran, 24–25 May 2012; pp. 1–5.
27. Naeiji, N.; Hamzeh, M.; Rahimi Kian, A. A modified model predictive control method for voltage control of an inverter in islanded microgrids. In Proceedings of the 2015 IEEE Power Electronics, Drives Systems & Technologies Conference (PEDSTC), Tehran, Iran, 3–4 February 2015; pp. 555–560.
28. Han, J.; Solanki, S.K.; Solanki, J. Coordinated predictive control of a wind-battery microgrid system. *IEEE J. Emerg. Sel. Top. Power Electron.* **2013**, *1*, 296–305. [CrossRef]
29. Rodríguez, J.; Cortés, P. *Predictive Control of Power Converter and Electrical Drives*; John Wiley & Sons: West Sussex, UK, 2012.
30. Xia, C.; Wang, M.; Song, Z.; Liu, T. Robust model predictive current control of three-phase voltage source PWM rectifier with online disturbance observation. *IEEE Trans. Ind. Informat.* **2012**, *8*, 459–471. [CrossRef]
31. Antoniewicz, P.; Kazmierkowski, M.P. Virtual-flux-based predictive direct power control of AC/DC converters with online inductance estimation. *IEEE Trans. Ind. Electron.* **2008**, *55*, 4381–4390. [CrossRef]
32. Rivera, M.; Yaramasu, V.; Rodriguez, J.; Wu, B. Model predictive current control of two-level four-leg inverters—Part II: Experimental implementation and validation. *IEEE Trans. Power Electron.* **2013**, *28*, 3469–3478. [CrossRef]
33. Natesan, C.; Ajithan, S.; Mani, S.; Kandhasamy, P. Applicability of droop regulation technique in microgrid—A survey. *Eng. J.* **2014**, *18*, 23–35. [CrossRef]
34. Wu, B.; Lang, Y.; Zargari, N.; Kouro, S. *Power Conversion and Control of Wind Energy Systems*, 1st ed.; Wiley-IEEE Press: Hoboken, NJ, USA, 2011; pp. 153–170.
35. Yoo, H.-J.; Kim, H.-M.; Song, C.-H. A coordinated frequency control of lead-acid BESS and Li-ion BESS during islanded microgrid operation. In Proceedings of the 2012 IEEE Vehicle Power and Propulsion Conference (VPPC), Seoul, Korea, 9–12 October 2012; pp. 1453–1456.

Article

Design and Field Tests of an Inverted Based Remote MicroGrid on a Korean Island

Woo-Kyu Chae [1], Hak-Ju Lee [1], Jong-Nam Won [1], Jung-Sung Park [1] and Jae-Eon Kim [2,*]

1 Research Institute, Korea Electric Power Corporation, Daejeon 305-760, Korea; wkchae@kepco.co.kr (W.-K.C.); juree@kepco.co.kr (H.-J.L.); jnwon@kepco.co.kr (J.-N.W.); jindulpa@kepco.co.kr (J.-S.P.)

2 School of Electrical Engineering, Chungbuk National University, Chungbuk 361-763, Korea

* Author to whom correspondence should be addressed; jekim@cbnu.ac.kr; Tel.: +82-432-612-423; Fax: +82-432-632-419.

Academic Editor: William Holderbaum
Received: 14 May 2015; Accepted: 27 July 2015; Published: 5 August 2015

Abstract: In this paper, we present the results of an economic feasibility study and propose a system structure to test and maintain electrical stability. In addition, we present real operation results after constructing a remote microgrid on an island in South Korea. To perform the economic feasibility study, a commercial tool called HOMER was used. The developed remote microgrid consists of a 400 kW wind turbine (WT) generator, 314 kW photovoltaic (PV) generator, 500 kVA $\times$ 2 grid forming inverter, 3 MWh lithium ion battery, and an energy management system (EMS). The predicted renewable energy fraction was 91% and real operation result was 82%. The frequency maintaining rate of the diesel power plants was 57% but the remote microgrid was 100%. To improve the operating efficiency of the remote microgrid, we investigated the output range of a diesel generator.

Keywords: microgrid; wind turbine; remote; island; hybrid power system; battery; HOMER; feasibility study

---

1. Introduction

Power supply in isolated regions far from land, including islands, is typically provided by small capacity diesel power plants. To overcome the high cost of diesel fuel in these small-capacity electrical power systems, and to prevent environmental pollution, a hybrid power system has begun to be applied, including in Alaska (USA). A hybrid power system is a diesel power plant system interconnected with a wind-turbine generator (WT) and photovoltaic (PV) array [1]. However, a restriction on renewable energy capacity that can be interconnected with a diesel power plant is still applied on account of the output variances of WTs and PVs. An attempt has been made to add a large capacity battery to the hybrid power system to solve the above problem owing to the sharp decline in battery prices in recent years. Such a system is called a remote microgrid (Figure 1) or hybrid microgrid [2–5]. However, renewable energy or batteries remain expensive, therefore, it is necessary to have an appropriate combination to construct an economically feasible system.

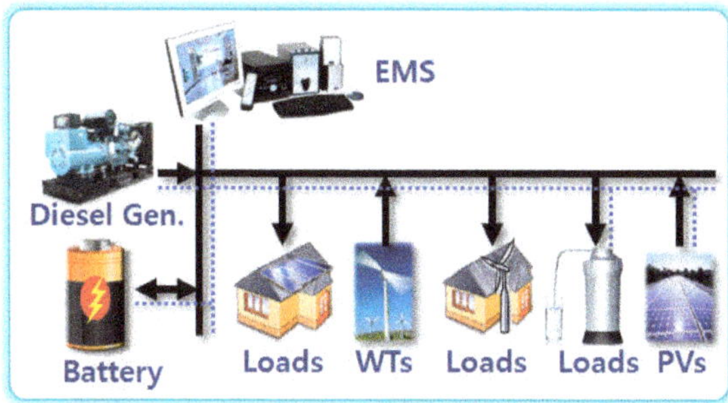

Figure 1. Example of a remote microgrid.

To effectively utilize the renewable energy cost, many researchers have studied algorithms to calculate the capacity of applicable generator units that can constitute a low cost reliable power system [6]. Li *et al.* [6] compared some proposed algorithms and presented a simple algorithm to determine the required number of generating units of a WT generator and PV array, as well as the associated storage capacity for a standalone hybrid microgrid. The algorithm is based on the observation that the battery charge state should be periodically invariant. Liang *et al.* [7] investigated the stochastic modeling and optimization tools for microgrids. Wang *et al.* [8] proposed an optimal capacity allocation method for standalone microgrids using a particle swarm optimization algorithm. Bansal *et al.* [9] analyzed the economic feasibility of a hybrid power system using the biogeography-based optimization (BBO) algorithm. Wies *et al.* [10] modeled load sharing between a diesel generator and PV using Simulink and presented how to analyze the economic feasibility. Xu *et al.* [11] proposed an optimal sizing method to achieve higher power supply reliability. However, Stiel *et al.* [12] and Yoo *et al.* [13] used a commercial feasibility study tool called HOMER instead of complex algorithms and thereby presented economic feasibility assessments results on a wind-diesel hybrid power system.

Purser *et al.* [14] presented the results of a technical and economic feasibility study of implementing a microgrid at Georgia Southern University using HOMER. Mizani *et al.* [15] presented the demonstration results for optimal design and operation of a grid-connected microgrid.

Fan *et al.* [16] presented design considerations to develop a standalone smart grid on Ubin Island (Pulau) as well as simulation results. Kojima *et al.* [17] proposed a structure and control method of an existing diesel power plant interconnected with a small capacity of WT, PV, and batteries. Prull *et al.* [18] proposed a design procedure, including a load study and meteorological information study, as well as design results to construct a remote microgrid on Necker Island (UK), but they didn't include any field test results. Fay *et al.* [1] presented the performance and economic analysis results for isolated wind-diesel systems in Alaska. Kaldellis *et al.* [19] presented cost-benefit analysis results of remote hybrid wind-diesel power stations. Ulleberg *et al.* [20] presented a remote renewable energy system for the Faroe Islands, but these wind-diesel systems usually didn't include the battery and/or photovoltaic generator. However, some studies presented algorithms that use only partial information to quickly calculate renewable energy and battery capacity. In particular, such algorithms do not consider the charge and discharge efficiency of the battery system and the depth of discharge (DoD) according to battery type. Some algorithms were only theoretically validated. Although some of the above studies proposed a design procedure for a remote microgrid, they provided no operational results from real sites. Some studies presented the comparison results between simulation and real implementation but they usually didn't include the design targets such like raising the system efficiency, stable operation, power quality, *etc.* To construct a remote microgrid, it is necessary to have an optimal system design that considers the power reserve ratio to both ensure the system economic feasibility and maintain the stable operation and rated voltage and frequency of the system. In addition to the above considerations, an appropriate system structure should be considered. The system should be constructed according to the design results;

moreover, the design procedure should be validated and fed back through comparisons with long-term operation results. However, no studies have provided such a series of procedures.

The present paper presents economic feasibility study results and a system structure to develop a grid-forming inverter-based remote microgrid for distant islands located on the west coast of South Korea. It presents design parameters for economical and reliable remote microgrid designs through comparisons between simulation results and real operation results. Moreover, this study aims to share the optimal design direction through real long-term operation results of the remote microgrid.

2. Isolated Power System

2.1. Penetration Level of Renewable Energy

When incorporating renewable-based technologies into isolated power systems, the amount of energy that will be obtained from the renewable sources will strongly influence the technical layout, performance and economics of the system. For this reason, it is necessary to explain two new parameters—the instantaneous and average power penetration of wind—as they help define system performance [21].

The average and peak penetration of renewable generation in a hybrid power system can be defined as shown in Equations (1) and (2) [18]:

$$\text{Average Penetration} = \frac{\text{Energy from Renewable Generation (kWh)}}{\text{Electrical Load (kWh)}} \tag{1}$$

$$\text{Peak Penetration} = \frac{\text{Peak Power from Renewable Generation (kW)}}{\text{Electrical Load (kW)}} \tag{2}$$

In [22], these definitions are used to categorize hybrid power systems into three classes: low, medium and high penetration, as shown for reference in Table 1.

Table 1. Hybrid power system penetration classifications.

System Class	Peak Penetration	Annual Average Penetration
Low	<50%	<20%
Medium	50%–100%	20%–50%
High	100%–400%	>50%

2.2. Hybrid Power System

The electric power production cost for diesel power plants located in islands or remote areas—depending on their scale and distance from land—can be up to 10 times higher that of the large-scale electrical power systems [23]. To reduce the electric power production cost, renewable energy systems such as wind turbine generators or photovoltaic generators are sometimes installed in parallel with the diesel power plant. This kind of system is called a hybrid power system [22]. Depending on its structure and the portion of renewable energy in the whole system, the diesel generators of the hybrid power system should maintain their voltage and frequency most of the time.

2.3. Inverter-Based Remote MicroGrid

Whereas the hybrid power system controls the voltage and frequency through its diesel generator, a remote microgrid has its Grid Forming Inverter (GFI) and individual distributed generators contribute to voltage/frequency control [18]. GFI maintains the system voltage/frequency through charging the battery if there is excess energy and *vice versa*. Coordination between GFI and WT is critical for stability performance. Using a large capacity GFI, a remote microgrid can usually utilize more renewable energy than a hybrid power system because a GFI can very quickly absorb the excess energy from the renewable energy using the battery [24].

3. Feasibility Study for Test Island

The remote microgrid referenced in this paper was designed to minimize power supply costs through an optimal combination of distributed power and the energy storage devices, and to consider reliable operations. In this section, the economic feasibility study results of the proposed remote microgrid are presented using HOMER. The economic feasibility study used real meteorological and load data from the actual site.

3.1. Feasibility Study Tool

To economically construct a remote microgrid, renewable energy type, battery type, and capacity should be accurately calculated. In addition, precise analysis of target area meteorological data is likewise needed. Such a series of processes is called a feasibility study, which can be performed by the most widely used commercial tool called HOMER.

HOMER is a computer simulation program designed by the National Renewable Energy Laboratory (NREL) in the United States. Coined as the Optimization Model for Distributed Power, HOMER allows the modelling of both grid and non-grid connected power systems consisting of conventional and renewable technologies. The program considers the economic and technical feasibility of desired power systems and delivers comprehensive reports covering a range of subjects from the net present capital cost of the system to the renewable penetration. It allows the input of renewable resources such as wind speeds, battery data, demand load data, capital and operation and maintenance (O&M) costs among others, as well as sensitivity analyses modelling the impact on the system to variations in any input [25].

HOMER uses the following equation to calculate the total net present cost:

$$C_{NPC} = \frac{C_{ann,tot}}{CRF(i, R_{proj})} \tag{3}$$

where $C_{ann,tot}$ is the total annualized cost, i the annual real interest rate (the discount rate), R_{proj} the project lifetime, and CRF (discount raterecovery factor), given by Equation (4):

$$CRF(i, N) = \frac{i(1+i)^N}{(1+i)^N - 1} \tag{4}$$

where i is the annual real interest rate and N is the number of years.

HOMER uses the following equation to calculate the levelized cost of energy:

$$COE = \frac{C_{ann,tot}}{E_{prim} + E_{def} + E_{grid,sales}} \tag{5}$$

where $C_{ann,tot}$ is the total annualized cost, E_{prim} and E_{def} are the total amounts of primary and deferrable load, respectively, that the system serves per year, and $E_{grid,sales}$ is the amount of energy sold to the grid per year [25].

3.2. Test Island in Korea

A distant target island was selected to validate the remote microgrid technology. Table 2 shows the information about the target island. The target island is located approximately 6 km from the mainland and currently has approximately 280 residents. There are three 100 kW diesel generators on the island; two generators operate in parallel during normal hours and one is for reserve. The distributed voltage is 6.6 kV, and there are two distribution lines. The heavy load supplied to the diesel power plant during normal operation hours is produced by an induction motor whose rated capacity is 11 kW. The mean wind velocity and solar radiation in 2013 were 5.1 m/s and 3.68 kWh/m^2·day, respectively, which represent typical data in offshore islands in South Korea [26].

Table 2. Overview of the Test Island [27].

Name	Gasado (Gasa Island)
Location	Southern part of South Korea
Area	6.4 km^2
Population	286 person (168 house)
Electrical System	Gen-set: 100kW × 3 (1992 year); Distribution Line (6.6 kV): 8 km; Main Transformer (380 V/6.9 kV): 300 kVA × 2
Fuel Consumption	285,000 L/year (2013 year)
Load (2013year)	Average: 113 kW; Peak: 210 kW; Minimum: 50 kW; Heavy load: 11 kW; Induction Motor × 2
Weather (Annual average)	Daily Radiation: 3.68 kWh/m^2·day; Wind Speed: 5.0 m/s @ 30 mL; Temperature: 13.4 °C

3.3. Results of Economic Feasibility Study

The data presented in Section 3.2 and the Appendix A were entered into HOMER to run the economic feasibility study simulation The simulation input data were set with consideration of the current diesel power plant operation results and current prices of renewable energy and batteries to be as realistic as possible.

The simulation results of the target island are summarized in Table 3, which shows a typical combination according to the renewable energy fraction. As shown in the table, the Cost of Energy (COE) of the diesel power plant in the target island was $0.992 and the annual operating cost was $869,000. The operation cost consisted of the diesel generator fuel cost, maintenance cost for the power generator and lines, personnel cost for operators, and other expenses. Power bills paid by customers were omitted.

Table 3. Simulation result for economic feasibility study.

Ren. Fraction (%)	COE ($)	NPC (1000$)	Operat. Cost (1000$)	Initial Capital (1000$)	Fuel (kL)	Photovoltaic (kW)	Wind Turbine (100 kW)	Battery (100 kWh)	Inverter (kVA)
0	0.913	11,258	903	0	311	0	0	0	0
10.9	0.833	10,277	735	1120	244	150	0	5	500
20.9	0.822	10,134	698	1435	217	250	0	5	500
30.8	0.834	10,285	671	1920	190	50	2	5	500
40.9	0.875	10,794	693	2160	177	300	1	5	500
50.5	0.879	10,849	653	2710	149	300	2	5	500
59.7	0.885	10,916	622	3165	121	350	2	10	500
69.8	0.901	11,111	587	3800	91	450	2	15	500
80.9	0.919	11,342	546	4540	57	600	2	20	500
89.9	1.051	12,963	553	6067	34	600	3	35	500
94.4	1.284	15,835	575	8668	18	600	6	50	500

Notes: Ren. Fraction: Renewable Fraction, COE: Cost of Energy, NPC: Net Present Cost, Operat. Cost: Operating Cost per Year.

The most economical system configuration can be formed with approximately 20% of renewable energy. However, the COE was lower or similar compared to that of the existing diesel power plant until a proportion of renewable energy in the remote microgrid reached approximately 70%. Furthermore, under the same COE, various combinations of capacities of renewable energy and batteries can be possible.

4. Remote Microgrid System Design

In this section, the design objectives for achieving the implementation and demonstration of an energy independent island using remote microgrid technology, as well as the results, are presented.

4.1. Design Target

Table 4 shows the design target and results for the test island. The energy independence of the test island was set to 99%, which means that 100% energy independence would not be pursued in consideration of the economic feasibility. That is, if required, diesel generators would be partially used. The daily average load in the test island was 113 kW. However, an average load was set to 120 kW to account for the supply-ready loads and future load increase rate. The required renewable energy capacity could be calculated as shown in Equation 6. The result is consistent with that in Table 3.

Table 4. Design target and design results for the test island [26].

Category	Design Target	Results
Renewable Energy Fraction	99% of power supplied by renewable energy	wind turbine (WT): 400 kW; photovoltaic (PV): 314 kW
Battery Capacity	Power supply by batteries only for one or more days	3 MWh
Inverter Capacity	-Total recharge of renewable energy output; -Inverter operation efficiency improvement	500 kVA × 2 ea, 250 kVA × 1ea; If CVCF inverter capacity is not sufficient, parallel inverter is running
System optimization	-Parallel operation of inverters and diesel generators; -Improvement of heat efficiency by fixed-speed operation of the diesel generator	-Inverter: CVCF (Constant Voltage Constant Frequency) operation; -Diesel generator : Droop operation
Electrical System	-Microgrid technology validation; -Commercial operation and cost minimization	-Construction of new distribution line; -Parallel interconnection with the main transformer
EMS (Energy Management System)	-Overall monitoring and control of the system; -System automatic operation	-SCADA + application; -Prediction of loads and renewable energy; -Direct load control

According to the renewable energy capacity, 400 kW of WT and 314 kW of PV were installed with comprehensive consideration of the geographical characteristics, installation site, and building permit issues. A design goal was to set the power supply for only one day with batteries when no renewable energy output was supplied to achieve an energy independent island. This was calculated with Equation (7). Additional Depth of Discharge (DoD) was not considered because we selected lithium batteries:

$$\text{Capacity of Renewable Energy} = \text{Daily Average Load} \div \text{Average Capacity Factor of Renewable Energy} = 120\ \text{kW} \div 17\% = 706\ \text{kW} \tag{6}$$

$$\text{Capacity of Battery} = 120\ \text{kW (Daily Average Load)} \times 24\ \text{h} = 2880\ \text{kWh} \tag{7}$$

Finally, we set the size of battery as 3 MWh considering the 5% margin.

The inverter capacity was calculated via Equation (8) in consideration of the maximum renewable energy output:

$$\begin{aligned}\text{Capacity of Inverter} &= \text{Maximum Output of Renewable Energy} - \text{Minimum Load} \\ &= 706\ \text{kW} - 60\ \text{kW} = 646\ \text{kW}\end{aligned} \quad (8)$$

To account for a possible commercial inverter purchase and inverter operation efficiency, two 500-kVA inverters and one 250-kVA inverter were set up. One 500-kVA inverter was used for backup. A 250-kVA inverter was set in the "off" state during normal hours and was automatically turned on if renewable energy output exceeded the threshold value. Through this, we can raise the operation efficiency of the system. The inverters could be charged and discharged bi-directionally for system optimization and could be run with the diesel generators in parallel. To employ the existing facilities in their current states, the diesel generator was operated using the droop method.

To validate the microgrid technology prior to the commercial operation, a test distribution line (D/L) was installed. Using the test D/L and main transformer for backup, the additional test system could be operated without power interruption. GFIs and the diesel generators are interconnected into the low voltage side of the main transformers to minimize the construction cost.

The overall system was monitored and controlled via an energy management system (EMS). To improve system efficiency, battery life management and diesel generator output were controlled using the prediction of loads and renewable energy power generation. For example, if the EMS predicts that the renewable energy supply will be enough within 24 h, the EMS calculates the output power and the stop time of the diesel generators.

4.2. Simulation Result for Test Island

Renewable energy and battery capacities determined by the design target were entered into HOMER to run the simulation of annual power production and loss. The results are shown in Figure 2 and Table 5. It was estimated that the PV produced 354,156 kWh per year, the WT produced 759,718 kWh per year, and the diesel power generator produced 180,137 kWh per year. A capacity factor of the PV was estimated less than that of other PV stations in South Korea, which was because the target area was an island. Although an energy independent island was designed as a goal, the renewal energy produced only 79.4% of the power required under the current construction conditions. Thus, the remainder of the power (20.6%) should be supplied by the diesel generator. Excess power supplied by the renewable energy was estimated as 379,987 kWh/year, which amounted to 29.4% of the total power production. This result was obtained because wind in South Korea is concentrated in the winter season on account of its geographical characteristic.

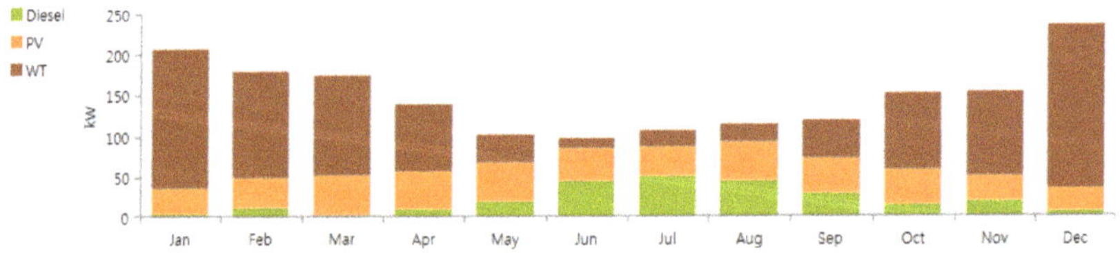

Figure 2. Expected monthly energy production by generator of the test island.

5. Results and Discussion

In this section, the result of a real remote microgrid construction in the test island and its operation results are presented. The results are compared with the simulation results in Section 3.2.

5.1. Implementation Results of the Field Test Site

Remote microgrid system was constructed on the test island in accordance with the design targets (Section 3.1). The system architecture is shown in Figure 3 and Table 6 shows functions and features of installed equipment. The remote microgrid on the test island is an inverter-based small power system. Usually, just one Grid Forming Inverter (GFI) controls the system voltage (V) and frequency (f) of Remote microgrid. The #2 inverter is for backup (if there is a fault at #1) or *vice versa*. It means that two GFIs don't operate in parallel because each GFI will control voltage & frequency and there be voltage oscillation called "hunting" [26].

Table 5. Simulation Results for Design Target.

Contents	Quantity	Value
PV	Rated Capacity	314 kW
	Total Production	355,124 kWh/year
	Mean Output	40.5 kW
	Capacity Factor	12.90%
	Penetration Rate	26.30%
WT	Rated Capacity	100 kW × 4 ea
	Total Production	733,790 kWh/year
	Mean Output	83.7 kW
	Capacity Factor	20.90%
	Penetration Rate	54.40%
Diesel Generator	Rated Capacity	100 kW × 3ea
	Total Production	260,207 kWh/year
	Mean Output	137.8 kW
	Fuel Consumption	78,927 L/year
	Hours of Operation	1888 h/year
	Penetration Rate	19.30%
Battery	Energy In	346,597 kWh/year
	Energy Out	328,682 kWh/year
	Losses	17,915 kWh/year
Energy Flows	Load	989,878 kWh/year
	Production	1,349,120 kWh/year
	Excess Electricity	318,316 kWh/year
	Renewable Fraction	73.70%

If the output power from WTs/PVs is bigger than the amount of load, GFI will charge the battery to maintain V&F. On the contrary, the GFI will discharge the electrical power from the battery to the load. If the surplus power is bigger than the GFI's rated power (500 kVA), the #3 inverter (250 kVA) will be operated automatically.

Sometimes, the diesel generators are interconnected to the GFI to charge the battery or to supply the power to the loads. We used the previously installed diesel generators to operate in parallel with the GFI without any functional modification. Because each diesel generators can operate in parallel with each other with its droop function. Voltage droop is the intentional loss in output voltage from a device as it drives a load [28]. Frequency droop allows synchronous generators to run in parallel, so that loads are shared among generators in proportion to their power rating [28]. In our test island, GFI will control system voltage & frequency and diesel generators will be operated in droop mode, so there is no concern for the voltage oscillation problem. WTs & PVs don't run without GFI, because the diesel generators cannot control the frequency properly if there is too much power from WTs & PVs. In this system, we impose the spinning reserve at the GFI because this system is the inverter-based remote microgrid. Through this, we can reduce the fuel consumption for the spinning reserve of the diesel generators.

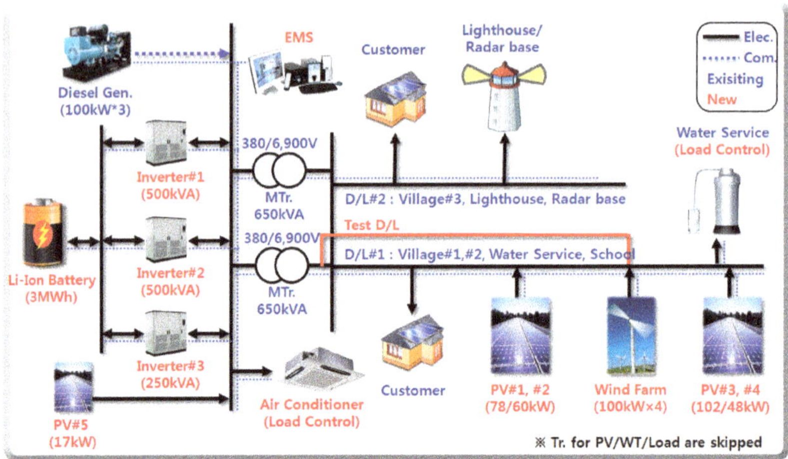

Figure 3. System architecture of the test island.

Test D/L is for the initial system test using GFI, WTs, PVs and load simulator, so normally Test D/L is not charged.

Table 6. Functions and Features of Installed Equipment.

Contents	Specification	Functions and Features
Energy Manage System	SCADA + Application	Battery SOC management, Forecasting of load and renewable energy, Direct load control, Automation
Grid Forming Inverter	500 kVA × 2, 250 kVA × 1	Frequency &voltage control, P/Q control 500 kVA #2: Backup, 250 kVA: for shortage of rating
Battery	3 MWh, Li-ion	Electrical energy storage, 1 C-rate, NMC type Three GFIs are connected to 3 MWh in parallel
WT	100 kW × 4	Permanent Magnet Synchronous Generator (PMSG) + Full converter, Power limitation, Power factor & Voltage control, LVRT, FRT
PV	314 kW (8 ea)	Power limitation, Monitoring of each module, Water floating PV system for limited site
Diesel Generator	100 kW × 3	Droop control, Remote on/off
Load	Water pump, Air conditioner	Water tank is used to energy storage. Battery room temperature control using surplus energy

Notes: LVRT: Low Voltage Ride Through, FRT: Fault Ride Through.

In the test island, a water supply tank was installed on a mountainside and water was supplied to households via natural water pressure. Thus, the water level in the water supply tank should be 70% or higher at all times; the water level was maintained by a water supply motor. If excess power in the renewable energy was used for charging the battery, the excess power was used to run the motor via the EMS for water pumping in advance instead of battery charging. In this way, the battery charging could be minimized to extend the battery life. Figures 4–7 are the installed equipment in the test island.

Figure 4. Wind and solar farm.

Figure 5. Grid forming inverters.

Figure 6. Li-ion battery.

Figure 7. Energy management system.

5.2. Analysis of Operation Results

Table 7 shows the operation results from October 2014 to March 2015 after the construction of the remote microgrid. On the test island, wind velocity was high and solar radiation was low in the winter season, whereas wind velocity was low and solar radiation was high in the summer season. The commercial operation of the remote microgrid began in October 2014. Because the WT capacity was greater than the PV capacity, power generation via WT was more than that of PV. During a six-month period, the average renewable energy fraction was approximately 82% and diesel generator fuel consumption was reduced by 80%.

Figure 8 shows a graph of the frequency comparison between the remote microgrid and the diesel power plant. The frequency-maintaining standard in South Korea is 59.8 to 60.2 Hz as an average value of a 30 min interval, which is displayed in the figure as a red dotted line. When supplying power with the diesel generator (green color solid line), the maintenance rate was 57% as the average value of a 0.2 s interval. However, when supplying power with the remote microgrid (blue solid line), the maintenance rate was 100% as the average value of a 0.2 s interval. That is, the power quality of the remote microgrid was better than that of the diesel power plant, despite the high renewable energy rate.

Table 7. Energy production and load of the test island after commission.

Generator	Unit	October	November	December	January	February	March	Average
Wind Turbine	kWh	33,301	42,107	56,577	55,700	50,200	31,400	44,881
Photovoltaic	kWh	27,659	23,537	15,005	21,064	23,705	35,940	24,485
Diesel Generator	kWh	20,300	16,839	21,061	12,672	6222	13,539	15,106
Total Production (=Load and Loss)	kWh	81,260	82,483	92,643	89,436	80,127	80,879	84,471
Renewable Fraction	%	75.0	79.6	77.3	85.8	92.2	83.3	82.0
Fuel Consumption (after commission)	Liter	6379	5272	6049	3703	1699	3600	4450
Fuel Consumption (before commission)	Liter	21,828	21,829	21,830	24,836	19,998	23,611	22,322

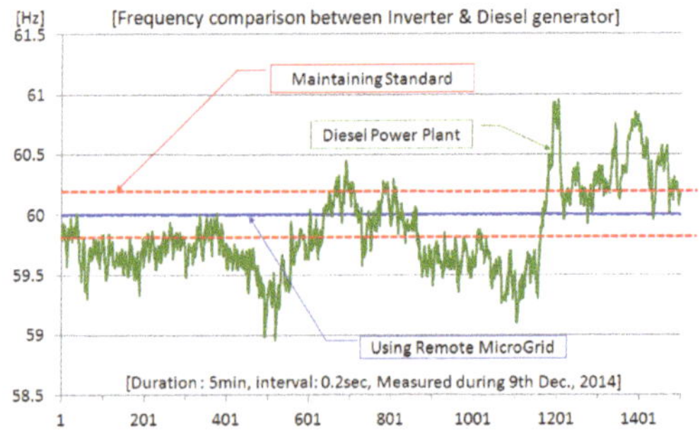

Figure 8. Frequency comparison between the remote microgrid and diesel power plant.

5.3. Operation Results Comparison with Economic Feasibility Study

Table 8 shows the expected energy production and load on the test island according to HOMER. The average load was 84,650 kWh, which was similar to that of 84,471 kWh in Table 5. That is, the loads were relatively well simulated. The power generation results via PV were 26,750 and 24,485 kWh, respectively, which showed an approximately 8.5% error, which was not significantly different compared to the expected power generation. However, the average WT power generation results were 96,987 and 44,881 kWh, respectively, which showed a difference by more than two-fold. This result was due to the significant WT power generation in the winter caused by high wind velocity; moreover, the battery was fully charged, thereby often stopping WT or enforcing output restriction. Another key reason for the error was the lower average wind velocity than in past years and several line fault accidents.

Table 8. Expected energy production and load of the test island by HOMER.

Generator	Unit	October	November	December	January	February	March	Average
Wind Turbine	kWh	67,760	71,989	146,867	123,287	83,624	88,398	96,987
Photovoltaic	kWh	32,788	21,948	20,403	23,634	25,319	36,411	26,750
Diesel Generator	kWh	16,988	20,411	8143	5894	12,964	4965	11,561
Total Production	kWh	117,536	114,347	175,413	152,815	121,907	129,773	135,299
Total Load Served	kWh	81,563	81,901	85,719	87,700	82,580	88,437	84,650
Renewable Fraction	%	85.5	82.2	95.4	96.1	89.4	96.2	91.0
Fuel Consumption	Liter	5124	6115	2462	1782	3911	1485	3480
Excess Energy and Loss	kWh	35,973	32,446	89,694	65,115	39,327	41,336	50,649

Accordingly, the average renewable fraction was also low at approximately 82%, which was unlike the initial expectation. Moreover, the average fuel consumption was also increased from 3480 to 4450 L. The average fuel consumption was reduced by 81% compared to that of the previous year.

5.4. Efficient Operation of Diesel Generators in the Remote Microgrid

If power is supplied to loads using only a diesel generator, the diesel generator would be entirely responsible for frequency control due to invariant loads. Under this circumstance, a fuel input amount can suddenly vary according to invariant loads so that fuel efficiency can be degraded. If a diesel generator is run in parallel with a grid forming inverter (GFI), the GFI handles frequency control due to invariant loads. Accordingly, the diesel generator can be run with constant output as well as improved fuel efficiency. Because parallel operation of two generators is not needed for reserve power, the diesel generator can be run in the highest efficiency region.

Table 9 shows a comparison of power generation efficiency between a power supply using only a diesel generator and a power supply using a diesel generator running in parallel with the GFI. This is the result of fuel consumption and power generation when real loads were supplied for 24 h each. According to data in Table 9, efficiency was improved by 14.2%. Operation efficiency of the inverter is normally 88% to 95%, and round-trip efficiency of the battery is 90% to 95%. Thus, when a diesel generator is run in the remote microgrid, it should be run in the highest efficiency section. The maximum output of the diesel generator should also not exceed the load to prevent a power loss due to charging and discharging of the battery. This would result in improvements in the overall system efficiency.

Table 9. Fuel consumption comparison of a diesel generator.

Operation Type	At Diesel Power Plant	At Remote Microgrid
	2 gen-set in parallel	one gen-set with grid forming inverter (GFI)
Fuel Consumption	766.2 L/24 h	562.7 L/24 h
Total Production	2319.3 kWh	1946.2 kWh
Average Power	96.6 kW	81 kW
Energy per Fuel	3.02 kWh/L	3.45 kWh/L
Fuel per Energy	0.3304 L/kWh	0.2892 L/kWh

6. Conclusions

In this paper, we presented a constructed energy independent island on a test island in South Korea using remote microgrid technology. We presented the operation results from a six-month period study. To construct a remote microgrid, an economic feasibility study was conducted, and the results were presented. The system structure required for reliably testing and operating the developed system was described. The power generation produced from the developed generators and predicted power generation using HOMER were compared. Loads and predicted results of PV had similar outputs with those of the real system. However, WT and diesel power generation showed a large difference, which was due to frequent stops of WT due to high wind velocity on the test island in the winter season. The power quality of the remote microgrid was also improved more than that of the diesel generator. When a diesel generator is run in the remote microgrid, it should be run in the highest efficiency section and the maximum output should not exceed the load to prevent a power loss.

7. Future Work and Contributions

Based on the analysis results to date, we will examine how to utilize excess power in the winter season. The commercial operation of the developed remote microgrid will continue. The demonstration results for one year or longer will be compared with HOMER simulation results. Furthermore, we are developing an EMS to efficiently run the remote microgrid and we will present the demonstration results. This paper contributed to the production of more accurate simulation results from economic feasibility studies for remote microgrids than previous studies. In addition, information in this study is expected to reduce trials and errors in real sites in when remote microgrid technology is implemented in the future.

Acknowledgments: This work was supported by Korea Institute of Energy Technology Evaluation and Planning (KETEP) grant funded by Korea government Ministry of Knowledge Economy (No. 20123010020080).

Author Contributions: W.-K.C and H.-J.L conceived and designed the experiments; W.-K.C and J.-N.W performed the simulation and the experiments; J.-S.P analyzed the data; W.-K.C and J.-E.K wrote the paper.

Conflicts of Interest: The authors declare no conflict of interest.

Appendix Input Data for Economic Feasibility Study

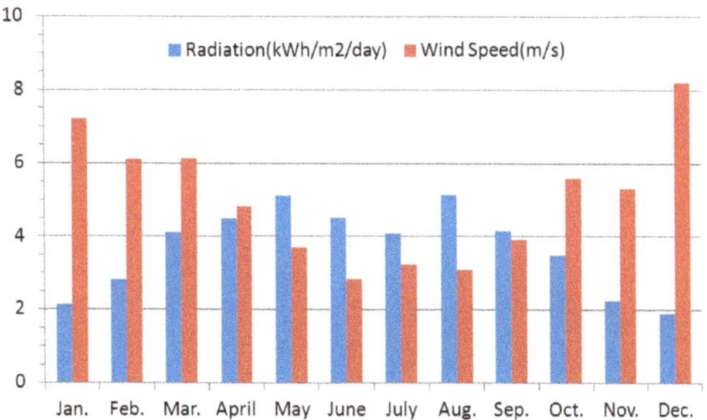

Figure A1. Weather Data of Test Island [26].

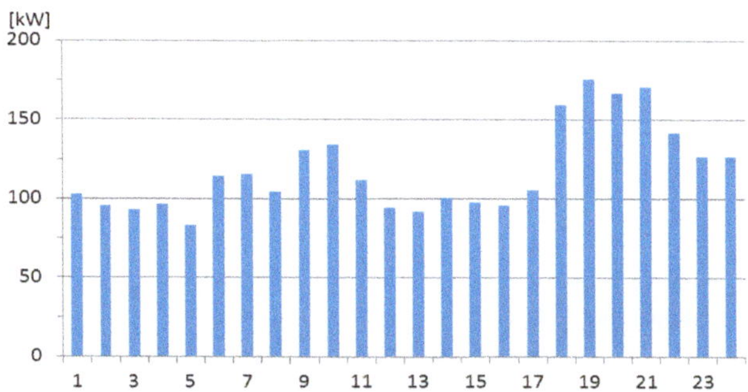

Figure A2. Daily Load Profile of Test Island (January) [26].

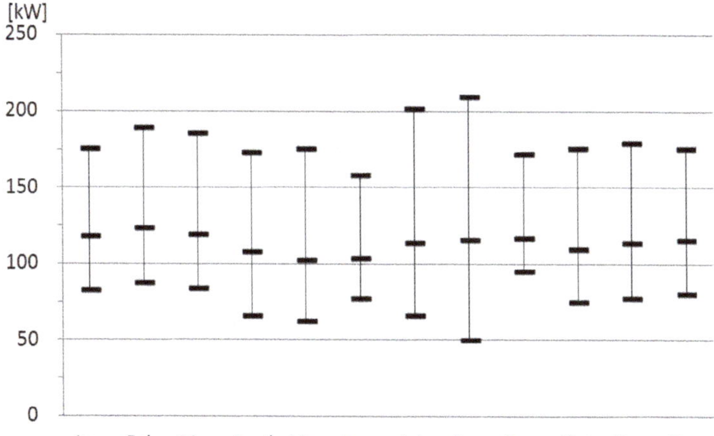

Figure A3. Annual Load Profile of Test Island (2013) [26].

Table A1. Economic Feasibility Study Input Data [26].

Contents	Item	Value
PV	Capital	3200 $/kW
	O&M	300 $/kW
	Life Time	20 years
WT	Capital	550,000 $/100 kW
	O&M	6000 $/100 kW
	Life Time	20 years
Inverter	Capital	280,000 $/500 kVA
	O&M	4000 $/500 kVA
	Life Time	20 years
Battery	Capital	63,000 $/100 kWh
	Replacement Cost	38,000 $/100 kWh
	O&M	600 $/100 kWh
	Life Time	10 years
Gen-set	Capital	Existing
	O&M	5.5 $/h/100 kW
	Life Time	175,200 h/100 kW
Economics	Interest rate	5%
	Project Lifetime	20 years
	Fixed O&M Cost	386,000 $/year

References

1. Fay, G.; Schwoerer, T.; Keith, K. *Alaska Isolated Wind-Diesel Systems: Performance and Economic Analysis*; University of Alaska Anchorage: Alaska, AK, USA, 2010.
2. Chae, W.K.; Lee, H.J.; Hwang, S.W.; Song, I.K.; Kim, J.E. Isolated microgrid's voltage and frequency characteristic with induction generator based wind turbine. *Smart Grid Renew. Energy* **2014**, *5*, 180–192. [CrossRef]
3. Martinez-Cid, R.; O'Neill-Carrillo, E. Sustainable Microgrids for Isolated Systems. In Proceedings of the Transmission and Distribution Conference and Exposition, New Orleans, LA, USA, 19–22 April 2010; pp. 1–7.
4. Muljadi, E.; McKenna, H.E. Power Quality Issues in a Hybrid Power System. In Proceedings of the Thirty-Sixth IAS Annual Meeting, Conference Record of the Industry Applications Conference, Chicago, IL, USA, 30 September–4 October 2001.
5. Senjyu, T.; Nakaji, T.; Uezato, K.; Funabashi, T. A hybrid power system using alternative energy facilities in isolated island. *IEEE Trans. Energy Convers.* **2005**, *20*, 406–414. [CrossRef]
6. Li, J.; Wei, W.; Xiang, J. A simple sizing algorithm for stand-alone PV/wind/battery hybrid microgrids. *Energies* **2012**, *5*, 5307–5323. [CrossRef]
7. Liang, H.; Zhuang, W. Stochastic modeling and optimization in a microgrid: A survey. *Energies* **2014**, *7*, 2027–2050. [CrossRef]
8. Wang, J.; Yang, F. Optimal capacity allocation of standalone wind/solar/battery hybrid power system based on improved particle swarm optimisation algorithm. *IET Renew. Power Gener.* **2013**, *7*, 443–448. [CrossRef]
9. Bansal, A.K.; Kumar, R.; Gupta, R.A. Economic analysis and power management of a small autonomous hybrid power system (SAHPS) using biogeography based optimization (BBO) algorithm. *IEEE Trans. Smart Grid* **2013**, *4*, 638–648. [CrossRef]
10. Wies, R.W.; Johnson, R.; Agrawal, A.N.; Chubb, T.J. Simulink model for economic analysis and environmental impacts of a PV with diesel-battery system for remote villages. *IEEE Trans. Power Syst.* **2005**, *20*, 692–700. [CrossRef]
11. Xu, L.; Ruan, X.; Mao, C.; Zhang, B.; Luo, Y. An improved optimal sizing method for wind-solar-battery hybrid power system. *IEEE Trans. Sustain. Energy* **2013**, *4*, 774–785.
12. Stiel, A.; Skyllas-Kazacos, M. Feasibility study of energy storage systems in wind/diesel applications using the HOMER model. *Appl. Sci.* **2012**, *2*, 726–737. [CrossRef]

13. Yoo, K.; Park, E.; Kim, H.; Ohm, J.Y.; Yang, T.; Kim, K.J.; Chang, H.J.; del Pobil, A.P. Optimized renewable and sustainable electricity generation systems for Ulleungdo Island in South Korea. *Sustainability* **2014**, *6*, 7883–7893. [CrossRef]
14. Purser, M.S. A Technical and Economic Feasibility Study of Implementing a Microgrid at Georgia Southern University. Master's Thesis, Georgia Southern University, Statesboro, GA, USA, May 2014.
15. Mizani, S.; Yazdani, A. Optimal Design and Operation of a Grid-Connected Microgrid. In Proceedings of the Electrical Power and Energy Conference (EPEC), Montreal, QC, Canada, 22–23 October 2009.
16. Fan, Y.; Rimali, V.; Tang, M.; Nayar, C. Design and Implementation of Stand-alone Smart Grid Employing Renewable Energy Resources on Pulau Ubin Island of Singapore. In Proceedings of the 2012 Asia-Pacific Symposium on Electromagnetic Compatibility (APEMC), Singapore, 21–24 May 2012; pp. 441–444.
17. Kojima, T.; Fukuya, Y. Microgrid system for isolated islands. *Fuji Electric Rev.* **2011**, *57*, 125–130.
18. Prull, D.S. Design and Integration of an Isolated Microgrid with a High Penetration of Renewable Generation. Ph.D. Thesis, University of California, Berkeley, CA, USA, May 2008.
19. Kaldellis, J.K.; Kavadias, K.A. Cost-benefit analysis of remote hybrid wind-diesel power stations: case study Aegean Sea islands. *Energy Policy* **2007**, *35*, 1525–1538. [CrossRef]
20. Ulleberg, Ø.; Rinnan, A. *Renewable Energy System Concepts for Nolsoy, the Faeore Islands*; Institute for Energy Technology: Haden, Norway, 2006.
21. Ackermann, T. *Wind Power in Power Systems*; John Wiley & Sons: Hoboken, NJ, USA, 2005.
22. Baring-Gould, I.; Dabo, M. Technology, Performance, and Market Report of Wind-Diesel Applications for Remote and Island Communities. In Proceedings of the WINDPOWER 2009 Conference and Exhibition, Chicago, IL, USA, 4–7 May 2009.
23. KEPCO (Korea Electric Power Corporation). *Report for the Prime Cost Report of Diesel Power Plant*; KEPCO: Seoul, Korea, 2012.
24. Brett, P. A New Isolated Grid Paradigm. In *International Wind-Diesel Workshop*; US Department of Energy: Alaska, AK, USA, 2011.
25. HOMER. Available online: http://www.homerenergy.com (accessed on 13 May 2015).
26. KEPRI (Korea Electric Power Research Institute). *2nd Year Report for Development of Convergence and Integration Technology for Renewable-Based Energy System and Its Grid Interconnection*; KEPCO (Korea Electric Power Corporation): Daejeon, Korea, 2014.
27. *Report for Operation Results for the Diesel Power Plant for Gasado*; Jindo-Gun (Local Government of Korea): Jindo-Gun, Korea, 2014.
28. Chowdhury, S.; Chowdhury, S.P.; Crossley, P. *Microgrids and Active Distribution Networks*; The Institution of Engineering and Technology: Herts, UK, 2009.

MDPI

Article

Distributed Energy Storage Using Residential Hot Water Heaters

Linas Gelažanskas * and Kelum A.A. Gamage

Engineering Department, Lancaster University, Bailrigg, Lancaster LA1 4YW, UK; k.gamage@lancaster.ac.uk
* Correspondence: linas@gelazanskas.lt; Tel.: +370-68210215

Academic Editor: William Holderbaum
Received: 28 December 2015; Accepted: 16 February 2016; Published: 25 February 2016

Abstract: This paper proposes and analyses a new demand response technique for renewable energy regulation using smart hot water heaters that forecast water consumption at an individual dwelling level. Distributed thermal energy storage has many advantages, including high overall efficiency, use of existing infrastructure and a distributed nature. In addition, the use of a smart thermostatic controller enables the prediction of required water amounts and keeps temperatures at a level that minimises user discomfort while reacting to variations in the electricity network. Three cases are compared in this paper, normal operation, operation with demand response and operation following the proposed demand response mechanism that uses consumption forecasts. The results show that this technique can produce both up and down regulation, as well as increase water heater efficiency. When controlling water heaters without consumption forecast, the users experience discomfort in the form of hot water shortage, but after the full technique is applied, the shortage level drops to nearly the starting point. The amount of regulation power from a single dwelling is also discussed in this paper.

Keywords: demand side management (DSM); distributed thermal storage; forecasting; water heater

---

1. Introduction

A distinctive characteristic of the electric power sector is that the amount of generated electricity has to be equal to the amount of consumed electricity at every single instance [1]. Unfortunately, there are peaks and valleys of total consumed electric energy, which do not always coincide with available generation patterns. People tend to have habits, including morning and evening rituals, that require large amounts of energy; thus, peaks are created. In addition, the generation side failures or other disruptions necessitate costly regulation ancillary services to match the demand with supply [2]. As a result, national transmission system operators (NTSO) constantly monitor the system and adjust the generation to meet the demand using ancillary services.

The increase of renewable energy generation attempts to solve problems associated with the conventional generation (such as emissions of greenhouse gasses), but creates power balancing issues [3]. Renewable energy is inherently intermittent and hard to control. As a result, its output is highly variable, and the electricity balancing problem becomes even more difficult [4]. Many researchers agree that wind generation introduces unprecedented amounts of uncertainty. The importance of demand side management (DSM) for long-term sustainable energy use in high renewable energy penetration areas is discussed in [5]. The power reserve limit needs to be increased when adding wind power to the system; otherwise, reliability is sacrificed [6]. It also makes unit commitment and economic dispatch problems more complicated, which are assessed in [7]. Studies show that in some countries in 2020, up to 13% of trade periods will require wind curtailment [8], indicating high wind generation uncertainty. According to [9], a forecasting horizon further than 4 h requires weather information to acquire better accuracy; therefore, wind generation forecasting results are highly dependant on the climate of a location. Wind power forecast uncertainty using probabilistic forecasting is described in [10], and in [6], the authors demonstrate the standard deviation of error of day-ahead forecast to be 0.22 per MW of installed power in Ireland. Up to now, traditional pumped hydro storage facilities primarily have served as part of the backup power, but this cannot meet the high rate of output change from renewable power plants [2,11]. In addition, centralised backup power requires energy to be transmitted back and forth; thus, transmission losses have to be accounted for, as well.

Energy storage fundamentally improves the way electricity is generated, transmitted and consumed [12]. It allows the decoupling of generation from consumption to a certain level [13]. Hence, more storage on the grid significantly reduces generation dependency on the consumption. In addition, storage devices would also help during power outages, caused by equipment failures/faults or accidents. Moreover, the transmission and distribution grid has capacity limits, which might be exceeded during peak electricity usage. Energy storage would also help the grid to smooth energy transportation, increase electricity throughput to its maximum and increase load factor [14]. This would significantly lower the infrastructure costs as the transmission and distribution equipment has to be designed for peak demand, which occurs less than 5% of the time [3]. Furthermore, it enables the potential of running generating units at their maximum efficiency point, thus eventually decreasing generation costs.

DSM is a broad set of means to alter the time and magnitude of end user's electricity consumption, one of which is load shifting. Load shifting techniques require storage capabilities, such as thermal storage devices. Water heaters are perfect candidates as demand responsive devices. In general, water heating accounts for 17% of all residential energy use in the United States [15]. Resistive hot water heaters are common in residential houses and make up 40% of all hot water heaters in the U.S. [15] and 12%–20% in the U.K. (depending on the season) [16], meaning the infrastructure is already established. They exhibit good thermal storage properties [17], possess high nominal power ratings and large thermal buffer capacities, as well as a fast response to load change [18–20]. Water has relatively high specific heat, which allows it to store large amounts of energy. Furthermore, in resistive water heaters, electricity is transferred to useful heat at 100% efficiency, and energy is lost only due to heat transfer through insulating walls.

Various hot water heater control techniques can be seen in the literature. The load commitment technique using real-time and forecasted pricing of electricity was researched by scientists in [21], whereas other researchers discussed a technique using timer switches for hot water load management [22]. Kepplinger *et al.* [23] demonstrate optimal control of hot water heaters using linear optimisation. The aggregate regulation service for renewable energy using thermally-stratified water heater model was analysed by Kondoh *et al.* [24]. The model is designed to have two heating elements, but only one is assigned for regulation services; thus, in essence, only one half of the thermal capacity is used for demand response (DR), and the other half is used to guarantee end users' comfort. Furthermore, there is an ongoing work to increase the efficiency of water heaters using baffles based on computational fluid dynamics [25]. Electric water heating control techniques to integrate wind power are compared by Fitzgerald *et al.* [26], whereas Finn *et al.* examines the impact of load scheduling on the adaption of wind generation [8]. Another study on the load balancing technique using an aggregate heating, ventilation and air conditioning (HVAC) system is presented by Lu in [27].

The widespread acceptance of DSM programs relies on minimal impact to the comfort of users [18]. This paper proposes a new strategy to control residential hot water heaters with minimal change in users' comfort levels. In this research, the focus was to eliminate the imbalance caused by wind power plants, although this technique is not limited to solving problems associated with renewable energy generation. It could help in cases of generation faults or it could be used as an ancillary service or by energy traders to profit from the fluctuating real-time price of electricity.

2. System Description and Methodology

This section describes the general methodology and techniques used in the design of the residential water heater-based distributed energy storage system. It also describes the data preparation, model design, evaluation and comparison of different scenarios.

2.1. Thermal Water Heater Model

The dynamic thermal water heater model was derived based on open system energy balance [21,23,28]. The amount of energy consumed by the electric heating element is added to the model as an input, whereas the outputs are (1) energy consumed by hot water usage and (2) thermal energy losses due to imperfect thermal insulation. The amount of water drawn from the tank is based on measurement data collected from individual dwellings [29]. The temperature of the inlet water and the specific heat of water at normal temperature and pressure (NTP) conditions were also taken into account. Thermal losses are calculated based on the temperature difference between water and ambient temperature and thermal conductivity. The model is fully mixed, unstratified, meaning water temperature is the same throughout the tank. The effect of temperature variation at the output is compensated by demanding more water in case the temperature is cooler than the setpoint and demanding less if the temperature is higher. According to [30], the fully-mixed model shows increased thermal energy losses, so heat transfer coefficients were adjusted to compensate for this. Figure 1 graphically depicts the energy conservation of the system.

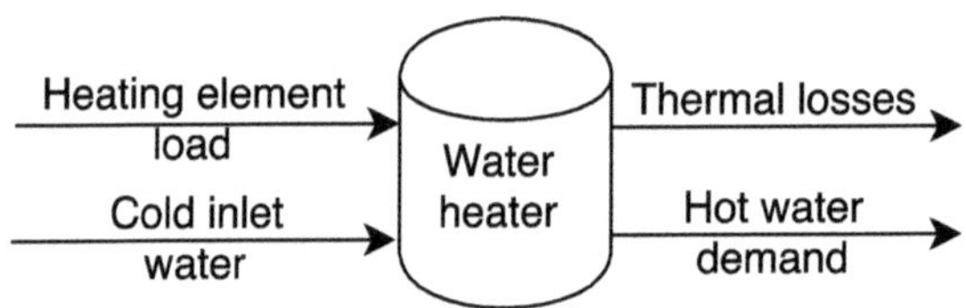

Figure 1. Thermal water heater diagram.

The mathematical model of the thermal system could be described as [21]:

$$Q_{t+1} = Q_t + \Delta t S_{0/1} K_{\mathrm{HE}} + C_{\mathrm{W}} D_t (T_{\mathrm{WH}} - T_{\mathrm{in}}) + \Delta t k (T_{\mathrm{WH}} - T_{\mathrm{amb}}) \quad (1)$$

$$T_{\mathrm{WH}} = \frac{Q_t}{m C_{\mathrm{W}}} \quad (2)$$

$$40\ ^\circ\mathrm{C} < T_{\mathrm{WH}} < 90\ ^\circ\mathrm{C} \quad (3)$$

where Q_t (J) is the thermal energy stored in the water tank (integrator); Δt (s) is the time step length; $S_{0/1}$ is the on/off state of the heating element (WH control); K_{HE} (W) is the heating element rating; C_{W} (J/kg°C) is the specific heat of water; m (kg) is the mass of water in a single device; D_t (kg) is the demand of hot water at time t; k (J/s°C) is the heat transfer coefficient for particular device and T_{in} (°C), T_{WH} (°C) and T_{amb} (°C) are inlet cold water, hot water and ambient temperatures respectively. The model was then implemented in the Matlab Simulink software environment which can be seen in Figure 2.

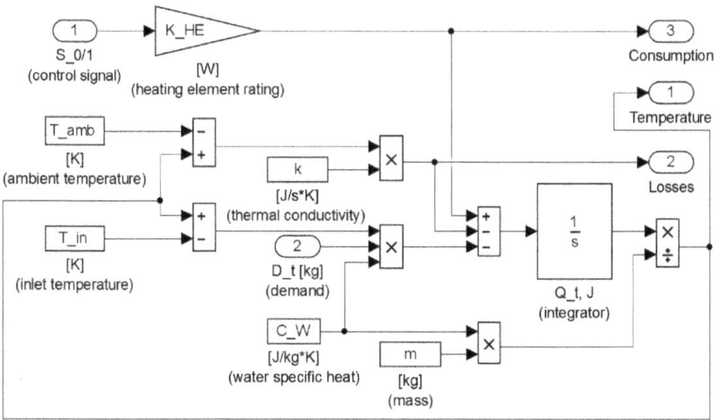

Figure 2. Thermal water heater block diagram model.

2.2. Smart Hot Water Heater Controller

The smart hot water heater controller in the proposed system controls the heating element according to the consumption forecasts and the signal sent from the smart grid. The controller is capable of locally forecasting hot water consumption of a particular dwelling. It contains an artificial neural network (ANN) model, which is trained based on the past hot water consumption information. The ANN model can compute short-term hot water usage forecasts tailored for the particular house. The controller also contains thermal model, so based on the consumption forecast, it can compute water temperature for the next 12 h period. It also receives a signal from the grid showing the requested duty cycle of the heating element. The signal is percentage-wise, where 0% means that the grid experiences a shortage of electricity, thus requesting to turn the heating element off, and 100% means a surplus of energy in the grid. The overall operation of the controller is described in Section 2.5.

The ANN model that is used in the proposed system is based on the results from previous research [31,32]. In particular, a neural network nonlinear autoregressive exogenous (NARX) model is used. The configuration is the same as in Case #8 in [31] (p. 414). The ANN comprises an input layer, a single hidden layer consisting of 10 neurons and an output layer. The external inputs are the average consumption profile, as well as weekday and weekend dummy variables. The outputs of the ANN are fed back as inputs using a certain delay. It uses the Levenberg–Marquardt training algorithm, and the data are divided into training (15%), validation (15%) and test (70%) datasets. The training algorithm uses mean square error as the performance function to terminate the training. The overall performance of the model is summarised in Table 1.

Table 1. Forecasting measures.

Measure	Wind Generation Forecast (per 1.5 kW)	Hot Water Consumption Forecast (kg)
Mean	0.557 kW	6.145 kg
Standard deviation	0.374 kW	9.269 kg
Mean error	−0.012 kW	0.042 kg
Standard deviation of error	0.137 kW	1.541 kg
Mean absolute error	0.099 kW	0.870 kg
Root mean square error	0.137 kW	1.548 kg
Normalised mean absolute error [32]	0.264	0.108
Normalised root mean square error [32]	0.368	0.192
Regression value R	0.938	0.981

The controller also implements temperature control. Despite any other factor, the controller attempts to maintain instantaneous temperature within the limits described in Equation (3). These are the upper and lower temperature safety bounds. If for any reason the temperature increased above 90 °C, it would disconnect the heating element until the temperature dropped below 88 °C. Similarly, if the temperature dropped below the critical 40 °C, it would turn on the heater regardless of the control signal from the system. This mechanism helps to ensure that the comfort level for the user is not impacted.

2.3. Wind Imbalance and Normal Consumption

The performance of the system is assessed using previously-measured and -forecasted wind power generation data (total wind generation forecasts, as well as actual wind generation shown in Figure 3) provided by the Lithuanian NTSO [33]. The overall goal of the newly-proposed DSM system is to create a backup power aggregator to cover forecasting error. The mismatch between forecast and actual generation can be either positive (surplus of energy) or negative (shortage of energy), and it is being referred to as the imbalance throughout the paper. Minimising imbalance enables renewable electricity sellers to supply the exact amount of electricity. The electricity that sells in the market can be delivered with high certainty, eliminating costly fines for under delivery of power or loss of income due to a lower price of unexpected energy generation (disconnection in the worst case). Table 1 contains statistical measures of the wind generation forecast data. The wind generation forecasts throughout the paper are based on the next day-ahead predictions to comprise the electricity day-ahead market. Furthermore, Table 1 presents hot water consumption forecast statistical information. It contains the arithmetic average of measures from all houses. These figures are calculated for one hour ahead forecasts.

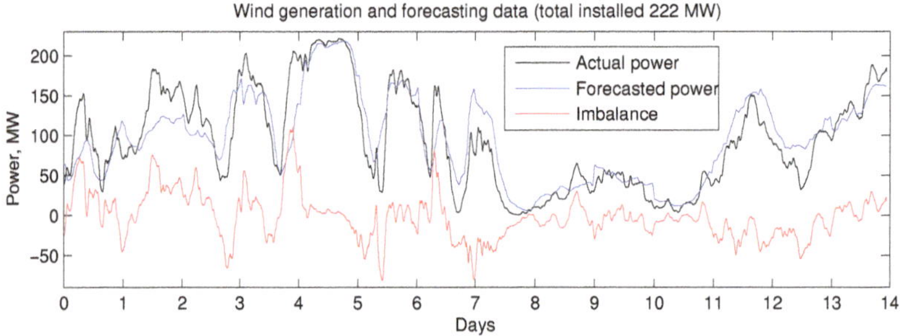

Figure 3. Total actual and forecasted wind power.

Figure 4 shows the normal electricity consumption of water heaters (per household) and the normalised wind power imbalance. The wind power imbalance is normalised by assigning 1.5 kW of installed power for every dwelling. The sum of the normal consumption and wind imbalance becomes the target total power consumption for hot water heaters participating in DSM. This way, the residential users can both shed the load (turn off the heating elements inside hot water heaters) or use more energy than they would normally use (turn on the heater, irrespective of the water setpoint temperature). This is particularly useful when compensating the negative imbalance in the system; the users would have to use less energy than they would normally use without DSM (regulation up). It should be noted that individual houses follow different loads specified by the smart controller, but the average target hot water heaters' consumption of electricity is shown in Figure 4.

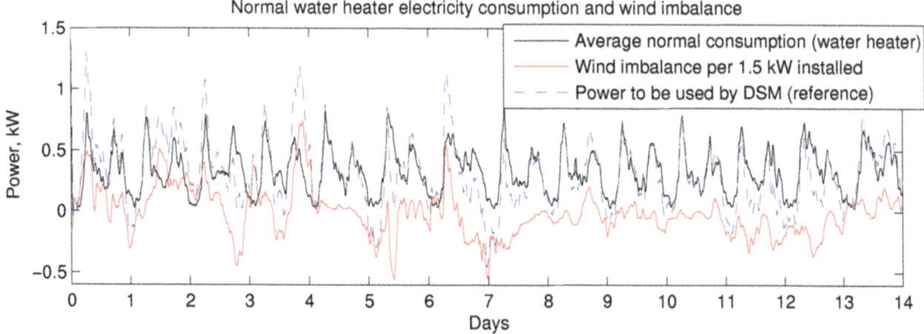

Figure 4. Average normal power consumption, wind power imbalance (fraction of 1.5 kW out of total 222 MW) and power to be used by the proposed demand side management (DSM) system per household.

2.4. Model Parameters and Assumptions

Modelling such a complex system required the careful selection of parameters, including temperature setpoints, sizes of the tanks, heating element ratings, ambient temperatures, thermal conductivity of the hot water tank, inlet water temperature, *etc.* One of the most important parameters in the context of energy accumulation is hot water tank volume. It describes how long a user can last without using electrical energy (in case of a shortage) or how much excessive electrical energy can be stored (in case of a surplus). In this paper, the hot water tanks were sized between 85 L and 200 L taking into account the average water consumption rate for a particular dwelling. Randomly-picked tank sizes from the chosen range were sorted in ascending order. The highest volume tank was matched to the dwelling with the most hot water consumption, and *vice versa*. Another crucial parameter of water heating devices is the rated power, where it defines how fast the electric energy is transferred to heat. From a demand response point of view, it is important during the times of energy surplus. The heating element power ratings were chosen to fall in a range from 1.5 kW to 2.5 kW [16]. The relationship of tank volume and heating elements can be seen in Figure 5.

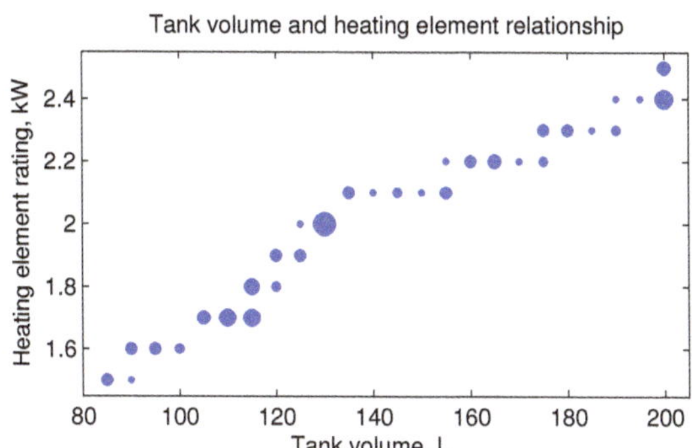

Figure 5. Water heater tank size and heating element power rating relationship.

The inlet water temperature was chosen to be slightly different for all households (between 9 °C and 11 °C) and was kept constant throughout the testing period. Similarly, the ambient air temperature surrounding the hot water tanks was chosen to be between 19 °C and 23 °C. The optimal setpoint temperature was set to be around 68 °C [26].

2.5. Proposed Demand Side Management System Overall Operation

The main goal of the proposed system is to compensate day-ahead wind generation forecast errors. It enables the supply of the exact amount of wind energy that was sold in the day-ahead market and avoids charges for costly regulation ancillary services. At first, the forecast error is calculated by subtracting the day-ahead forecast from the actual wind generation. This is the power to be regulated using DR. Since water heaters can only consume electricity (regulate down), the imbalance is added on top of the predicted normal consumption to enable up regulation. The predicted normal water heater consumption information can be taken from the distribution system operator or, in this paper, it is modelled by the same ANN. Secondly, the actual electricity usage is aggregated and subtracted from the reference load. It is then used by the demand response controller to compute the request signal for the water heaters, which in turn decides whether to participate in the DR or not. Every 5 min, the controller forecasts individual demand for the next 12 h and computes the ability to participate in the demand response. It is only necessary to forecast 12 h ahead, because it takes about the same amount of time to raise the temperature by 50 degrees for a 200 L tank using a 1.5 kW heating element. Then, the controller computes the worst case scenario and checks whether the temperature is maintained in between the boundaries of comfort. The worst case scenario is achieved by turning the heater off for 5 min and when leaving it to work according to the thermostat. In the case of participation, the water heater reacts to the request signal and alters the energy use accordingly. As a result, the wind forecast error ends up balanced.

The simulation framework comprises 95 dwellings equipped with resistive hot water heater models of different sizes and power ratings, as well as 95 ANN models for every dwelling. The overall system diagram can be seen in Figure 6.

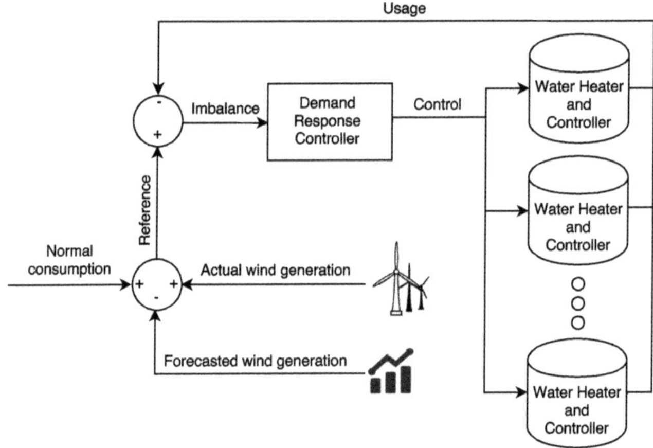

Figure 6. Overall diagram of the system.

3. Results and Discussion

The simulations were split into three different cases. Each case adds DSM capabilities step by step. Table 2 summarises the performance of five different scenarios. Case #1 represents the normal use of hot water heaters without DSM. Case #2 involves DSM, but excludes forecasting of hot water consumption, *i.e.*, it does not look ahead to how much water is to be potentially needed during the next 12 h. In this case, users' comfort is not taken into account and might be compromised. Power in brackets next to the case number in Table 2 shows the amount of installed wind power that is on average assigned to every dwelling. It demonstrates the backup power capability of a single unit using the DSM technique.

This case involves three different scenarios, −1 kW, 1.5 kW and 2 kW. Finally, Case #3 depicts the proposed DSM with forecasting and the method of looking ahead. All values are per household.

Table 2. Performance measures.

Case	Mean Power Consumption (W)	Mean Absolute Final Imbalance (W)	Mean Losses (W)	Mean Temperature (°C)	Shortage (% of Time)	Participation, %
#1 (N/A)	325.7	144.3	49.4	67.5	0.11	(N/A)
#2 (1.0 kW)	309.4	26.6	54.4	73.1	1.19	100.0
#2 (1.5 kW)	298.7	47.1	52.6	71.4	1.95	100.0
#2 (2.0 kW)	290.0	72.3	51.2	70.0	2.74	100.0
#3 (1.5 kW)	313.9	52.1	46.9	65.9	0.30	94.0

Performance measures used in Table 2 can be summarised as follows:

- Mean power consumption is calculated by simply taking the arithmetic mean of the consumption profile from all dwellings.
- Mean absolute final imbalance is the arithmetic average of final absolute imbalance values. Figures are scaled to be per household per 1.5 kW of installed wind power.
- Mean losses: arithmetic average of thermal losses per hot water heater.
- Mean temperature: arithmetic average of water temperature inside tanks.
- Shortage: average percentage of time the demanded water temperature was not supplied.
- Participation: the average percentage of time that each water heater was participating in DSM. The only time they are not participating is when there is expected high future consumption of hot water; thus, the temperature was expected to drop below critical, so the controller disconnects the particular water heater from DSM (therefore, increasing/maintaining user comfort).

Figure 7 shows the relation between the time of shortage of hot water and the tank volume (Case #3). Most dwellings have not experienced any hot water shortage during the simulated period. Houses that suffered from the lack of hot water at some point in time show no correlation between their the tank size. As a result, it can be concluded that the installed tank size does not dictate how suitable the house is for DSM participation.

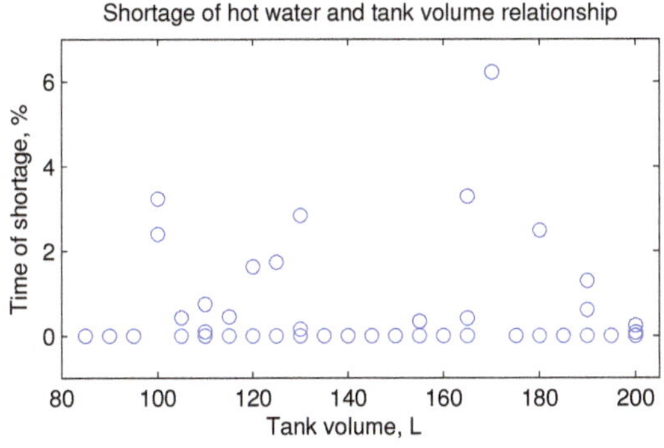

Figure 7. The relationship of water heater tank size and the percentage of time the users experienced a shortage of hot water.

The results provide evidence that the proposed DSM technique is capable of (1) lowering the energy requirements for hot water preparation and (2) supplying an ancillary service (power regulation) to the grid with a minor change in user comfort. The average energy required to supply the same amount of hot water is decreased due to increased efficiency. Contrary to the traditional temperature control, when the temperature is kept at a constant level and the amount of prepared hot water is inadequate for the amount that is actually needed, the proposed look ahead mechanism forecasts the required amount of hot water and controls temperature in a more efficient way. The temperature inside the water reservoir is decreased during energy shortages, whereas at the times of surplus energy, the temperature is increased to store energy. In fact, user comfort was affected in Case #2, but after demand forecasting was applied, it got restored to nearly the same level (shortage in Table 2). Ancillary balancing services become available at virtually no cost, because the users do not notice any major difference in hot water supply due to the correct amounts of hot water that are prepared using forecasting.

3.1. Limitations

The fact that a negative imbalance can only be compensated by shedding the load leads to a certain limitation. The maximum power that can be shed is equal to the cumulative power the residences would normally use minus the power needed to maintain critically low water temperatures. In this particular case, the hot water consumption profile has very distinctive daily and weekly patterns. The consumption profile does not always coincide with the wind generation imbalance, thus during the valleys of normal energy consumption, there might be insufficient energy to be shed. Clearly, it can be expected that the proposed DSM mechanism will work best during peak hot water consumption periods and, hence, reduce the energy demand from the network. Figure 8 demonstrates the average weekly consumption profile.

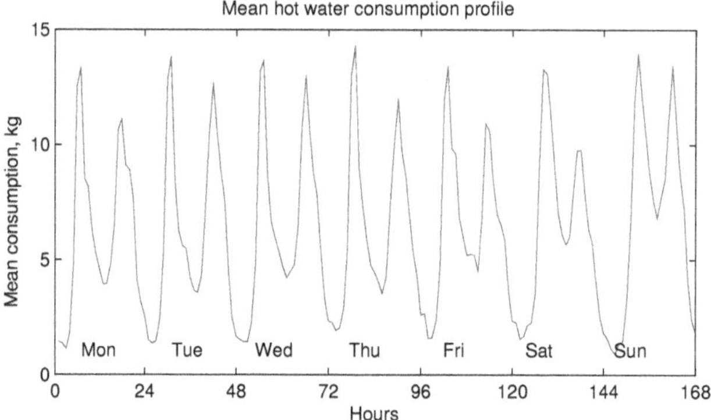

Figure 8. Weekly mean hot water consumption pattern [31].

This hypothesis is confirmed by the scatter plot in Figure 9. The scatter plot depicts the relationship between normal power consumption (x axis) and the absolute final power imbalance (y axis). As can be seen, the system is capable of reaching a more accurate final balance during times of higher normal consumption, *i.e.*, when the DSM mechanism has a wider margin for error. Figure 10 also confirms this fact. It can be seen that during the times around midnight, the normal energy consumption is low. By subtracting the shortage of energy (caused by negative wind balance), the reference power curve is moved below zero. Obviously, water heaters cannot work in reverse; thus, wind power energy is not fully balanced, and negative dips of final system balance can be seen during these hours.

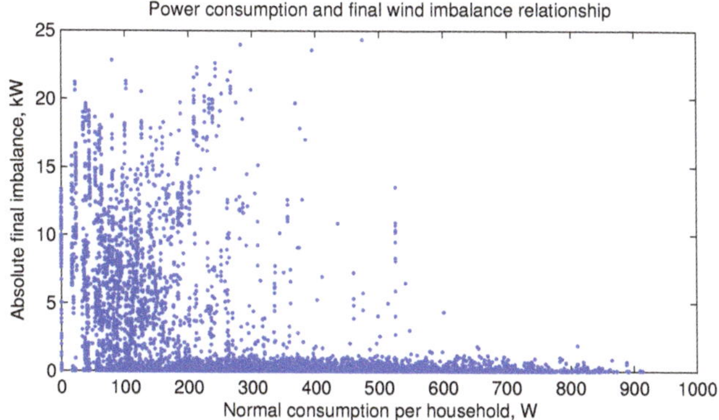

Figure 9. Scatter diagram showing the relationship between normal consumption and final power imbalance.

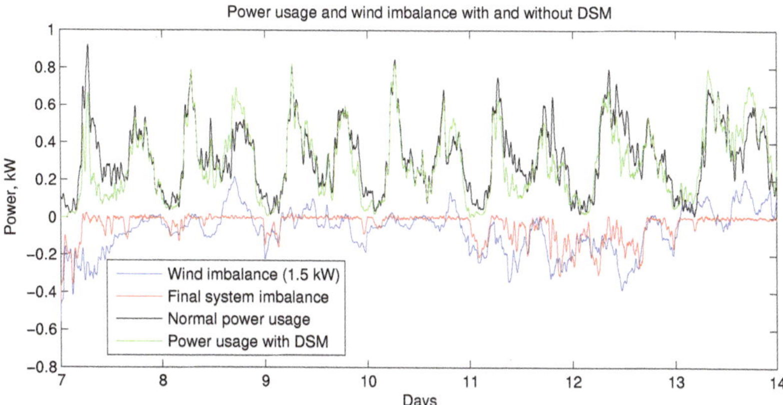

Figure 10. Sample time plot showing alterations in power consumption and wind imbalance. The plot depicts results from Cases #1 and #3.

Another limitation is for the surplus energy, *i.e.*, the maximum positive power imbalance the system can compensate. It is equal to the summed power rating of responsive water heaters (the ones with water temperatures below critically high) minus the forecasted normal consumption. In this paper, the normal consumption forecasts are computed using the same ANN models. As a result, every single dwelling cannot backup more installed power than its maximum rating, hence the chosen 1.5 kW value to be backed up by each dwelling. In addition, once the heater is fully charged (critical temperature reached) it is forced to the off state and cannot participate in DR. This creates vulnerability for long periods of surplus energy.

3.2. Temperatures

Clearly, it is expected that during the normal operation of the hot water heater (no DSM), the temperature does not go above the setpoint. The heater is simply turned off after a certain temperature is reached and turns on when the temperature is dropping. During the high hot water demand periods, the temperature might drop below the given setpoint. Theoretically, the heater should be sized such that it always satisfies users' demands.

In case of hot water heater control using the DSM technique, without look ahead, there might be a situation where the temperature drops below a critical level. Such a situation occurs when an electricity shortage period is followed by substantial demand for hot water. The heating element is simply not capable of transferring heat at the same rate the water is drawn (otherwise, there would be no need for an accumulation tank). This case depicts a situation where the grid is satisfied by sacrificing user comfort (Case #2).

To overcome this problem, a control technique is added, which looks 12 h ahead and takes into account the forecasted consumption at every dwelling. Figure 11 depicts the average temperatures of normal consumption (*i.e.*, the setpoint does not change), three DSM scenarios using different amounts of installed wind power to be balanced (per household) and average temperatures using the proposed DSM technique. It can be seen that using the traditional method, the temperature fluctuates around setpoint. In Case #2, three different amounts of backup power force the temperatures to swing in higher amplitudes, respectively. Finally, the mean temperature in Case #3 shows a different pattern, as there is a participation factor introduced to the system, which allows users to choose whether to participate in the DSM or not.

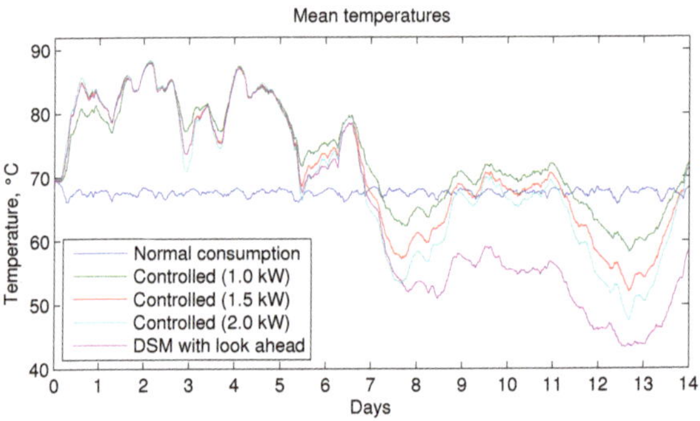

Figure 11. Sample average temperature time plot showing different simulation scenarios.

3.3. Losses

Thermal losses depend on the thermal conductivity coefficient of the tank walls and the difference in water and air temperatures. Since the thermal conductivity coefficient is constant and room temperature is also fairly constant, losses are mainly a function of temperature. Greater losses are experienced when water temperature is kept high. Therefore, in the event of shifting energy use into the future (delay raising the temperature), the heater exerts less heat waste, and *vice versa*.

3.4. Energy Balance

Figure 10 illustrates the exemplar time plot of energy balancing results from the simulation. It shows the normal consumption and wind power imbalance without DSM. The same figure also depicts the power consumption of Case #3, as well as the final balance that was achieved using DSM with look ahead. Table 2 compares the performance measures of the chosen simulation cases. It can be seen that mean power consumption has decreased by about 5% when the DSM technique was applied. The decrease in energy consumption was caused by a higher system efficiency (lower thermal losses), lower final average water temperature and overall negative wind power imbalance. The results suggest that users experienced some hot water shortage in Case #2 due to the fact that 100% of the users were forced to alter their energy use (see sixth and seventh columns in Table 2). On the other hand, in Case #3, the look ahead forecasting mechanism allowed the users to decide the most suitable times to participate in order to prevent their comfort violation. It can be seen that using the proposed DSM technique and the current setup, the average of about 94% of users were able to participate. The other

6% were notified by the tailored forecasting models that in case of participation there is a high chance of a hot water shortage. Therefore, user satisfaction was restored and the shortage percentage decreased. At the same time, Cases #2 and #3 demonstrate a decrease in final wind imbalance, *i.e.*, wind generation variation was successfully backed up by the DSM technology. It should also be noticed that mean absolute final imbalance varied in Case #2 due to different amounts of installed wind power per household. The 1.5 kW per household of installed wind power has been observed to be optimal, as higher values cause the system to saturate and increase the final imbalance, which contradicts the key objective of this paper.

4. Conclusions

Due to the increased number of renewable energy sources, the electricity system requires more ancillary backup services every day. DSM techniques, such as distributed thermal energy storage using individual hot water heaters, can be utilised to tackle this problem. Forecasting hot water consumption at an individual level unveils each users needs; thus, the control can be applied such that the comfort is maintained at almost the same level. By having precise consumption forecasts, it is possible to prepare more accurate amounts of hot water compared to the functioning of a conventional water heater. At the same time, there is a wider margin for DSM operations. Using the proposed technique, time of water shortage increases from 0.11% to 0.3%. Compared to the results of Case #2 (1.95%), the increase in Case #3 is negligible. At the same time, the mean absolute final imbalance decreased by about 64%. The results confirm the initial hypothesis, that using such a DSM technique, it is possible to (1) lower the energy requirements for hot water preparation and (2) supply an ancillary service to the grid with minimal change in user comfort.

Acknowledgments: The authors would like to acknowledge the financial support of the Department of Engineering and Faculty of Science and Technology, Lancaster University, U.K., as well as the Energy Saving Trust for providing the necessary hot water consumption data.

Author Contributions: Linas Gelažanskas designed the models, performed simulations, and wrote the paper. Linas Gelažanskas and Kelum A.A. Gamage analyzed the data and corrected the paper.

Conflicts of Interest: The authors declare no conflict of interest.

References

1. Kundur, P. *Power System Stability and Control*; McGraw-Hill: New York, NY, USA, 1994.
2. Baranauskas, A.; Gelažanskas, L.; Ažubalis, M.; Gamage, K. Control strategy for balancing wind power using hydro power and flow batteries. In Proceedings of the 2014 IEEE International Energy Conference (ENERGYCON), Cavtat, Croatia, 13–16 May 2014; pp. 352–357.
3. Gelažanskas, L.; Gamage, K.A. Demand side management in smart grid: A review and proposals for future direction. *Sustain. Cities Soc.* **2014**, *11*, 22–30.
4. Lund, P.D.; Lindgren, J.; Mikkola, J.; Salpakari, J. Review of energy system flexibility measures to enable high levels of variable renewable electricity. *Renew. Sustain. Energy Rev.* **2015**, *45*, 785–807.
5. Pina, A.; Silva, C.; Ferrao, P. The impact of demand side management strategies in the penetration of renewable electricity. *Energy* **2012**, *41*, 128–137.
6. Doherty, R.; O'Malley, M. A new approach to quantify reserve demand in systems with significant installed wind capacity. *IEEE Trans. Power Syst.* **2005**, *20*, 587–595.
7. Wang, J.; Botterud, A.; Bessa, R.; Keko, H.; Carvalho, L.; Issicaba, D.; Sumaili, J.; Miranda, V. Wind power forecasting uncertainty and unit commitment. *Appl. Energy* **2011**, *88*, 4014–4023.
8. Finn, P.; Fitzpatrick, C.; Connolly, D.; Leahy, M.; Relihan, L. Facilitation of renewable electricity using price based appliance control in Ireland's electricity market. *Energy* **2011**, *36*, 2952–2960.
9. Black, M.; Strbac, G. Value of Bulk Energy Storage for Managing Wind Power Fluctuations. *IEEE Trans. Energy Convers.* **2007**, *22*, 197–205.
10. Matos, M.; Bessa, R. Setting the Operating Reserve Using Probabilistic Wind Power Forecasts. *IEEE Trans. Power Syst.* **2011**, *26*, 594–603.

11. Gelažanskas, L.; Baranauskas, A.; Gamage, K.A.; Ažubalis, M. Hybrid wind power balance control strategy using thermal power, hydro power and flow batteries. *Int. J. Electr. Power Energy Syst.* **2016**, *74*, 310–321.
12. Divya, K.; Østergaard, J. Battery energy storage technology for power systems—An overview. *Electr. Power Syst. Res.* **2009**, *79*, 511–520.
13. Chen, H.; Cong, T.N.; Yang, W.; Tan, C.; Li, Y.; Ding, Y. Progress in electrical energy storage system: A critical review. *Prog. Nat. Sci.* **2009**, *19*, 291–312.
14. Denholm, P.; Sioshansi, R. The value of compressed air energy storage with wind in transmission-constrained electric power systems. *Energy Policy* **2009**, *37*, 3149–3158.
15. Hepbasli, A.; Kalinci, Y. A review of heat pump water heating systems. *Renew. Sustain. Energy Rev.* **2009**, *13*, 1211–1229.
16. Newborough, M.; Augood, P. Demand-side management opportunities for the UK domestic sector. *IEE Proc. Gener. Transm. Distrib.* **1999**, *146*, 283–293.
17. Armstrong, P.; Ager, D.; Thompson, I.; McCulloch, M. Improving the energy storage capability of hot water tanks through wall material specification. *Energy* **2014**, *78*, 128–140.
18. Paull, L.; Li, H.; Chang, L. A novel domestic electric water heater model for a multi-objective demand side management program. *Electr. Power Syst. Res.* **2010**, *80*, 1446–1451.
19. Ericson, T. Direct load control of residential water heaters. *Energy Policy* **2009**, *37*, 3502–3512.
20. Vanthournout, K.; D'hulst, R.; Geysen, D.; Jacobs, G. A Smart Domestic Hot Water Buffer. *IEEE Trans. Smart Grid* **2012**, *3*, 2121–2127.
21. Du, P.; Lu, N. Appliance Commitment for Household Load Scheduling. *IEEE Trans. Smart Grid* **2011**, *2*, 411–419.
22. Atikol, U. A simple peak shifting DSM (demand-side management) strategy for residential water heaters. *Energy* **2013**, *62*, 435–440.
23. Kepplinger, P.; Huber, G.; Petrasch, J. Autonomous optimal control for demand side management with resistive domestic hot water heaters using linear optimization. *Energy Build.* **2015**, *100*, 50–55.
24. Kondoh, J.; Lu, N.; Hammerstrom, D. An evaluation of the water heater load potential for providing regulation service. In Proceedings of the 2011 IEEE Power and Energy Society General Meeting, San Diego, CA, USA, 24–29 July 2011; pp. 1–8.
25. Sedeh, M.M.; Khodadadi, J. Energy efficiency improvement and fuel savings in water heaters using baffles. *Appl. Energy* **2013**, *102*, 520–533.
26. Fitzgerald, N.; Foley, A.M.; McKeogh, E. Integrating wind power using intelligent electric water heating. *Energy* **2012**, *48*, 135–143.
27. Lu, N. An Evaluation of the HVAC Load Potential for Providing Load Balancing Service. *IEEE Trans. Smart Grid* **2012**, *3*, 1263–1270.
28. Nehrir, M.; Jia, R.; Pierre, D.; Hammerstrom, D. Power Management of Aggregate Electric Water Heater Loads by Voltage Control. In Proceedings of the 2007 IEEE Power Engineering Society General Meeting, Tampa, FL, USA, 24–28 June 2007; pp. 1–6.
29. *Measurement of Domestic Hot Water Consumption in Dwellings;* Technical Report; Energy Saving Trust: London, UK, 2008.
30. Celador, A.C.; Odriozola, M.; Sala, J. Implications of the modelling of stratified hot water storage tanks in the simulation of *CHP* plants. *Energy Convers. Manag.* **2011**, *52*, 3018–3026.
31. Gelažanskas, L.; Gamage, K. Forecasting hot water consumption in dwellings using artificial neural networks. In Proceedings of the 2015 IEEE 5th International Conference on Power Engineering, Energy and Electrical Drives (POWERENG), Riga, Latvia, 11–13 May 2015; pp. 410–415.
32. Gelažanskas, L.; Gamage, K.A.A. Forecasting Hot Water Consumption in Residential Houses. *Energies* **2015**, *8*, 12702–12717.
33. Gelažanskas, L.; Gamage, K.A. *Wind Generation;* Data catalogue, Lancaster University Library: Lancaster, UK, 2015.

Article

The Demand Side Management Potential to Balance a Highly Renewable European Power System

Alexander Kies [1,2,*], Bruno U. Schyska [1,2] and Lueder von Bremen [1]

[1] ForWind, Center for Wind Energy Research, Küpkersweg 70, 26129 Oldenburg, Germany; bruno.schyska@forwind.de (B.U.S.); lueder.von.bremen@forwind.de (L.v.B.)

[2] Institute of Physics, University of Oldenburg, Ammerländer Heerstr. 114, 26129 Oldenburg, Germany

* Correspondence: alexander.kies@uni-oldenburg.de

Academic Editor: William Holderbaum

Received: 03 July 2016; Accepted: 08 November 2016; Published: 15 November 2016

Abstract: Shares of renewables continue to grow in the European power system. A fully renewable European power system will primarily depend on the renewable power sources of wind and photovoltaics (PV), which are not dispatchable but intermittent and therefore pose a challenge to the balancing of the power system. To overcome this issue, several solutions have been proposed and investigated in the past, including storage, backup power, reinforcement of the transmission grid, and demand side management (DSM). In this paper, we investigate the potential of DSM to balance a simplified, fully renewable European power system. For this purpose, we use ten years of weather and historical load data, a power-flow model and the implementation of demand side management as a storage equivalent, to investigate the impact of DSM on the need for backup energy. We show that DSM has the potential to reduce the need for backup energy in Europe by up to one third and can cover the need for backup up to a renewable share of 67%. Finally, it is demonstrated that the optimal mix of wind and PV is shifted by the utilisation of DSM towards a higher share of PV, from 19% to 36%.

Keywords: demand side management; renewable energy systems; European power system; energy system modelling; wind energy; solar energy

---

1. Introduction

Aiming at sustainability and reduced CO_2 emissions, shares of renewable generation are on the rise all across Europe. This is in line with the 2015 United Nations Climate Change Conference (CMP 11) commitments. However, the integration of intermittent renewable generation from wind and photovoltaics (PV) into energy systems poses severe balancing challenges [1]. Unlike conventional generation (e.g., nuclear, fossil, etc.), renewable generation is driven by the weather and can not be reliably dispatched and therefore not directly adopted to follow the demand. Several possible approaches to this issue have been investigated: (i) Optimising the mix of different renewable sources [2–8]; (ii) Storage to shift generation in time [9,10]; (iii) Backup [11–13]; (iv) Or the reinforcement of the transmission grid to shift generation in space [14,15]. In addition to these approaches on the generation side, part of the need for balancing might also be covered by the modification of the demand for power on the consumer side with the objective of increasing its manageability (e.g., to match renewable generation in time [16,17]). One definition of demand side management (DSM) is "the planning, implementation, and monitoring of utility activities designed to encourage consumers to modify patterns of electricity usage, including the timing and level of energy demand." [18]. The interest in different aspects of DSM has risen in recent years, along with the rising general interest in renewable power systems. DSM storage strategies for end-users were investigated in [19]. Large-scale industrial processes might be able to provide up to 50% of backup capacity need by 2020 [20,21]. Furthermore, DSM can significantly reduce the need for conventional generation [22] even on a residential scale. It is expected that most contributions to a flexible demand side will be provided by industry. In Germany, for instance, a major contributor might be the automotive industry [23]. Klobasa et al. [24] investigate the interplay of load management and wind power forecasts in Germany and concludes that load management can reduce balancing costs by up to 20% and is mostly economically useful. Lund et al. [25] investigate the potential impact of electric cars in a vehicle-to-grid (V2G) system on a renewable Danish power system. Stadler et al. [26] conclude that different DSM branches can complement each other fairly well in Germany to provide seasonally independent power and that the biggest fraction of demand side management can be provided by storage heating and combined heat and power (CHP). Moura et al. [27] use a heuristic approach to show a reduction of Portuguse peak loads of more than 10% through the use of DSM. A similar heuristic approach to characterise the potential of DSM is applied in [28]. A broad overview of different DSM types is given in [29]. In [30], an energy system model is used to estimate the impact of DSM in the EU-27. However, it only inhibits a low renewable share in the system and focuses on congestion. This model is methodologically extended in [31].

There are many studies that investigate aspects of the potential of demand side management in Europe, and most of them conclude that DSM can contribute significantly to a reliable renewable energy supply, but they only provide rough estimates of the benefit [32,33]. However, we are not aware of any studies that show the overall impact of DSM on the need for backup energy by a large scale integration into a fully renewable European power system in a systematic way as we did in this paper.

In this paper, the novel framework described in [34] is used to implement DSM into a model of a fully renewable European power system with country-level resolution. All of the DSM potentials of a single country are treated as one large storage-equivalent with time dependent constraints. Weather data is used to model feed-in from the renewable sources of wind and PV. Together with historical load data and a power flow model, the need for backup energy was calculated and the impact of DSM was investigated.

We do not include hydropower into the simulations, although it already contributes approximately 10% to the European electricity mix today. This is because European hydropower, with its seasonal storage characteristics, would likely be used after DSM with its daily storage characteristics, and, therefore, has little effect on the results, but instead replaces a large share of the need for backup energy.

This paper focuses on the following objectives: (i) What is the potential of DSM to reduce the need for backup energy in a fully renewable European power system? (ii) Until which amount of renewable generation can DSM replace backup? and (iii) How does the successful integration of DSM affect the optimal mix of wind and PV generation?

This paper is structured in the following way: first, the model is described in detail. This includes a description of the main components renewable generation and load, the transmission model, and the incorporation of demand side management. Second, the potential benefit of DSM for two scenarios of generation capacity distribution and two scenarios of transmission grid strength is investigated. Third, the impact of the full utilisation of DSM on the optimal share of wind and PV is studied. For this, all generation, load, and DSM potentials are aggregated into a single European node (copper plate approximation). Finally, the potential reduction of the need for backup energy in dependency of the renewable share is investigated.

2. Model Description

A highly renewable European power system covering 33 countries was simulated. Every country was aggregated into a single node, and the countries are interconnected via transmission links (Figure 1). Every node n has a generation time series $G_n(t)$ from the renewable sources of wind and PV, derived from weather data, and a load time series $L_n(t)$, which consists of historical data. We used ten years of data ranging from 2003–2012 for all of the following computations. The time series of the mismatch between generation and load of country n is given by

$$\Delta_n(t) = G_n(t) - L_n(t) \tag{1}$$

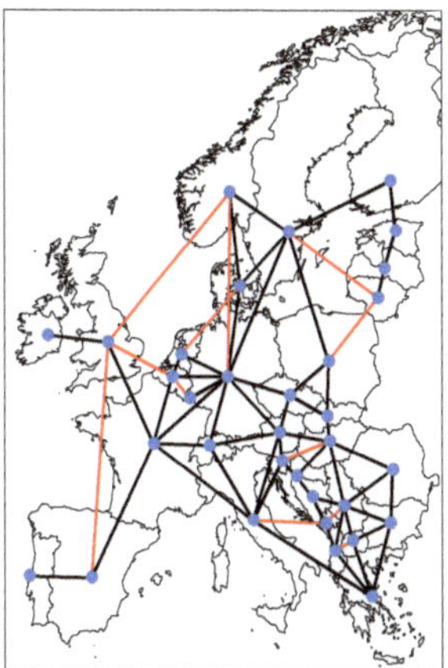

Figure 1. Topology of the investigated simplified European power system. Countries are modeled as nodes and connected by inter-country transmission links. **Black** links are existing connections, and **red** ones are either planned or under construction.

At each node and at all times, the power system must be balanced. This is expressed in the nodal balancing equation

$$G_n(t) - L_n(t) = \Phi_n(t) - B_n(t) + C_n(t) + S_n(t) \tag{2}$$

$B_n(t)$ is the time series of backup, $C_n(t)$ is the excess energy that is curtailed and $\Phi_n(t)$ is the injection pattern (Exports-Imports). $S_n(t)$ is the interaction (charge/discharge) with the storage-equivalent DSM. After transmission and DSM, the remaining residual mismatch is handled by backup, which is assumed to be perfectly flexible, i.e., neither subject to ramping nor must-run constraints. Thus, the backup time series is calculated as

$$B_n(t) = \max\left(\{0, L_n(t) - G_n(t) + \Phi_n(t) + S_n(t)\}\right) \tag{3}$$

Consequently, the backup energy need in a given period of time T is given by

$$B_n^E = \int_T B_n(t)dt \tag{4}$$

In reality, backup energy could, for example, be provided by dispatchable gas power plants. The time series for curtailment is given by

$$C_n(t) = \max\left(\{0, G_n(t) - L_n(t) - \Phi_n(t) - S_n(t)\}\right) \tag{5}$$

Hence, either backup (if $\Delta_n(t) < 0$) or curtailment ($\Delta_n(t) > 0$) occurs at a node n. For example, curtailment can be realised by feathering wind turbine blades. The share of renewable generation of a node is denoted as α_n and is defined via

$$\alpha_n = \frac{\langle G_n(t)\rangle}{\langle L_n(t)\rangle} \tag{6}$$

Equivalently, the share of renewable generation of the whole system consisting of multiple nodes is given by

$$\alpha = \sum_n \alpha_n \frac{\langle L_n\rangle}{\langle L\rangle} \tag{7}$$

Throughout this paper, the terminology "dispatchable generation" refers to power from sources that could be dispatched (e.g., gas). "Backup" refers to the fraction of the dispatchable generation that is needed to cover the intermittency of the renewable generation. For example, if $\alpha = 0.7$, we have an average renewable share of 70% and at least 30% from dispatchable sources.. The part of dispatchable generation that is needed in addition is referred to as "backup" or "backup energy".

2.1. Generation and Load Data

Feed-in from the renewable sources of wind and PV was simulated using a ten-year weather database with a spatial resolution of 7 × 7 km and an hourly temporal resolution. Wind speed (and 2 m temperature) was downscaled from MERRA reanalysis [35] and converted to wind power through the use of an Enercon E-126 power curve with 5% plain losses. Surface irradiance was calculated using the Heliosat method [36,37] from satellite pictures (Meteosat First Generation, Meteosat Second Generation). To obtain irradiation on the tilted modules, the Klucher model was applied [38]. Detailed information on the database is given in [39]. Finally, generation was aggregated from the 7 × 7 km grid to the country level.

For load time series of all considered European countries, historical data provided by the *European Network of Transmission System Operators for Electricity* (ENTSO-E) was used. This data was split into different load categories and modified to account for expected future changes caused by the increased use of heat pumps and e-mobility within the RESTORE 2050 project.

2.2. Demand Side Management

To incorporate demand side management into our model, the methodology of [34] is adopted and described in this section. Equations (8)–(15) are taken from [34]. In addition, a simple example is given in that paper. In this methodology, DSM is treated like storage with time dependent charging and energy constraints. The load time series can be split into different categories, i.e., it is composed of load time series of different categories $L_n^c(t)$, $L_n(t) = \sum_c L_n^c(t) + L_n^{\text{stat.}}(t)$. This load is referred to in the following as scheduled load. $L_n^{\text{stat.}}$ refers to the part of the load that can not be shifted. It implicitly contains the DSM utilisation shares of different DSM categories, which are given in Table 1. Without DSM, each country would simply have one (scheduled) load time series $L_n(t)$.

DSM allows for replacing a scheduled load $L_n^c(t)$ by a realized load $R_n^c(t)$. The difference is the charging or discharging rate of the storage-equivalent DSM buffer

$$P_n^c[R_n^c(t)](t) = R_n^c(t) - L_n^c(t) \tag{8}$$

The square brackets indicate that it takes a function, i.e., the realized load, as an argument. Because we assume no impact of DSM usage on overall energy consumption, a DSM storage filling level can be calculated as the temporal integral over the charging rate

$$E_n^c[R_n^c(t)](t) = \int_0^t P_n^c[R_n^c(t')](t')dt' \tag{9}$$

This storage-equivalent differs from a classical storage (e.g., a battery system or pumped hydro plant) by the time dependency of its filling level and charging limits. These time dependent constraints of charging and filling level of the DSM buffer are defined as

$$E_n^{c,+}(t) = \int_t^{t+\Delta t^c} L_n^c(t')dt' \tag{10}$$

$$E_n^{c,-}(t) = -\int_{t-\Delta t^c}^{t} L_n^c(t')dt' \tag{11}$$

$$P_n^{c,+}(t) = \Lambda_n^c(t) - L_n^c(t) \tag{12}$$

$$P_n^{c,-}(t) = -L_n^c(t) \tag{13}$$

Upper limits are indicated by the index $^+$ and lower limits indicated by the index $^-$. Δt^c is the time frame of management of a category and Λ_n^c the maximal realisable load. All realised loads within the imposed constraints are valid:

$$E_n^{c,+}(t) \geq E[R_n^c(t)](t) \geq E_n^{c,-}(t) \tag{14}$$

$$P_n^{c,+}(t) \geq P[R_n^c(t)](t) \geq P_n^{c,-}(t) \tag{15}$$

Time series of the DSM constraints of different categories for the different countries were developed within RESTORE 2050 and are described in [40]. Five categories were defined with individual time frames of management and utilisation shares (Table 1). Utilisation shares determine what share of a category is available for DSM.

Table 1. Load categories defined for demand side management [40].

Category	Δt^c (h)	Utilisation (%)
Industrial bandload	4	25
Cooling	1	12
Households	12	10
Heat pumps	24	100
E mobility	6	80

How are the charging rates and thus the usage of DSM determined in our model? Since rescheduling of loads is assumed to leave the total energy demand unchanged, DSM does neither cause losses or gains. Hence, the simple assumption is made that local excess energy after transmission is used to charge and local deficits used to discharge the DSM buffer within the constraints. However, in a real-world power market, backup and curtailment might be a more convenient option than DSM. Then again, this study aims at the theoretical potential of DSM. Whether it can be fully exploited, depends strongly on the market conditions.

The algorithm to distribute the charging/discharging among categories consists of the following five steps for time t and node n:

(i) Compute the ratio of the power limit to remaining energy storage for each category

$$\text{ratio} = \begin{cases} \frac{P_n^{c,-}(t)}{E_n^c(t)-E_n^{c,-}}, \text{ if } (G_n(t) - L_n(t) - \Phi_n(t)) < 0 \\ \frac{P_n^{c,+}(t)}{E_n^{c,+}-E_n^c(t)}, \text{ if } (G_n(t) - L_n(t) - \Phi_n(t)) > 0 \end{cases} \tag{16}$$

(ii) Compute the DSM charging rate $P_n^c(t)$ of category c with the lowest ratio

$$\tilde{P}_n^c(t) = G_n(t) - L_n(t) - \Phi_n(t) - \tilde{S}_n^c(t) \tag{17}$$

$$P_n^c(t) = \min\left\{P_n^{c,+}(t), \max\left\{P_n^{c,-}(t), \tilde{P}_n^c(t)\right\}\right\} \tag{18}$$

$\tilde{S}_n^c = \sum_{c'} P_n^{c'}$ is the sum of charging rates of all categories with a lower ratio than c. For the category with the lowest ratio, it equals zero.

(iii) Compute the storage filling level of category c for the next time step

$$E_n^c(t+\Delta t) = \begin{cases} \min\left\{E_n^c(t) + P_n^c(t)\Delta t, E_n^{c,+}(t+\Delta t)\right\}, \text{if } P_n^c(t) \geq 0 \\ \max\left\{E_n^c(t) + P_n^c(t)\Delta t, E_n^{c,-}(t+\Delta t)\right\}, \text{if } P_n^c(t) < 0 \end{cases} \tag{19}$$

(iv) Repeat steps (ii) and (iii) for the category with the next lowest ratio until the storage filling level for all categories was computed.

(v) Finally, compute the total interaction with the DSM storage at node n, $S_n(t)$, via

$$S_n(t) = \sum_n P_n^c(t)$$

Thus, charging and discharging rates are computed for the DSM categories in time dependent ascending order of the ratios of power capacity to remaining energy capacity. However, we believe that the choice of the distribution of charging among categories has little effect on the results because loads of all categories can only be shifted by up to one day. This section has summarised the most relevant details of the DSM approach from [34,40].

Figure 2 exemplarily shows the usage of the DSM in Germany for three days. Excess energy is stored in the DSM buffer, whereas deficits are covered from it. In the evening of the second day shown, the maximum of the energy capacity is reached. In the following hours, the energy capacity maximum is reduced, thereby forcing the DSM storage to discharge. Furthermore, the minimum charging rates are reached on day one and two.

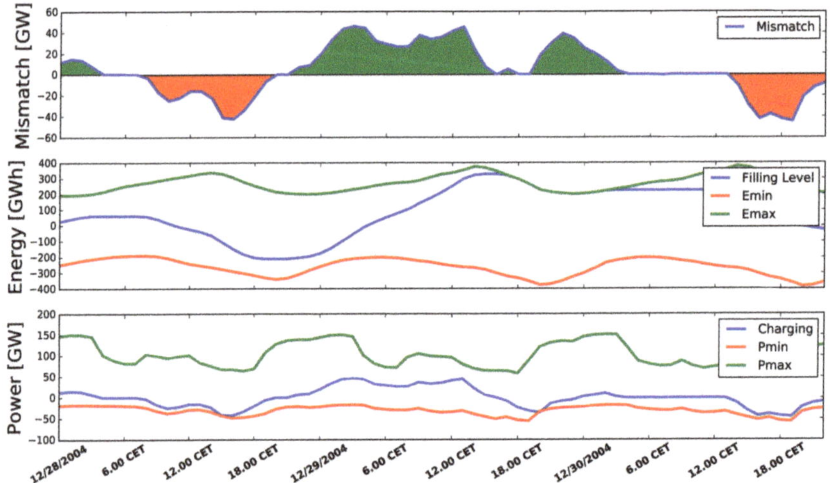

Figure 2. Mismatch (renewable generation-load-injection pattern), DSM filling level, and DSM charging rates for three exemplary days in Germany.

2.3. Transmission

Countries in our model are connected via inter-country transmission links. Hence, nodes can exchange excess energy and partially balance their mismatches. For transmission, the equations of a full electric power-flow in an alternating current (AC) electricity network are used in a common linear approximation [41] (occasionally referred to as DC approximation because the structure of the obtained equations is similar). Transmission is used prior to DSM, backup and curtailment and formulated as an optimisation problem consisting of two steps. The first step minimises the overall need for backup energy and the second step the dissipation by transmission. This can be interpreted as a cost-optimal dispatch strategy, if all nodes are assumed to have no limits on dispatchable generation and the same marginal cost, i.e., the same conventional generation technology is serving the load that cannot be supplied by renewable generation at any node in the system. The first step reads

$$\underset{\Phi(t)}{\text{minimise}} \quad \sum_n B_n(t) =: B^{\min}(t) \tag{20}$$

$$\text{subject to} \quad \sum_n \Phi_n(t) = 0 \tag{21}$$

$$F_l^- < \left[K^T L^+ \Phi(t)\right]_l \leq F_l^+ \tag{22}$$

where L^+ is the Moore–Penrose pseudo-inverse of the Laplacian, and $F_l^{\pm}$ are the limits imposed on the flow of link l in both direction, which can, for example, be thermal limits.

The resulting need for backup energy is fixed for the second step. A second step is necessary because the solution Φ is generally not unique. This second step ensures the uniquety of the solution by minimising the dissipation of the flows ($\propto F^2$ in a resistor network). It reads

$$\underset{\Phi(t)}{\text{minimise}} \quad \sum_l \left[K^T L^+ \Phi(t)\right]_l^2 \tag{23}$$

$$\text{subject to} \quad \sum_n \Phi_n(t) = 0 \tag{24}$$

$$F_l^- < \left[K^T L^+ \Phi(t)\right]_l \leq F_l^+ \tag{25}$$

$$\sum_n B_n(t) = B^{\min}(t) \tag{26}$$

The result is the injection pattern $\Phi(t)$ as the unique solution of the optimisation problem. The incidence matrix K is defined as

$$K_{nl} = \begin{cases} 1 & \text{if link l begins at node n} \\ -1 & \text{if link l ends at node n} \\ 0 & \text{otherwise} \end{cases} \tag{27}$$

and the Laplace Matrix L is given by

$$L_{nm} = \begin{cases} -1 & \text{if node } m \text{ and } n \text{ are connected by a link} \\ \deg(v_n) & \text{if } n = m \\ 0 & \text{otherwise} \end{cases} \tag{28}$$

If the injection pattern $\Phi(t)$ is known, the flows can be computed via

$$F = K^T L^+ \Phi \tag{29}$$

$K^T L^+$ is often referred to as the PTDF (Power Transfer Distribution Factors) matrix. This transmission methodology is described in more detail in [42]. An equivalent formulation is used in [11,14,43–46].

3. The Impact of DSM on Backup Energy Need

We quantified the possible reduction of the need for backup energy in a fully renewable Europe ($\alpha = 1.0$) for two scenarios of transmission, which we refer to as *vision 2030* (Vis.) and *unlimited* (Unl.), and two scenarios of capacity distribution, entitled *homogeneous* (Hom.) and *inhomogeneous* (Inh.). For *unlimited* transmission, no limits are imposed on transmission links (Equations (22) and (25)).

Vision 2030 refers to the capacities as envisioned by ENTSO-E for 2030 [47]. Hence, these capacities are assigned as the imposed link constraints $F_l^{\pm}$. In addition, symmetry is assumed: if $F_l^+ \neq F_l^-$ in [47], both are set to $F_l^{\pm} = \max\left\{F_l^+, F_l^-\right\}$.

The scenarios of capacity distribution differ by the distribution of the shares of renewables. In both cases, the mix of wind and PV generation capacities for each country is adopted from [48]. In the homogeneous scenario, however, each country covers its own load on average (i.e., $\alpha_n = 1$), whereas in the inhomogeneous scenario, the ratio of generation capacities to load for each country are also taken from [48]. Shares of renewables from all countries are depicted in Figure 3. High shares of renewables can be observed in the *inhomogeneous* scenario for countries on the shores of the North Sea, such as Denmark or Great Britain.

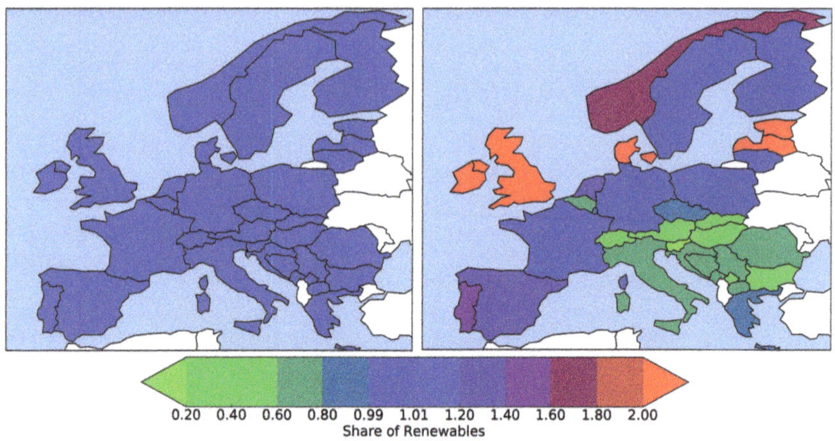

Figure 3. Shares of renewables (α_n, average renewable generation over consumption) for single countries in the homogeneous (**left**) and inhomogeneous (**right**) scenarios.

Figure 4 shows the need for backup energy and the possible reduction by DSM for all four scenarios. The complete bar shows the respective need for backup energy without DSM (e.g., ca. 16% of the consumption in the case of *homogeneous* capacity distribution with *unlimited* transmission capacities). The blue component of each bar shows the corresponding reduction of the backup energy need by fully utilised DSM. If transmission is unlimited, backup energy need equals approximately 15%–16% of the total consumption in both scenarios. However, because DSM is not interacting with the inter-country transmission system, its potential is strongly reduced in the case of an inhomogeneous capacity distribution. In the inhomogeneous scenario, the need for backup energy can be reduced by only 15% compared to ca. one third in the homogeneous scenario because some countries produce more than they consume on average, and others produce less. Those with overproduction have a DSM buffer, which is full most of the time; those with little production relative to their consumption have a DSM buffer that is mostly empty. Thus, the uneven distribution of surpluses and deficits makes the usage of DSM less optimal in the *inhomogeneous* scenario.

If the transmission capacities are limited (*vision 2030*), two situations can be observed: first, the overall need for backup is increased by 45% in the *homogeneous* scenario and by 75% in the *inhomogeneous*. The reduction of the backup energy need grows as well (*homogeneous*: +42%, *inhomogeneous*: +44%), but not as much as the need for backup energy. Especially in the *inhomogeneous* scenario, transmission limitations hamper energy exchange to account for partial mismatches and thereby intensify the contrast between exporters and importers.

The frequency distribution of DSM energy filling levels (Figure 5) shows that DSM buffer is negative on average for all four scenarios. This is likely due to the higher number of hours with a negative mismatch (54% of hours for both capacity distributions). In addition, high levels of DSM storage filling of more than 1500 GWh are never reached, if capacities are distributed *inhomogeneously*. This is caused by a large proportion of countries generating small amounts of renewable energy in this case, which is not sufficient to fill the buffer. On the lowest end of the scale, scenarios do not differ significantly. Below −500 GWh, all scenarios show similar frequency distributions, which is likely due to the same occurrence of lasting periods without significant renewable generation. In these periods, neither the generation capacity distribution nor the transmission grid matters.

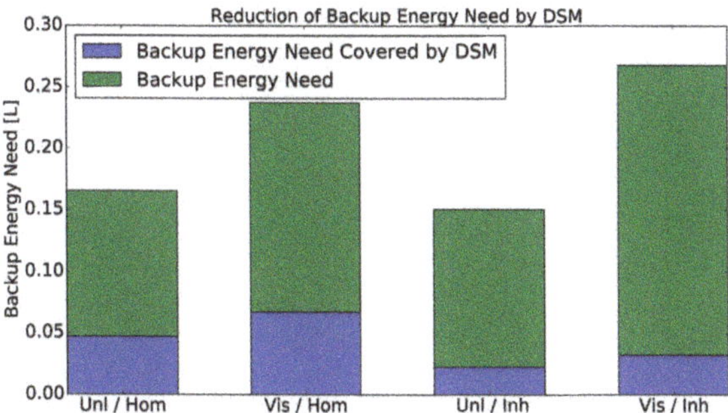

Figure 4. Need for backup energy and possible reduction by DSM. Entire bars show the need for backup energy without DSM. **Blue** components show the possible reduction by DSM. The backup energy need is measured in units of the overall consumption.

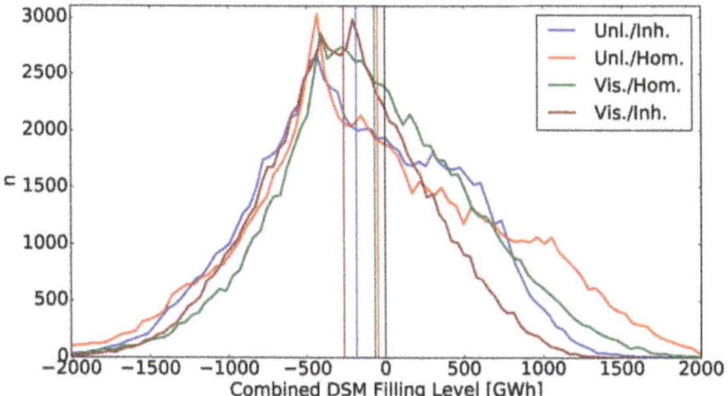

Figure 5. Frequency distribution of combined DSM storage filling levels. Vertical lines indicate the averages.

4. Influence of DSM on the Optimal Mix of Wind and PV

The optimal mix of different renewable sources in Europe has been investigated in different papers. In [3], the seasonal optimal mix is specified to be 40% PV and 60% wind with respect to the monthly standard deviation, and, in [5], it is calculated at 20% PV and 80% wind with respect to the backup energy need (note: we use the same generation and similar load data as [5]).

To calculate the optimal mix of wind and PV with and without DSM, we make major simplifications within this section: first, Europe is treated as a copper plate with homogeneous capacity distribution having only one time series of generation $G(t) = \sum_n G_n(t)$ and load $L(t) = \sum_n L_n(t)$. Second, DSM is treated like one large storage equivalent with time dependent constraints:

$$E^+(t) = \sum_{n,c} E_n^{c,+}(t) \tag{30}$$

$$E^-(t) = \sum_{n,c} E_n^{c,-}(t) \tag{31}$$

$$P^+(t) = \sum_{n,c} P_n^{c,+}(t) \tag{32}$$

$$P^-(t) = \sum_{n,c} P_n^{c,-}(t) \tag{33}$$

This simplification might violate the constraints imposed in Equations (10)–(13) and can therefore not be fully justified by assuming unlimited transmission between nodes and the possibility of exporting/importing energy to charge/discharge DSM storages at a different node. However, violations should rarely, if ever, occur, and the results can be interpreted as an upper limit of the potential benefit.

Figure 6 shows the need for backup energy with and without DSM. Without DSM, the optimal European mix with respect to the need for backup energy is 19% PV and 81% wind. If DSM is fully utilised, this changes to an optimal mix of 36% PV and 64% wind. PV profits much more from DSM than wind. If the entire generation side is comprised of PV power, DSM can reduce the need for backup by ca. 40%, whereas for a wind-only scenario, the possible reduction of the backup energy need is below 20%. The reason for this is straightforward: Compared to wind, PV power has a deterministic diurnal cycle and therefore uses the DSM storage, which has properties that can be characterized as "daily storage" due to limited storage reservoir capacity more efficiently than wind. The more efficient usage of PV is also partly reflected by the seasonal variability of feed-in being the least if the solar generation share is higher than the wind share [49]. The red curve in Figure 6 shows the difference between the backup energy need with and without fully utilised DSM and thus the potential benefit from DSM. Three phases can be seen: first, it remains flat to a solar share of 20%. Second, it increases steadily up to a solar share of ca. 60%, and, third, it remains nearly unaltered up to a solar share of 100%. Above a solar share of 60%, the potential for DSM to reduce the need for backup energy is fully exploited.

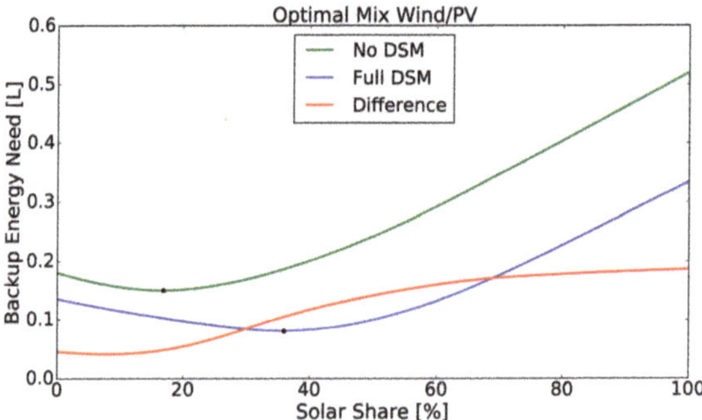

Figure 6. Backup energy need in dependency of the mix of wind and PV in Europe. Europe is treated like a copper plate. **Black** dots indicate the minima. The **red** line shows the difference between both curves. The backup energy need is measured in units of the overall consumption.

5. DSM Potential vs. Share of Renewables

The last investigated question of this paper is: until which share of renewables can DSM virtually cover the whole need for backup energy? We assume that the remaining share $1-\alpha$ is covered by a perfectly flexible dispatchable generation. This means that the dispatchable generation in our model has no must-run or ramping constraints. This question is important because the share of renewables in the European electricity mix has barely reached 25%, but it continues growing and is expected to reach values of more than 80% by 2050.

The need for backup energy in dependency of the share of renewables α is shown in Figure 7. Generation capacities are distributed *homogeneously* ($\alpha_n = \alpha \forall n$) to reduce the need for backup energy among countries, and transmission is assumed to be *unlimited* between nodes. However, DSM is only used, like backup, locally after transmission and cannot be exchanged between countries. This means, for example, that after transmission, excess energy at node a can not be transferred to node b to charge the DSM buffer there. The inter-country transmission system is solely used to cover residual loads and does not interact with DSM. This can be justified because planned reinforcement measures of inter-country transmission links focus on the exchange of renewable energy and not on backup or storage energy. For up to a share of renewables of 67%, all need for backup energy can be covered by DSM. Without DSM, the need for backup energy at a renewable share of 65% equals 2% of the yearly consumption (approximately 70 TWh).

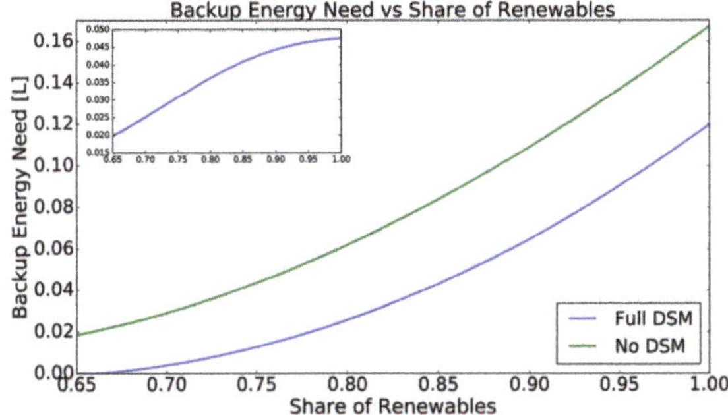

Figure 7. Need for backup energy in dependency of the renewable share in a European power system. Transmission capacities are assumed to be unlimited. The enclosed figure shows the difference between the two curves. The backup energy need is measured in units of the overall consumption.

Figure 7 (subfigure) also shows the reduction in backup energy need by DSM in units of the consumption. For $\alpha = 0.65$, ca. 2% relative to the consumption could be replaced. This value continues to increase steadily over the interval $\alpha \in [0.65, 1.0]$ and reaches its maximum at 4.8% for $\alpha = 1$.

6. Conclusions

We have simulated a simplified highly renewable European power system to investigate the possible impact of demand side management on the need for backup energy. We have defined two scenarios of capacity distribution. In the *homogeneous* scenario, each country on average produces as much from renewable sources as it consumes, and, in the *inhomogeneous* scenario, installed capacities are distributed unevenly among European countries. First, it was shown that DSM can reduce the need for backup energy in a fully renewable European power system in dependency of the scenario by up to one third. Second, the optimal mix of wind and PV was found to be shifted from 19% PV, and 81% wind without DSM, to 36% PV and 64 % wind, if DSM is fully utilised. The beneficial interaction of wind and PV is also reflected by the fact that a wind-only Europe could reduce its backup need through the use of DSM by merely 20%, whereas this number doubles in a PV-only scenario. Therefore, it can be concluded that the importance and economic performance of PV can be substantially increased, if the role of DSM within the energy market becomes vital.

It was also shown that DSM can theoretically cover all needs for backup energy up to a renewable share of ca. 67%. This is still far from the current share of renewables in the European electricity mix of around 25%. It is comparable to the reduction by a lossless storage in a fully renewable copper plate Europe with a storage size equal to three hours of the average load (ca. 1.2 TWh) [50]. This raises the question of what the interplay of DSM with its "daily storage" characteristics and European hydropower, with its large storage reservoir capacities and the seasonal dependency of its natural inflow, would look like. We believe that this is an interesting extension of this existing work.

DSM was shown to be an appropriate means of compensating the variable nature of renewable power sources such as wind and photovoltaics. Therefore, we conclude that demand side management has the potential to contribute significantly to issues that arise with the energy transition, which are currently in sight. This requires the implementation of a proper market environment and an appropriate understanding of risks and benefits by industry and policy makers.

Acknowledgments: The work is part of the RESTORE 2050 project (Wuppertal Institute, Next Energy, University of Oldenburg) that is financed by the Federal Ministry of Education and Research (BMBF, Fkz. 03SFF0439A). The work was also supported by the Ministry for Science and Culture of Lower Saxony (Hanover, Germany) in the 'ventus efficiens' project. We thank our project partners for helpful discussions and the provision of load and DSM data. Furthermore, we thank Martin Greiner (Aarhus) for helpful suggestions. We thank the editor and two anonymous reviewers for their constructive comments and suggestions, which helped us to improve the manuscript.

Author Contributions: All authors have discussed the findings, contributed to the writing of this article and approved the final manuscript.

Conflicts of Interest: The authors declare no conflict of interest.

Abbreviations

$n \in N$	European country/node of the network
$G_n(t)$	generation time series from renewable sources of wind and PV of country n
$L_n(t)$	scheduled load time series of country n
$\Delta_n(t)$	generation-load mismatch time series of country n
$\Phi_n(t)$	injection pattern time series of country n
$B_n(t)$	backup time series of country n
$C_n(t)$	curtailment time series of country n
$S_n(t)$	DSM charging/discharging time series of country n
$P_n^c(t)$	DSM charging rate of country n and load category c
B_n^E	backup energy need of country n
α_n	share of renewables of country n
$R_n^c(t)$	realised load time series of country n
$E_n^{c,\pm}(t)$	time series of DSM storage energy limits
$P_n^{c,\pm}(t)$	time series of DSM power limits
$E_n^c(t)$	time series of DSM energy filling level
K	incidence matrix of the network
L	Laplace matrix of the network
$F_l^{\pm}$	link transmission constraints in both directions

References

1. Lund, H. Large-scale integration of wind power into different energy systems. *Energy* **2005**, *30*, 2402–2412.
2. Lund, H. Large-scale integration of optimal combinations of PV, wind and wave power into the electricity supply. *Renew. Energy* **2006**, *31*, 503–515.
3. Heide, D.; von Bremen, L.; Greiner, M.; Hoffmann, C.; Speckmann, M.; Bofinger, S. Seasonal optimal mix of wind and solar power in a future, highly renewable Europe. *Renew. Energy* **2010**, *35*, 2483–2489.
4. François, B.; Hingray, B.; Raynaud, D.; Borga, M.; Creutin, J. Increasing climate-related-energy penetration by integrating run-of-the river hydropower to wind/solar mix. *Renew. Energy* **2016**, *87*, 686–696.
5. Kies, A.; Nag, K.; von Bremen, L.; Lorenz, E.; Heinemann, D. Investigation of balancing effects in long term renewable energy feed-in with respect to the transmission grid. *Adv. Sci. Res.* **2015**, *12*, 91–95.
6. Santos-Alamillos, F.; Pozo-Vázquez, D.; Ruiz-Arias, J.; Von Bremen, L.; Tovar-Pescador, J. Combining wind farms with concentrating solar plants to provide stable renewable power. *Renew. Energy* **2015**, *76*, 539–550.

7. Tafarte, P.; Das, S.; Eichhorn, M.; Thrän, D. Small adaptations, big impacts: Options for an optimized mix of variable renewable energy sources. *Energy* **2014**, *72*, 80–92.
8. Kies, A.; Schyska, B.U.; von Bremen, L. The optimal share of wave power in a highly renewable power system on the Iberian Peninsula. *Energy Rep.* **2016**, *2*, 221–228.
9. Weitemeyer, S.; Kleinhans, D.; Vogt, T.; Agert, C. Integration of renewable energy sources in future power systems: The role of storage. *Renew. Energy* **2015**, *75*, 14–20.
10. Weitemeyer, S.; Kleinhans, D.; Wienholt, L.; Vogt, T.; Agert, C. A European perspective: Potential of grid and storage for balancing renewable power systems. *Energy Technol.* **2015**, *4*, 114–122.
11. Heide, D.; Greiner, M.; von Bremen, L.; Hoffmann, C. Reduced storage and balancing needs in a fully renewable European power system with excess wind and solar power generation. *Renew. Energy* **2011**, *36*, 2515–2523.
12. Huber, M.; Dimkova, D.; Hamacher, T. Integration of wind and solar power in Europe: Assessment of flexibility requirements. *Energy* **2014**, *69*, 236–246.
13. Schlachtberger, D.; Becker, S.; Schramm, S.; Greiner, M. Backup flexibility classes in emerging large-scale renewable electricity systems. *Energy Convers. Manag.* **2016**, *125*, 336–346.
14. Becker, S.; Rodriguez, R.; Andresen, G.; Schramm, S.; Greiner, M. Transmission grid extensions during the build-up of a fully renewable Pan-European electricity supply. *Energy* **2014**, *64*, 404–418.
15. Brown, T. Transmission network loading in Europe with high shares of renewables. *IET Renew. Power Gener.* **2015**, *9*, 57–65.
16. Rabl, V.A.; Gellings, C.W. The concept of demand-side management. In *Demand-Side Management and Electricity End-Use Efficiency*; Springer: Heidelberg, Germany, 1988; pp. 99–112.
17. Strbac, G. Demand side management: Benefits and challenges. *Energy Policy* **2008**, *36*, 4419–4426.
18. Diamond, A.P. *Energy Glossary*; Nova Science Publishers: New York, NY, USA, 2002.
19. Atzeni, I.; Ordóñez, L.G.; Scutari, G.; Palomar, D.P.; Fonollosa, J.R. Demand-side management via distributed energy generation and storage optimization. *IEEE Trans. Smart Grid* **2013**, *4*, 866–876.
20. Stötzer, M.; Gronstedt, P.; Styczynski, Z. Demand side management potential-A case study for Germany. In Proceedings of the 21st International Conference on Electricity Distribution, Frankfurt, Germany, 6–9 June 2011.
21. Paulus, M.; Borggrefe, F. The potential of demand-side management in energy-intensive industries for electricity markets in Germany. *Appl. Energy* **2011**, *88*, 432–441.
22. Mesarić, P.; Krajcar, S. Home demand side management integrated with electric vehicles and renewable energy sources. *Energy Build.* **2015**, *108*, 1–9.
23. Emec, S.; Kuschke, M.; Chemnitz, M.; Strunz, K. Potential for demand side management in automotive manufacturing. In Proceedings of the 2013 4th IEEE/PES Innovative Smart Grid Technologies Europe (ISGT EUROPE), Lyngby, Denmark, 6–9 October 2013; IEEE: Piscataway, NJ, USA; pp. 1–5.
24. Klobasa, M. Analysis of demand response and wind integration in Germany's electricity market. *IET Renew. Power Gener.* **2010**, *4*, 55–63.
25. Lund, H.; Kempton, W. Integration of renewable energy into the transport and electricity sectors through V2G. *Energy Policy* **2008**, *36*, 3578–3587.
26. Stadler, I. Power grid balancing of energy systems with high renewable energy penetration by demand response. *Util. Policy* **2008**, *16*, 90–98.
27. Moura, P.S.; De Almeida, A.T. The role of demand-side management in the grid integration of wind power. *Appl. Energy* **2010**, *87*, 2581–2588.
28. Pollhammer, K.; Kupzog, F.; Gamauf, T.; Kremen, M. Modeling of demand side shifting potentials for smart power grids. In Proceedings of the 2011 AFRICON, Livingstone, Zambia, 13–15 September 2011; IEEE: Piscataway, NJ, USA, 2011; pp. 1–5.
29. Palensky, P.; Dietrich, D. Demand side management: Demand response, intelligent energy systems, and smart loads. *IEEE Trans. Ind. Inform.* **2011**, *7*, 381–388.
30. Göransson, L.; Goop, J.; Unger, T.; Odenberger, M.; Johnsson, F. Linkages between demand-side management and congestion in the European electricity transmission system. *Energy* **2014**, *69*, 860–872.
31. Zerrahn, A.; Schill, W.P. On the representation of demand-side management in power system models. *Energy* **2015**, *84*, 840–845.

32. Tröster, E.; Kuwahata, R.; Ackermann, T. *European Grid Study 2030/2050;* Greenpeace International: Langen, Germany, 18 January 2011.
33. Kohler, S.; Agricola, A.; Seidl, H. *Dena Grid Study IIe—Integration of Renewable Energy Sources in the German Power Supply System from 2015–2020 with an Outlook to 2025;* Deutsche Energie-Agentur (dena), German Energy Agency: Berlin, Germany, 2010.
34. Kleinhans, D. Towards a systematic characterization of the potential of demand side management. 2014, arXiv:1401.4121.
35. Rienecker, M.M.; Suarez, M.J.; Gelaro, R.; Todling, R.; Bacmeister, J.; Liu, E.; Bosilovich, M.G.; Schubert, S.D.; Takacs, L.; Kim, G.K.; et al. MERRA: NASA's modern-era retrospective analysis for research and applications. *J. Clim.* **2011**, *24*, 3624–3648.
36. Cano, D.; Monget, J.; Albuisson, M.; Guillard, H.; Regas, N.; Wald, L. A method for the determination of the global solar radiation from meteorological satellite data. *Sol. Energy* **1986**, *37*, 31–39.
37. Hammer, A.; Heinemann, D.; Westerhellweg, A. Derivation of daylight and solar irradiance data from satellite observations. In Proceedings of the 9th Conference on Satellite Meteorology and Oceanography, Paris, France, 25–29 May 1998; pp. 747–750.
38. Klucher, T. Evaluation of models to predict insolation on tilted surfaces. *Sol. Energy* **1979**, *23*, 111–114.
39. Kies, A.; Chattopadhyay, K.; von Bremen, L.; Lorenz, E.; Heinemann, D. Simulation of renewable feed-in for power system studies; In *RESTORE 2050 Project Report;* Technical Report; NEXT ENERGY-EWE-Forschungszentrum für Energietechnologie: Oldenburg, Germany, 2016.
40. Meyer, K.; Kleinhans, D. Arbeitspaket 5: Lastmanagement charakterisierung und quantifizierung des lastmanagementpotentials fuer Europa. In *Restore 2050;* Technical Report; NEXT ENERGY—EWE-Forschungszentrum für Energietechnologie: Oldenburg, Germany, 2015.
41. Oeding, D.; Oswald, B.R. *Elektrische Kraftwerke und Netze;* Springer: Berlin/Heidelberg, Germany, 2004; Volume 6.
42. Heide, D. Statistical Physics of Power Flows on Networks with a High Share of Fluctuating Renewable Generation. Ph.D. Thesis, Johann Wolfgang Goethe-Universitaet Frankfurt am Main, Frankfurt, Germany, 2010.
43. Rodriguez, R.A.; Becker, S.; Greiner, M. Cost-optimal design of a simplified, highly renewable pan-European electricity system. *Energy* **2015**, *83*, 658–668.
44. Kies, A.; von Bremen, L.; Chattopadhyay, K.; Lorenz, E.; Heinemann, D. Backup, storage and transmission estimates of a supra-European electricity grid with high shares of renewables. In Proceedings of the 14th Wind Integration Workshop, Brussels, Belgium, 20–22 October 2015.
45. Rodriguez, R.A.; Becker, S.; Andresen, G.B.; Heide, D.; Greiner, M. Transmission needs across a fully renewable European power system. *Renew. Energy* **2014**, *63*, 467–476.
46. Kies, A.; Schyska, B.; von Bremen, L. Curtailment in a highly renewable power system and its effect on capacity factors. *Energies* **2016**, *9*, 510.
47. European Network of Transmission System Operators for Electricity. *Ten-Year Network Development Plan 2016;* European Network of Transmission System Operators for Electricity: Brussels, Belgium, 2016.
48. Pfluger, B.; Sensfuß, F.; Schubert, G.; Leisentritt, J. *Tangible Ways towards Climate Protection in the European Union (EU Long-Term Scenarios 2050);* Fraunhofer Institut für System-und Innovationsforschung ISI: Karlsruhe, Germany, 2011.
49. Bett, P.E.; Thornton, H.E. The climatological relationships between wind and solar energy supply in Britain. *Renew. Energy* **2016**, *87*, 96–110.
50. Rasmussen, M.G.; Andresen, G.B.; Greiner, M. Storage and balancing synergies in a fully or highly renewable pan-European power system. *Energy Policy* **2012**, *51*, 642–651.

MDPI

Article

A Distributed Control Strategy for Frequency Regulation in Smart Grids Based on the Consensus Protocol

Rong Fu *, Yingjun Wu, Hailong Wang and Jun Xie

College of Automation, Nanjing University of Posts and Telecommunications, Nanjing 210023, China; ywu_njupt@163.com (Y.W.); hailongwang1988@126.com (H.W.); jxie@njupt.edu.cn (J.X.)

* Author to whom correspondence should be addressed; furong@njupt.edu.cn; Tel.: +86-25-85-866-500.

Academic Editor: William Holderbaum

Received: 29 April 2015; Accepted: 23 July 2015; Published: 31 July 2015

Abstract: This paper considers the problem of distributed frequency regulation based on the consensus control protocol in smart grids. In this problem, each system component is coordinated to collectively provide active power for the provision of ancillary frequency regulation service. Firstly, an approximate model is proposed for the frequency dynamic process. A distributed control algorithm is investigated, while each agent exchanges information with neighboring agents and performs behaviors based on communication interactions. The objective of each agent is to converge to a common state considering different dynamic load characteristics, and distributed frequency control strategy is developed to enable the agents to provide active power support. Then, the distributed proportional integral controllers with the state feedback are designed considering the consensus protocol with topology $\mathcal{G}$. The theory of distributed consensus protocol isfurther developed to prove the stability of the proposed control algorithm. Whenproperly controlled, the controllers can provide grid support services in a distributed manner that turn out the grid balanced globally. Finally, simulations of the proposed distributed control algorithm are tested to validate the availability of the proposed approach and the performance in the electrical networks.

Keywords: distributed control; consensus protocol; multi-agent system; dynamic loads; frequency regulation

1. Introduction

Inspired by the Smart Grid, electrical power systems are undergoing a global transformation in structure and functionality to increase efficiency and reliability. Such transformations are expanded by the introduction of new technologies such as advanced communication and control, integration of new flexible loads and new electricity generation sources. In smart electrical power networks, proper coordination and control of generation and load resources provide flexible frequency regulation services to enhance efficiency and reliability in smart grids. The distributed control strategies for coordination of distributed energy and load resources are proposed to provide active power for the provision of ancillary frequency regulations.

Traditionally, centralized frequency control is implemented and operated at different timescales in dispatching centering [1]. Automatic generation control (AGC) and governor control are adjusted to maintain the system frequency tightly around the nominal value when these distributed energy and load resources fluctuate uncertainly. The objective of the frequency controller is to keep the system frequency and the inter-area power transmission to the scheduled values during normal conditions, and when the system is subject to disturbances or sudden changes [2]. The primary frequency control operates at a timescale in minutes or so, and adjusts the operating points of governors in a centralized mode to drive the frequency back to its reasonable and secure value. Kothari *et al.* [3] have proposed an optimal PI controller by using area control error (ACE) stability control techniques. Malik *et al.* [4] have developed a generalized approach based on dual-mode discontinuous control and variable structure systems. Moon *et al.* [5] have devised a PID frequency controller to realize noise-tolerable differential control problems in power systems. Adaptive PI controllers are also proposed to regulate the power supply based on the self-tuning regulator. Yamashita *et al.* [6] have devised a method of designing a multi-variable self tuning-regulator for frequency problem on load demand. Khodabakhshian *et al.* [7] have proposed a new designed PID controller for automatic generation control in power systems.

Furthermore, decentralized control techniques have been used to deal with frequency control problems on the generation side. The robust decentralized controllers are designed independently, mainly based on the uses of a reduction model observer and a PI/PID controller. Yang *et al.* [8] have transformed the decentralized frequency controller design problem into an equivalent problem of controller design for a multi-port control system. Liu *et al.* [9] have proposed a new nonlinear constraint predictive control algorithm to guarantee the frequency dynamic stability. The design and operation of each local controller requires only its local states, and the errors between the outputs of two physical connected controllers are used to adaptively correct for the interactions from a global approximation model. Ilic *et al.* [10] have investigated a decentralized multi-agent frequency control system based on power communication technology. A robust load frequency controller is proposed to use genetic algorithms and linear matrix inequalities [11]. Nowadays, model free and data processing techniques also have been studied in control power generations to dump oscillations [12]. Some relative works are also proposed in literature [13] on the techniques of intelligent frequency control methods. It shows that decentralized control methods might provide efficient control with self-healing characters [14].

Recently, the distributed cooperative control manner for multi-agent systems has attracted increasing attention due to their flexibility and networked computational efficiency in many areas such as mobile robots, vehicle and traffic control. One kind of basic and challenging problem in distributed cooperative control is the consensus problem for multi-agent systems. The coordination and synchronization process necessitates that each agent could exchange information with neighboring agents according to some restricted communication protocols and distributed algorithms [15,16]. Ilic *et al.* [17] have proposed a fully distributed frequency control algorithm for electrical power systems. A push-sum algorithm is used to adapt to the demand [18], and a modified consensus algorithm including weights in the network is proposed in distributed control [19]. Andreasson *et al.* [20] have studied the consensus algorithm for frequency control considering agents with system dynamics. Zhao *et al.* [21] have designed continuous distributed load control for primary frequency regulation and the Lyapunov function method is used to prove the convergence of the analytic model. It shows that distributed control methods might provide efficient control with self-healing characters. If all the agents on a network converge to a common state, we could make a decision that the consensus problem has been solved and the common state is called the consensus state of the agents.

In smart grids, new hierarchical model of frequency adjustment and the distributed control techniques are proposed considering distributed communication protocols [2]. The cooperative frequency control strategy is executed to achieve a primary and secondary frequency recovery using the optimized average consensus algorithm. Furthermore, utilizing load side control is an appealing alternative to control the system frequency on the demand side, which can reduce the dependency of grids on the expensive generation side controllers. Remarkably, it emphasizes that such frequency adaptive loads would allow the system to accept more readily a stochastically fluctuating energy source. It can be seen that the proposed distributed control has possible benefits over centralized frequency control [22].

Compared with the traditional centralized power regulation strategies, the distributed controllers have the following features: (i) The pressure of communication becomes more distributed between various distributed controller devices; (ii) The distributed resources can take decisions collectively from the network to achieve better quality and efficiency. While there are many studies and discussions in frequency control, there is not much analytic study that relates the behavior of the distributed frequency controllers with the dynamic behavior of the loads in smart grids.

In this paper, we further focuses on the distributed primary frequency control on these energy resources and demand responses in smart grids. Considering the model of the power system dynamics by the swing equations, we apply and further develop the theory of distributed consensus to realize the stability of the proposed algorithm. The agents can reach an agreement on certain frequency deviations by sharing information locally with their neighbors. The analytic model and simulations exhibit that the proposed consensus protocol can attenuate the time-varying oscillations and lead the agents to steady state values for frequencies after disturbances.

The main structure and the content of this paper are organized as follows. In Section 2 (Consensus for Agents by Distributed Integral Action), we introduce the mathematical notation about the consensus protocol for the agents. In Section 3 (Network Model with Load Dynamics), we analyze the formulation of the frequency dynamic model and the consensus for agents. In Section 4 (Distributed Frequency Control Algorithm), based on the developed consensus protocol, we propose a distributed control algorithm for frequency control of electrical power systems, and compare the performance with traditional control algorithms. Simulation results of a two-area four-machine system using the proposed distributed control scheme are presented and discussed in Section 5 (Simulations). Finally, the conclusion of the distributed frequency control regulation is given in Section 6 (Conclusions).

2. Consensus for Agents by Distributed Integral Action

One basic and challenging problem in cooperative control is the consensus problem. It is assumed that there are multiple agents on a network. This network is usually modeled by a graph consisting of nodes (representing the agents) and edges (representing the interactions between agents). If all the agents on a network converge to a common state, we say that the multi-agent system solves a consensus problem or has a consensus property, and the common state is called group decision value or consensus state. This network is usually modeled by a graph consisting of nodes (representing the agents) and edges (representing the interactions between agents).

The proposed control architecture is illustrated including the distributed coordination frequency controller and the meshed electrical network describing the exchange of information among the multi-agents. For the application of frequency control in the electrical network, a connected and undirected graph $\mathcal{G}$ of order $n$is considered with the set of nodes $\mathcal{V} = \{1, ..., n\}$, set of edges $\mathcal{E} \subseteq \mathcal{V} \times \mathcal{V}$, and a weighted adjacency matrix $A = [a_{ij}]$ with nonnegative adjacency elements a_{ij}. In the undirected graph $\mathcal{G}$, the Laplacian matrix $\mathcal{L}$ holds that $\mathcal{L} = \mathcal{B}(\mathcal{G})\mathcal{B}^T(\mathcal{G})$, where $\mathcal{B}(\mathcal{G})$ means the vertex-edge adjacency matrix of $\mathcal{G}$. I_n denotes the identity matrix of dimension n.

Consider the agents with second-order dynamics:

$$\begin{cases} \dot{r}_i = v_i \\ \dot{v}_i = u_i \\ u_i = -\sum\limits_{j \epsilon \mathcal{N}_i} (\beta(r_i - r_j) + \alpha(v_i - v_j)) + d_i \end{cases} \tag{1}$$

where $r_i \in \mathbb{R}^n$ and $v_i \in \mathbb{R}^n$ are the position and velocity states of the ith agent(node), u_i is the system input,$\alpha \in \mathbb{R}^+$ and $\beta \in \mathbb{R}^+$ are fixed parameters, and $d_i \in \mathbb{R}$ is a disturbance. $\mathcal{N}_i$ denotes the set formed by all agent nodes connected to the node i.

Consider the linear coordinate change $z = \hat{S}^T v$,$w = \hat{S}^T r$, where$\hat{S} = \left[\frac{1}{\sqrt{n}} 1^{n\times 1}\ S\right]$, S is a matrix such that$\hat{S}$ is an orthonormal matrix. Thus the system dynamic (1) can be rewritten as:

$$\begin{gathered} \dot{w} = z \\ \dot{z} = \begin{bmatrix} 0 & 0_{1\times(n-1)} \\ 0_{(n-1)\times 1} & -\beta S^T L S \end{bmatrix} w + \begin{bmatrix} 0 & 0_{1\times(n-1)} \\ 0_{(n-1)\times 1} & -\alpha S^T L S \end{bmatrix} z + \begin{bmatrix} \frac{1}{n} 1_{1\times(n)} \\ S^T \end{bmatrix} d \end{gathered} \tag{2}$$

We can eliminate the uncontrollable state w_1 and z_1, thus obtaining the realization of the dynamic system by defining the new coordinates $w' = [w_2, \ldots, w_n]^T$ and $z' = [z_2, \ldots, z_n]^T$:

$$\begin{bmatrix} \dot{w}\prime \\ \dot{z}\prime \end{bmatrix} = \begin{bmatrix} 0_{(n-1)\times(n-1)} & I_{(n-1)} \\ -\beta S^T LS & -\alpha S^T LS \end{bmatrix} \begin{bmatrix} w\prime \\ z\prime \end{bmatrix} + \begin{bmatrix} 0_{(n-1)\times 1} \\ S^T d \end{bmatrix} \tag{3}$$

Since $S^T LS$ is invertible, the states w'' and $z\prime\prime$ can be defined to consider the disturbance matrix transform.

$$\begin{bmatrix} w\prime\prime \\ z\prime\prime \end{bmatrix} = \begin{bmatrix} w\prime \\ z\prime \end{bmatrix} - \begin{bmatrix} 0_{(n-1)\times 1} \\ \frac{1}{\alpha}(S^T LS)^{-1} S^T d \end{bmatrix} \tag{4}$$

so in the new coordinates the system dynamics become:

$$\begin{bmatrix} \dot{w}\prime\prime \\ \dot{z}\prime\prime \end{bmatrix} = \underbrace{\begin{bmatrix} 0_{(n-1)\times(n-1)} & I_{(n-1)} \\ -\beta S^T LS & -\alpha S^T LS \end{bmatrix}}_{\triangleq A''} \begin{bmatrix} w\prime\prime \\ z\prime\prime \end{bmatrix} \tag{5}$$

The characteristic polynomial of A'' is given by $\det(\rho^2 I_{(n-1)} + (\alpha\rho + \beta) S^T LS)$.Compared with the characteristic polynomial $\det(sI + S^T LS)$, we note that the eigenvalues satisfy with solutions $-s_i < 0$ by lemma 10 in [23]. Since $S^T LS$ is full-rank, we obtain that the eigenvalues of A'' could satisfydet$(\rho^2 + \alpha\, \rho s_i + \beta s_i) = 0$ with solutions $\rho \in \mathbb{C}^-$ by the Routh-Hurwitz stability criterion. Using the coordinates' shifts, it shows that the agents can converge to a common state and the consensus is reached for any $\alpha, \beta \in \mathbb{R}^+$.

3. Network Model with Load Dynamics

In the smart grid, the power system can be modeled by a graph $\mathcal{G} = (\mathcal{V},)$. There are two typical kinds of buses in the network: generator buses and load buses. The generator buses can convert the mechanical power into electric power and transmit them along the network. Then, the frequency dynamics on an ith synchronous generator can be modeled as follows:

$$\begin{cases} \dot{\delta}_i = \omega_i - \omega_{ref} \\ T_i \dot{\omega}_i = P_{mi} - P_{ei} - P_{di} - D_i \omega_i + u_i \end{cases} \quad \forall i \in \mathcal{V} \tag{6}$$

where δ_i is the phase angle of bus i, ω_i is the angular velocity of bus i, T_i and D_i are the inertia and damping coefficient, P_{mi} is the power injection at bus i, P_{ei} is the outputactive power of the generator i, P_{di} is the load at bus i, u_i is the mechanical input from frequency controller i. Let P_{mi}^0, P_{ei}^0, P_{di}^0 denote the initial uncontrolled operating point where$P_{mi}^0 - P_{ei}^0 - P_{di}^0 - D_i \omega_i^0 = 0$.

In general, load dynamics may diverge with the bus voltage magnitude (which is assumed fixed) and frequency. We distinguish between three types of loads, static controllable loads, frequency sensitive dynamic loads and uncontrollable loads. We assume that the frequency sensitive dynamic loads may increase linearly with frequency oscillations, and model these loads by $P_{di}(t) = P_{di}^0 + \Delta P_{di}(t) = P_{di}^0 + K_{di} \Delta\omega_i$, where K_{di} represents the load consumption due to the frequency deviation. Considering $P_{mi}(t) = P_{mi}^0 + \Delta P_{mi}(t)$, $P_{ei}(t) = P_{ei}^0 + \Delta P_{ei}(t)$, the deviation $\Delta P_{ei}(t)$ from the adjacent branch flows follows the linearized dynamic $\Delta P_{ei}(t) = \sum_{j \epsilon \mathcal{N}_i} B_{ij} cos\left(\Delta\delta_i^0 - \Delta\delta_j^0\right)(\Delta\delta_i - \Delta\delta_j)$, where $B_{ij} = \frac{|V_i||V_j|}{x_{ij}}$ is a constant determined by the operating bus voltages and the line reactance. We assume that the frequency deviations are small for all the buses $i \in \mathcal{V}$ and the differences between phase angle deviations are small across all the links in ε. Then, the deviation satisfy:

$$\begin{cases} \Delta\dot{\delta}_i = \Delta\omega_i \\ T_i\Delta\dot{\omega}_i = \Delta P_{mi} - \sum_{j\epsilon\mathcal{N}_i} B_{ij}cos\left(\Delta\delta_i^0 - \Delta\delta_j^0\right)(\Delta\delta_i - \Delta\delta_j) - K_{di}\Delta\omega_i - D_i\Delta\omega_i + u_i\forall i \in V \end{cases} \tag{7}$$

Let us consider the power system model by a graph $\mathcal{G} = (\mathcal{V}, \varepsilon)$. Each energy resource node here is denoted by each agent, which is assumed to obey the linearized swing equation. The phased angle and the angular velocity of the agent i is δ_i and ω_i. By defining the state vectors $\delta = [\delta_1, \ldots, \delta_n]$ and $\omega = \dot{\delta} = [\omega_1, \ldots, \omega_n]$, we may rewrite (7) in state-space form as

$$\begin{bmatrix} \Delta\dot{\delta} \\ \Delta\dot{\omega} \end{bmatrix} = \begin{bmatrix} 0_{n\times n} & I_n \\ -M\mathcal{L}_k & -MK_d - MD \end{bmatrix}\begin{bmatrix} \Delta\delta \\ \Delta\omega \end{bmatrix} + \begin{bmatrix} 0_{n\times 1} \\ M\Delta P_m \end{bmatrix} + \begin{bmatrix} 0_{n\times 1} \\ Mu \end{bmatrix} \tag{8}$$

where $M = diag\left(\frac{1}{T_1}, \ldots, \frac{1}{T_n}\right) = diag(M_1, \ldots, M_n)$, $K_d = diag(K_{d1}, \ldots, K_{dn})$, $D = diag(D_1, \ldots, D_n)$, $\mathcal{L}_k$ is the weighted Laplacian with agents edge weightsk_{ij}, $k_{ij} = B_{ij}cos\left(\Delta\delta_i^0 - \Delta\delta_j^0\right)$, $\Delta P_m = [\Delta P_{m1}, \ldots, \Delta P_{mn}]^T$, $u = [u_1, \ldots, u_n]^T$. The model (8) illustrates the power system dynamic behaviors. The system operates in an equilibrium point state where all frequency deviations are constant over time.

4. Distributed Frequency Control Algorithm

In this section, we design a distributed frequency controller based on the second-order consensus algorithm, where each agent measures its neighbors state information and integrates the relative differences. Compared with the traditional central controller, the distributed frequency controller solves the frequency control problem by several agents cooperatively, which results in better performance when an islanding network occurs or central signals are unavailable.

To control the agents reaching the consensus states, the controller of agent i from its adjacent agent j is assumed to be given by:

$$u_{ij} = -\alpha(\Delta\omega_i - \Delta\omega_j) - \beta(\Delta\delta_i - \Delta\delta_j) \tag{9}$$

We obtain a state feedback u_i which is designed based on a distributed protocolwith topology $\mathcal{G}$. The distributed frequency controller can be designed as follows:

$$u_i = -\sum_{j\epsilon\mathcal{N}_i}(\alpha(\Delta\omega_i - \Delta\omega_j) + \beta(\Delta\delta_i - \Delta\delta_j))\forall i \in \mathcal{V} \tag{10}$$

The protocol asymptotically solves the consensus problem when there exists an asymptotically stable equilibrium for all agent nodes. Then Equation (8) under the distributed Equation (10) can be given as:

$$\begin{bmatrix} \Delta\dot{\delta} \\ \Delta\dot{\omega} \end{bmatrix} = \underbrace{\begin{bmatrix} 0_{n\times n} & I_n \\ -M\mathcal{L}_k - M\beta I_n\mathcal{L}_k & -MK_d - MD - M\alpha I_n\mathcal{L}_k \end{bmatrix}}_{\triangleq A}\begin{bmatrix} \Delta\delta \\ \Delta\omega \end{bmatrix} + \begin{bmatrix} 0_{n\times 1} \\ M\Delta P_m \end{bmatrix} \tag{11}$$

It is easy to see that the characteristic equation of A can be given by $0 = \det((s^2 + MK_ds + MDs)I_n + (\alpha sM + M + \beta M)\mathcal{L}_k)$. Let $\overline{m} = \min_i M_i$, $\overline{k} = \min_i K_{di}$ and $\overline{d} = \min_i D_i$. We may rewrite: $MK_d = \overline{m}\overline{k}I_n + K'$, and $MD = \overline{m}\overline{d}I_n + D'$, where K',D' are diagonal matrix with positive entries respectively. We now define the matrix $A' \triangleq \begin{bmatrix} 0_{n\times n} & I_n \\ -\overline{m}\mathcal{L}_k - \overline{m}\beta I_n\mathcal{L}_k & -\overline{m}\overline{k}I_n - \overline{m}\overline{d}I_n - \overline{m}\alpha I_n\mathcal{L}_k \end{bmatrix}$. The eigenvalues of

A' are given by $\det\left(\left(s^2+s\overline{m}\overline{k}+s\overline{m}\overline{d}\right)I_n+(s\overline{m}\alpha+\overline{m}+\overline{m}\beta)\mathcal{L}_k\right)$. By noticing that the characteristic equation of $\mathcal{L}_k$: 0=det($\mathcal{L}_k-\lambda_i I_n$), where $\lambda_i \geq 0$, we can obtain the equation$s^2+(s\overline{m}\alpha+\overline{m}+\overline{m}\beta)\lambda_i+s\overline{m}\left(\overline{k}+\overline{d}\right)=0$. It shows that s must satisfy this equation for each λ_i. Considering $\alpha>0,\ \beta>0$, the above equation has all its solutions $s<0$. In the translated coordinates, it follows that the matrix A is also Hurwitz. By simple calculation under the Routh-Hurwitz stability criterion, it can be seen that the aforementioned equation has its solutions $s\in\mathbb{C}^-$if $\alpha,\beta\in\mathbb{R}^+$.

Given an initial position $\omega(0)=\omega_0$ under the dynamics (11), the power system is proved to be stable, hence, the consensus of the frequency $lim_{t\to\infty}|\omega_i(t)-\omega_j(t)|=0,\ \forall i,j\in\mathcal{V}$ is obtained. With the adjustment of the controllable loads, their frequency regulation functions are distributed to each node agent. For the case in which $(\mathcal{V},\varepsilon)$ is a undirected and connected network, it guarantees that every trajectory converges to a compact set as $t\to\infty$ and $\omega(t)$ converges to an optimal point ω^* for the distributed frequency control.

So, several important features are illustrated:

(1) Distributed Control. Each agent can make local decisions according to the local frequency and distributed coordination of power deviations. It allows a completely distributed solution and decreases the communication messages among the agents. A distributed control turns out to be gradually optimal with the coordination control of agents.

(2) Equilibrium Frequency Objective. The frequency deviations $\omega(t)$ of agents are synchronized to ω^* no matter the transient dynamic difference. The new common frequency may be different from the initial frequency point when different disturbances occur. Mechanical power supplies and frequency-sensitive dynamic load consumptions are illustrated to drive the new system frequency regulation. Thus, an equilibrium frequency objective is proposed to converge to an optimal value.

(3) Solution Optimization. The consensus algorithm by distributed integral dynamic action is developed to prove the stability of the frequency control problem. In an undirected and connected network, the consensus of the agents can be realized with distributed proportional-integral controllers. It illustrates that the trajectory of each agent can converge to the optimal frequency point ω^* to rebalance power flows after a disturbance.

5. Simulations

As a test system, a two-area four-machine system is provided to test the distributed frequency control algorithm [1]. The single line diagram of this system is given in Figure 1a. It consists of two areas and each area has two equivalent generators. The topology of communication network describing the exchange of information between generator agents is given in Figure 1b, where agents $A1$, $A2$, $A3$ and $A4$ represent generators G_1, G_2, G_3 and G_4. In this network, there are three pairs of agents, namely $A1$ and $A2$, $A2$ and $A4$, $A4$ and $A3$, and the agents in each pair share information with each other.

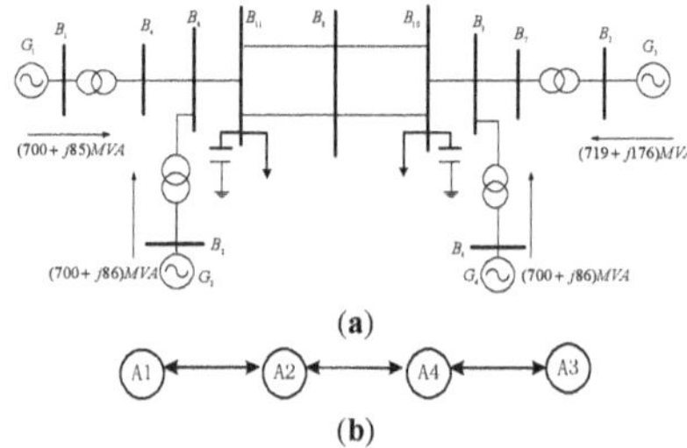

Figure 1. Two-area four-machine system. **(a)** Single line diagram; **(b)** The topology of communication network.

In our proposed method, generators are coordinated to collectively provide active power for the provision of ancillary frequency regulation service. We select loads in Bus 11 as controllable dynamic loads to perform load characteristic. The proportion of regulating frequency is −3%–3% of the average loads in Bus 11. So $K_d = 0.2$ is defined to these controllable loads. These loads are controlled in frequency ancillary regulation and power damping.

In the simulation, we use the Power System Toolbox in MATLAB/SIMULINK to test closed-loop responses of controlled nonlinear systems. The simulation step size is 0.001 s. Unlike the proposed analytic model, the simulation model is much more detailed and realistic, including two-axis transient generator model, AC nonlinear power flows, and non-zero line resistances. The simulation would show whether our analytic model and control algorithm is a suitable approximation of the simulation model.

Considering multi-agent system with dynamics (8) and the communication topology given in Figure 1b, the Laplacian matrix is shown as $\mathcal{L}_k = \begin{bmatrix} -0.3557 & 0.3557 & 0 & 0 \\ 0.3557 & -0.4228 & 0 & 0.0671 \\ 0 & 0 & -0.3744 & 0.3744 \\ 0 & 0.0706 & 0.3744 & -0.4450 \end{bmatrix}$. The simulations are conducted for different cases including the small signal disturbances and the short circuit faults. In addition, the system with various parameters for each of the cases are simulated. These simulations aim to check the robustness of distributed controllers obtained with parameter variations. Meanwhile, simulations for the system with different operating conditions are implemented for illustrating the effectiveness of the controller. For all simulations, detailed dynamic responses are considered. For evaluating the performance of the proposed controller, the integral of absolute frequency deviation, $\mathrm{J} = \frac{1}{\mathrm{N}} \sum_{\mathrm{i}=1}^{\mathrm{N}} \int_0^{\tau} \Delta \mathrm{f}_{\mathrm{i}}^2(\mathrm{t})\mathrm{dt}$, is selected as a performance index. Nis the total bus node number. The total time internal of τ for all simulations are taken as 20 s.

5.1. A Small Signal Disturbance and Stability Analysis

In this case, a small disturbance has occurred in the load demand. The system operates stability before the time of $t = 0$ s. At $t = 0$ s, there is a 1% step increase of the total demand. The frequency curve of the system for the case of without control is given in Figure 2. It shows that after the instant of load increase, the frequency of the system oscillates and the amplitude increases. Therefore, the system is prone to lose stability when small disturbance happens. In order to guarantee the system's stability, proper control schemes are required.

The proposed distributed frequency controllers are tested in the system operation. The power transmitted from area1 to area 2 through the tie-line for the cases of with and without the distributed frequency controller (DFC) are given in Figure 3. Without the DFC, the power oscillates and the amplitude increases largely. If frequency-sensitive load shedding control strategy is used in Bus 11, the curtailment of the loads is varied according to the frequency drop, which is fluctuated nearly from 16 MW to 10 MW around. The curve shows that the trajectories of the tie-line power continue oscillating at a long time. With the distributed frequency control and no dynamic loads, thepower deviations converge to zero in less than 3 s. Considering DFC and dynamic load characteristics, the power oscillations could be damped faster. It can been seen that the performance of DFCs is more beneficial than load shedding control strategy. It also shows that considering dynamic loads' characters for frequency regulations, the power transportation and frequency values can increase more than distributed frequency regulations without dynamic loads.

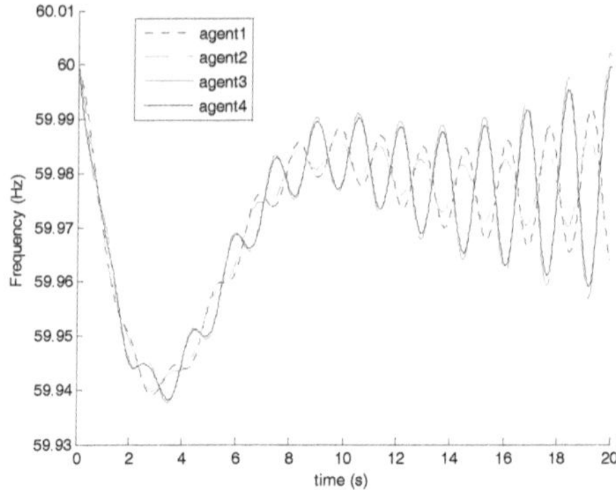

Figure 2. The frequency deviations without control in a small disturbance.

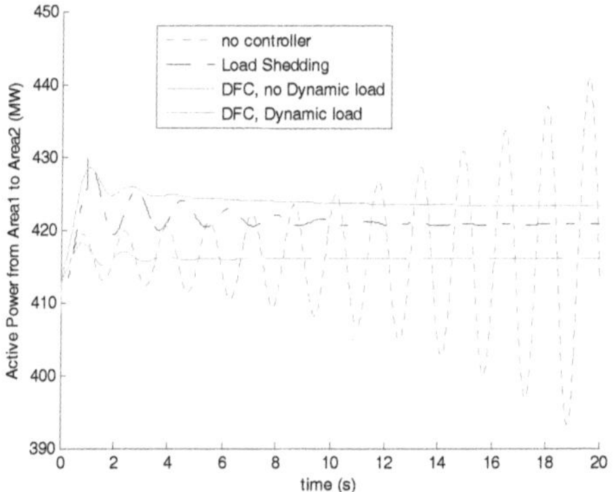

Figure 3. The tie-line active power (MW) *versus* time (s) with a small disturbance for cases (i) no controller; (ii) load shedding control; (iii) DFC without dynamic loads; (iv) DFC considering dynamic loads.

5.2. The Effects of Control Parameters of DFC on Frequency Stability

Previous simulation results indicate that the proposed method is capable to reduce the oscillation. In this section, we will discuss the effects of the control parameters, α and β, on power system frequency stability.

We first set β as zero to observe the effects of parameters α with various values. In fact, $\beta = 0$ means that the DFC cannot regulate the agents' angular acceleration. Figure 4a gives the frequency curves of system with different controller parameter α. From the figure, we found that the increase of parameter α brings a positive effect on damping frequency oscillations. Obviously, the control with the ability to regulate agents' angular acceleration would be more effective to damping frequency oscillations. Here we set $\alpha = 20$ to observe the effects of parameters β with various values. The simulation results are given in Figure 4b. It shows that the system frequency oscillation is suppressed. In addition, a smaller value of β indicates a better control performance.

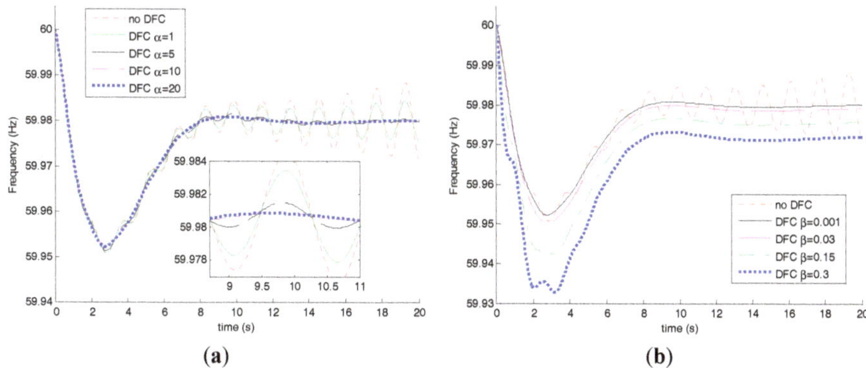

(a) (b)

Figure 4. Comparisons of the frequency deviations under different controller parameters in a small disturbance (**a**) different α (β = 0); (**b**) different β (α = 20).

The performance index of the system with different controller parameter α and β = 0 are calculated. The values of the performance index J and $max\{\Delta f_1(t)\}$ are given in Table 1. In addition, the performance index of the system with different controller parameter β and α = 20 are calculated. In Table 2, the values of J gradually increase with the increase of β. This is consistent with the observation that the largerβ has a larger $max\{\Delta f_1(t)\}$.

Table 1. The performance index under different controller parameters with the small disturbance (β = 0).

Different Controllers	J	max$\{\lvert\Delta f_1(t)\rvert\}$
NoDFC	0.0145	0.0492
DFC,α = 1	0.0144	0.0491
DFC, α = 5	0.0143	0.0485
DFC, α = 10	0.0143	0.0481
DFC, α = 20	0.0143	0.0477

Table 2. Different performances using different controller parameters with the small disturbance (α = 20).

Different Controllers	J	max$\{\lvert\Delta f_1(t)\rvert\}$
NoDFC	0.0145	0.0492
DFC, β = 0.001	0.0144	0.0477
DFC, β = 0.03	0.0157	0.0497
DFC, β = 0.15	0.0215	0.0577
DFC, β = 0.3	0.0287	0.0673

It can be seen that frequency-sensitive loads have shown their frequency regulations in power networks. We also present simulation results below with different kinds of frequency-sensitive loads. Let K_d denote different kinds of frequency-sensitive loads, different consumption of the frequency-sensitive loads are add on Bus 11. Thus, different kinds of frequency-sensitive load control performance are shown in Figure 5. In Figure 5a, the system has no controller, so that the different values of the parameter K_d cause different undamped oscillations. In Figure 5b, when the distributed frequency controllers are added, different values of the parameter K_d can contribute in the power balance and frequency regulation. It shows that these linear frequency-sensitive dynamic loads can rebalance power and resynchronize frequency after a disturbance. If K_d is larger, the load-side control time is often faster because of little time constants and evident load regulation solutions.

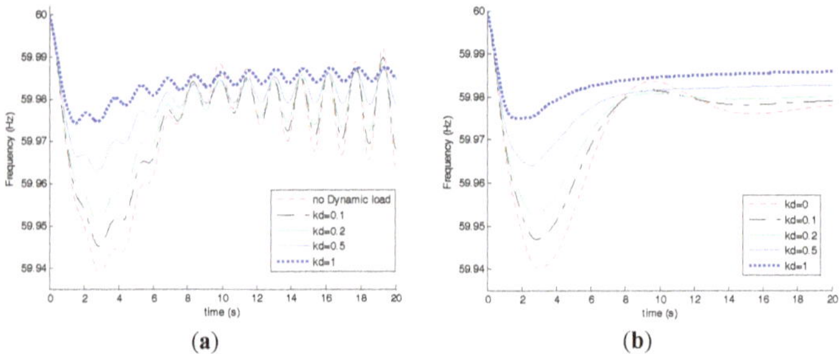

Figure 5. Comparisons of the frequency deviations under different dynamic load characteristics in a small disturbance. (**a**) No DFC; (**b**) with DFC.

5.3. A Short-Circuit Fault and Stability Analysis

In order to obtain more information to investigate the distributed controller performance in improving the system stability, transient stability simulations are tested to evaluate the results. A three-phase short-circuit fault is applied at bus8 at $t = 1$ s, which is cleared after $t = 200$ ms.

The power transmitted from area1 and area 2 after the fault is observed. As shown in Figure 6, with the distributed control, less fluctuation appears in the active power from area1 to area 2, and after a relatively short time (around 6 s) the power oscillation disappeared. It also shows that considering frequency-sensitive load regulation in a short circuit fault, the power transportation value could increase more than controllers without dynamic loads to improve system stability.

We also observed the frequency oscillation after the short circuit fault cleared. The curves of frequency oscillation of the system for the case with various α and $\beta = 0$ are given in Figure 7a. From the figure, it can be found that the system could keep stable when α is larger than 5. This means that the proposed DFC improves the transient stability if proper α is selected.

In Figure 7b, the curves of frequency oscillation after the short-circuit fault cleared for case with various β and $\alpha = 20$ are provided. Observing these curves, we saw that the systems keep stable for all combination patterns of α and β. For different combination patterns of α and β, the performance of the system are quitesimilar, which indicate that parameter β plays a less important role compared to parameter α. If α is larger, the damping control effect is more significant. In addition, a smaller value of β performs a better control effect in the long time.

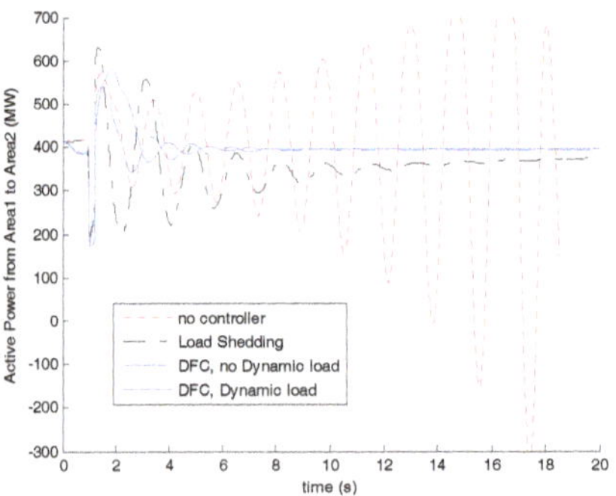

Figure 6. The tie-line active power (MW) from area1 to area2 when a three-phase short-circuit fault occurs (i) no DFC; (ii) load shedding control; (iii) DFC without dynamic loads; (iv) DFC considering dynamic loads.

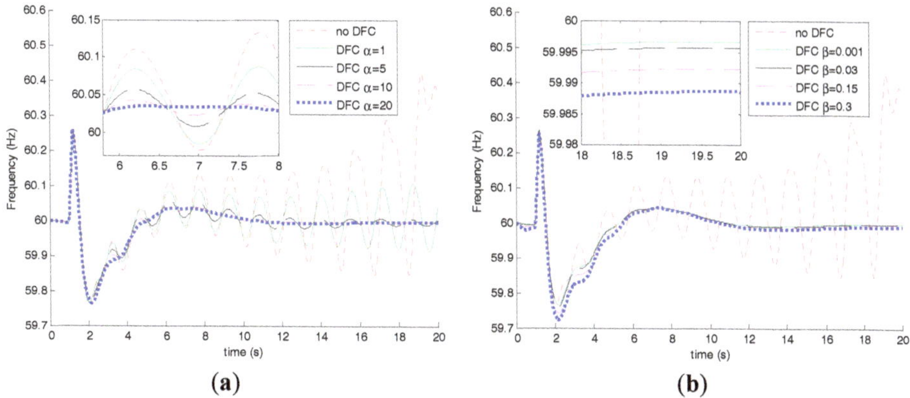

Figure 7. Comparisons of the frequency deviations under different controller parameters with a short-circuit fault (**a**) different α; (**b**) different β (α = 20).

Tables 3 and 4 provide the values of the performance indices for the cases of various combination patterns of parameters. These performance indices give the quantitative control evaluation values under different controller parameters.

Table 3. The performance index under different controller parameters with a short-circuit fault (β = 0).

Different Controllers	J	$\max\{\lvert\Delta f_1(t)\rvert\}$
NoDFC	0.3875	0.4230
DFC, α = 1	0.1132	0.2591
DFC, α = 5	0.0771	0.2600
DFC, α = 10	0.0765	0.2607
DFC, α = 20	0.0799	0.2616

Table 4. Theperformance index using different controller parameters with a short-circuit fault disturbance (α = 20).

Different Controllers	J	$\max\{\lvert\Delta f_1(t)\rvert\}$
NoDFC	0.3875	0.4230
DFC, β = 0.001	0.0866	0.2645
DFC, β = 0.03	0.0829	0.2602
DFC, β = 0.15	0.0949	0.2558
DFC, β = 0.3	0.1096	0.2730

Figure 8a shows the phase plane diagram of frequency oscillation for the case without DFC. The agents' frequencies oscillate and finally the system turns into being in an unstable state. Then, we install DFC to each of these agents. The phase plane diagram of frequency oscillation is given in Figure 8b. In the figure, the state of each agent deviates from its original state, and after a certain time, goes back to the original state. This means that the DFC can keep the frequencies stable in the case of short-circuit fault occurred.

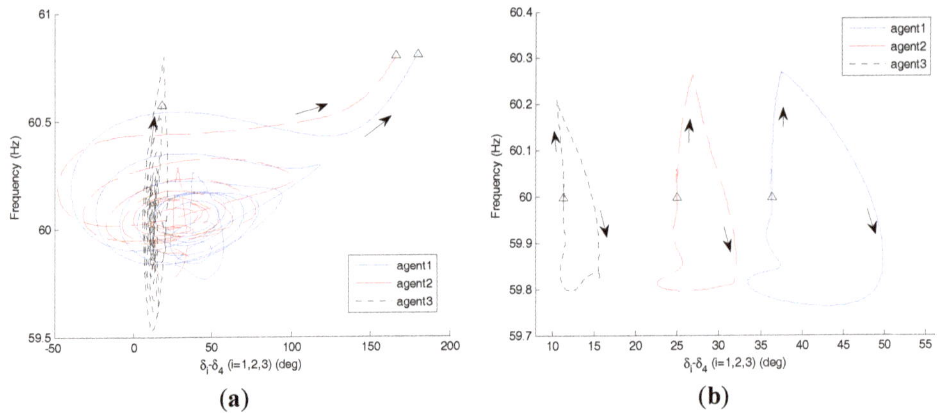

Figure 8. The phase plane diagram under (**a**) no DFC; (**b**) DFC with a short-circuit fault disturbance.

In summary, it can be concluded that the distributed frequency control can provide a guarantee of more reliable and stable power supply in the power system. The controller improves both the steady-state and transient performance of frequency. Compared with the central frequency control, the distributed frequency control method is more secure and efficient.

6. Conclusions

This paper develops a distributed control method to decide the active injection of the frequency regulation in the electrical network. Each distributed resource in the network computes the amount of active power that it needs to provide. A distributed frequency controller is designed considering the dynamic load charateristics, where each bus controls its own frequency based on local measurements and information from neighbouring places. For the purpose of designing the coordination controller, some dynamic assumptions are made, *i.e.*, the difference between phase angles of buses are small, and the frequency sensitivities with respect to changes in the operating point do not change much for different operating points. The proposed consensus protocol can provide grid support services in a distributed manner and achieve the frequency regulation consensus. The simulations illustrate the availability to regulate frequency oscillations during the power dynamic process.

Acknowledgments: This work has been partially funded by the State Grid Corporation of China Project: Study on Key Technologies for Power and Frequency Control of System with 'Source-Grid-Load' Interactions, and the Six Talent Summit Project in Jiangsu Province.

Author Contributions: Rong Fu and Yingjun Wu contributed in developing the ideas of this research and the consensus algorithm in the paper. Rong Fu, Hailong Wang and Jun Xie performed this research and simulated the power system operations, all the authors involved in preparing this manuscript.

Conflicts of Interest: The authors declare no conflict of interest.

References

1. Kundur, P. *Power System Stability and Control*, 2nd ed.; McGraw Hill: New York, NY, USA, 2008.
2. Zhao, C.; Topcu, U.; Low, S.H. Optimal load control via frequency measurement and neighborhood area communication. *IEEE Trans. Power Syst.* **2013**, *28*, 3576–3587. [CrossRef]
3. Kothari, M.L.; Nanda, J.; Kothari, D.P.; Das, D. Discrete-mode automatic generation control of a two-area reheat thermal system with new area control error. *IEEE Trans. Power Syst.* **1989**, *4*, 730–738. [CrossRef]
4. Malik, O.P.; Kumar, A.; Hopeg, S. A load frequency control algorithm based on a generalized approach. *IEEE Trans. Power Syst.* **1988**, *3*, 375–382. [CrossRef]

5. Moon, Y.H.; Ryu, H.S.; Lee, J.G.; Kim, S. Power system load frequency control using noise-tolerable PID feedback. In Proceedings of the IEEE International Symposium on Industrial Electronics, Pusan, Korea, 12–16 June 2001; Volume 3, pp. 1714–1718.
6. Yamashita, K.; Miyagi, H. Multi-variable self-tuning regulator for load frequency control system with interaction of voltage on load demand. *IEEE Proc. Control Theory Appl.* **1995**, *138*, 177–183. [CrossRef]
7. Khodabakhshian, A.; Hooshmand, R. A new PID controller design for automatic generation control of hydro power systems. *Int. J. Electr. Power Energy Syst.* **2010**, *32*, 375–382. [CrossRef]
8. Yang, T.C.; Cimen, H. Applying structured singular values and a new LQR design to robust decentralized power system load frequency control. In Proceedings of the IEEE International Conference on Industrial Technology, Shanghai, China, 2–6 December 1996; pp. 880–884.
9. Liu, X.; Kong, X.; Deng, X. Power system model predictive load frequency control. In Proceedings of the American Control Conference, Montreal, QC, Canada, 27–29 June 2012; pp. 6602–6607.
10. Ilic, M.D.; Xie, L.; Khan, U.A.; Moura, J.M. Modeling of future cyber-physical energy systems for distributed sensing and control. *IEEE Trans. Syst. Man Cybern.* **2010**, *40*, 825–838. [CrossRef]
11. Rerkpreedapong, D.; Hasanovic, A.; Feliachi, A. Robust load frequency control using genetic algorithms and linear matrix inequalities. *IEEE Trans. Power Syst.* **2003**, *18*, 855–861. [CrossRef]
12. Yin, S.; Li, X.; Gao, H.; Kaynak, O. Data-based techniques focused on modern industry: An overview. *IEEE Trans. Ind. Electron.* **2014**, *62*, 657–667. [CrossRef]
13. Kocaarslan, I.; Çam, E. Fuzzy logic controller in interconnected electrical power systems for load-frequency control. *Int. J. Electr. Power Energy Syst.* **2005**, *27*, 542–549. [CrossRef]
14. Chidambaram, I.A.; Velusami, S. Design of decentralized biased controllers for load-frequency control of interconnected power systems. *Electr. Power Compon. Syst.* **2005**, *33*, 1313–1331. [CrossRef]
15. Carli, R.; Chiuso, A.; Schenato, L.; Zampieri, S. Distributed Kalman filtering based on consensus strategies. *IEEE J. Sel. Areas Commun.* **2008**, *26*, 622–633. [CrossRef]
16. Liu, W.; Gu, W.; Sheng, W.; Meng, X.; Wu, Z.; Chen, W. Decentralized multi-agent system-based cooperative frequency control for autonomous microgrids with communication constraints. *IEEE Trans. Sustain. Energy* **2014**, *5*, 446–456. [CrossRef]
17. Ilic, M.D.; Liu, Q. Toward sensing, communications and control architectures for frequency regulation in systems with highly variable resources. In *Control and Optimization Methods for Electric Smart Grids*; Springer: New York, NY, USA, 2012; Volume 3, pp. 3–33.
18. Molina-Garcia, A.; Bouffard, F.; Kirschen, D.S. Decentralized demand-side contribution to primary frequency control. *IEEE Trans. Power Syst.* **2011**, *26*, 411–419. [CrossRef]
19. Pedroche, F.; Rebollo, M.; Carrascosa, C.; Palomares, A. On the convergence of weighted-average consensus. *Int. J. Elec. Power Energy Syst.* **2013**. arXiv:1307.7562v1.
20. Andreasson, M.; Sandberg, H.; Dimarogonas, D.V.; Johansson, K.H. Distributed integral action: Stability analysis and frequency control of power systems. In Proceedings of the IEEE 51st Annual Conference on Decision and Control, Maui, HI, USA, 10–13 December 2012; pp. 2077–2083.
21. Zhao, C.; Topcu, U.; Li, N.; Low, S. Design and stability of load-side primary frequency control in power systems. *IEEE Trans. Autom. Control* **2014**, *59*, 1177–1189. [CrossRef]
22. Yu, W.; Chen, G.; Cao, M. Some necessary and sufficient conditions for second-order consensus in multi-agent dynamical systems. *Automatica* **2010**, *46*, 1089–1095. [CrossRef]
23. Freeman, R.A.; Yang, P.; Lynch, K.M. Stability and convergence properties of dynamic average consensus estimators. In Proceedings of the 45th IEEE Conference on Decision and Control, San Diego, CA, USA, 13–15 December 2006; pp. 2398–2403.

Article

Application of a $LiFePO_4$ Battery Energy Storage System to Primary Frequency Control: Simulations and Experimental Results

Fabio Massimo Gatta [1], Alberto Geri [1], Regina Lamedica [1], Stefano Lauria [1], Marco Maccioni [1,*], Francesco Palone [2], Massimo Rebolini [2] and Alessandro Ruvio [1]

1. Department of Astronautics, Electric and Energy Engineering, Sapienza University of Rome, Rome 00184, Italy; fabiomassimo.gatta@uniroma1.it (F.M.G.); alberto.geri@uniroma1.it (A.G.); regina.lamedica@uniroma1.it (R.L.); stefano.lauria@uniroma1.it (S.L.); alessandro.ruvio@uniroma1.it (A.R.)
2. Terna S.p.A., Rome 00156, Italy; francesco.palone@terna.it (F.P.); massimo.rebolini@terna.it (M.R.)

* Correspondence: marco.maccioni@uniroma1.it; Tel.: +39-06-44585540

Academic Editor: Peter J. S. Foot
Received: 29 July 2016; Accepted: 25 October 2016; Published: 29 October 2016

Abstract: This paper presents an experimental application of $LiFePO_4$ battery energy storage systems (BESSs) to primary frequency control, currently being performed by Terna, the Italian transmission system operator (TSO). BESS performance in the primary frequency control role was evaluated by means of a simplified electrical-thermal circuit model, taking into account also the BESS auxiliary consumptions, coupled with a cycle-life model, in order to assess the expected life of the BESS. Numerical simulations have been carried out considering the system response to real frequency measurements taken in Italy, spanning a whole year; a parametric study taking into account different values of governor droop and of BESS charge/discharge rates (*C-rates*) was also performed. Simulations, fully validated by experimental results obtained thus far, evidenced a severe trade-off between expected lifetime and overall efficiency, which significantly restricts the choice of operating parameters for frequency control.

Keywords: battery energy storage system (BESS); $LiFePO_4$ battery; primary frequency control

---

1. Introduction

The "smart grid" paradigm envisages a massive presence of non-programmable renewable energy sources: in this context, battery energy storage systems (BESSs) are liable to play a key role at both distribution and transmission level, given their potential ability to fulfill roles such as load shifting, peak shaving, frequency and also voltage control [1–6]. Moreover, BESSs have also been proposed in integration to electric power systems supplying traction and mobility systems, with the aim to maximize the energy efficiency [7–10]. In principle, the study of BESS impact on the electric power system should include environmental aspects, optimal siting, as well as power quality issues and harmonic disturbances, in accordance with existing standards [11–18]. Generally speaking, technical features such as battery size (in terms of both rated power and energy), efficiency, transient performance, cycling and lifetime depend on the specific application. In order to evaluate the economic return of each application, local energy market rules must be taken into account.

This paper deals with the application of a BESS based on lithium iron phosphate ($LiFePO_4$) batteries to primary frequency control (PFC). Since conventional power plants are increasingly displaced by (mostly non-dispatchable) generation from renewable energy sources, transmission system operators (TSOs) are looking for new PFC providers to preserve frequency quality. Li-ion BESSs are being evaluated for the PFC role [19,20], which entails exacting requirements such as fast response, high number of charge/discharge cycles and wide depth-of-discharge (*DOD*); notably, $LiFePO_4$ batteries look very promising, due to their chemical and thermal stability which could ensure a long lifetime under the PFC cycling conditions at a relatively low cost [21–23].

However, to date there is not enough operating experience confirming the PFC applicability and the performances (expected lifetime, round trip efficiency) of $LiFePO_4$ batteries. To this end, a coupled electrical-thermal model of a $LiFePO_4$ battery has been developed and validated against experimental tests by Terna (the Italian TSO). The model has been used to simulate PFC operation of a 1-MW/1-MWh $LiFePO_4$ BESS deployed by Terna, considering different values of droop and discharge rate (*C-rate*). The paper is organized as follows. Section 2 briefly recalls the Terna experimental BESS system, while Section 3 details the proposed PFC application. Section 4 deals with BESS modelling; experimental test results are shown in Section 5 and PFC simulation results are reported in Section 6.

2. The Terna Experimental $LiFePO_4$ Battery Energy Storage System

There are a significant number of manufacturers of $LiFePO_4$ batteries, since they use readily available raw materials and are thermally and chemically stable, thus ensuring safety as well as long service life. Moreover, the high power-to-energy ratio makes $LiFePO_4$ batteries attractive for BESS applications. $LiFePO_4$ have a lower nominal cell voltage (3.2 V) than other Li-ion batteries. The normal voltage for grid (stationary) application ranges between 2.8 and 3.6 V, in order to increase battery life avoiding operation at extreme values of the state-of-charge (SOC) near full charge or full discharge. The maximum continuous discharge rate of presently available MWh-sized systems can vary from 0.2C [19] to 4C [20], depending on the manufacturing technology and module thermal design. *C-rate* range requirements vary widely with the specific application: renewable energy sources balancing typically requires *C-rates* ranging between 0.2C and 1C, whereas PFC might involve *C-rates* ranging between 1C and 4C. Several $LiFePO_4$ BESS projects have been recently commissioned in China, America and Europe; in Italy, two $LiFePO_4$ systems have been recently installed by Terna as part of the wider "Storage Lab" [24,25] experimental BESS project. Terna's $LiFePO_4$ BESS is based on prismatic cells (Figure 1) suitable for stationary applications, located in an aluminum case.

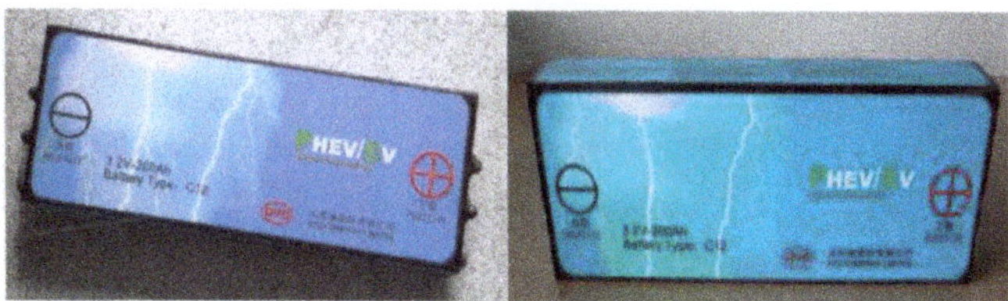

Figure 1. Prismatic $LiFePO_4$ cells.

The series connection of four such cells forms a battery module with a 12.8 V-2.37 kWh rating. The battery module is sealed to prevent moisture ingress and to avoid leaks in case of battery failure: as a consequence, the thermal behavior of the cells inside the module differs substantially from that of free-standing cells. Each module is provided with its own battery management system (BMS) for cell balancing and battery monitoring. Modules are series-connected to form battery strings (mounted on racks designed with sufficient spacing for proper ventilation and cooling [24]), which are paralleled inside the air-conditioned battery container (Figure 2). Each of Terna's 1 MW/1 MWh BESS includes a battery container and a dual-stage power conditioning system (PCS) [26]. The PCS is connected from the low voltage (LV) level to a 20 kV medium voltage (MV) busbar via an integrated MV/LV transformer; the whole system is in turn connected to the Terna 150 kV high voltage (HV) sub-transmission network through a HV/MV transformer.

Extensive testing was carried out on the above described $LiFePO_4$ batteries, focusing on safety requirements [27–29] and battery performance. The latter tests involved intensive cycling of single battery modules (12.8 V-185 Ah) and performance tests on battery string specimens (256 V-185 Ah), with the aim of verifying the expected battery life in normal operation. Moreover, PFC performance was evaluated by cycling a string specimen with a power profile emulating the response of a virtual governor to an actual ENTSO-E (acronym for European Network of Transmission System Operators for Electricity) measured frequency pattern.

Figure 2. Racks for battery strings, installed inside an air-conditioned cabinet.

The experimental setup included an electronic variable load (model ZS4206, H&H, Konzell, Germany), a controllable dc power supply (SM 15-400, Delta Elektronika, Zierikzee, Netherlands) and a measurement/monitoring system (cFP2220, cFP-AI-118, cFP-TC-120, National Instruments, Austin, TX, USA). This allowed to measure fundamental battery state variables such as current, individual cell voltages and temperatures, subsequently used to estimate battery parameters (resistance, thermal inertia) and performances (round-trip efficiency, battery life). Given the modular design of the BESS, results of tests performed on individual battery strings can be straightforwardly extended to the whole 1 MW/1 MWh system.

3. Application of the $LiFePO_4$ Battery Energy Storage System to Primary Frequency Control

3.1. Short Review of Battery Energy Storage System Applications to Primary Frequency Control

PFC is the most important task for the stability of the electrical power system. The first utility-scale BESS (based on lead-acid batteries) in Europe used for PFC was deployed in the 1980s in West Berlin [30], where for political reasons the supply system was not connected to the East Germany's national grid. Another relevant PFC application is the 1 MW Li-ion BESS operated by Elektrizitätswerke des Kantons Zürich (EKZ), the Canton of Zürich utility, in Dietikon, Switzerland [31]. Three applications (PFC, peak shaving and islanded operation) of the BESS are discussed and preliminary results regarding PFC application are supplied, showing the suitability of BESS for such tasks. Finally, many large-scale projects involving BESSs for PFC application have been recently deployed: these are recorded in the US Department of Energy database, together with hundreds projects involving other applications [32].

3.2. Primary Frequency Control in the ENTSO-E European Synchronous System

ENTSO-E is the association of 41 TSOs from 34 countries in Europe, accounting for three interconnections (namely the continental synchronous power system which links most European countries, plus the "Nordic" and "Baltic" interconnections), as well as the British—Irish and Sardinia—Corsica asynchronous power systems. The Continental Europe Operation Handbook [33] summarizes technical requirements and procedures for operation, control and security of the "continental" grid, in which frequency control (primary, secondary and tertiary control) obviously plays a paramount role. In particular, primary frequency control "[...] stabilizes the system frequency at a stationary value after a disturbance or incident in the time-frame of seconds" [33]. PFC is carried out by proportional regulators, so that a quasi-steady-state frequency deviation Δf (defined as $\Delta f = f - f_n$, being f_n the nominal frequency of the interconnected grid), caused by an unbalance ΔP_a between demand and generation, will cause all generators participating in PFC to change their output according to the Equation (1):

$$\Delta P_G = -\frac{\Delta f}{f_n} \cdot \frac{P_{Gn}}{s} \cdot 100, \tag{1}$$

where ΔP_G (MW) is the variation of the active power output of the generation unit, P_{Gn} (MW) is the rated active power output of the generation unit and s (%) is the governor droop. No change in output is required if Δf does not exceed ± 10 mHz, to cater for the combined effects of frequency response insensitivity and governor dead band. Terna mandates 4% droop for hydroelectric and 5% droop for thermal power plants, respectively [34], whereas no prescription regarding BESSs is reported at present (2016).

3.3. Frequency Profile Used in Simulations

The PFC tests were carried out offline, by feeding the PCS of a 50 kW–50 kWh battery string (20 series-connected modules) with a power command directly proportional to a frequency deviation signal obtained from the actual ENTSO-E real-time frequency recording for the year 2014, in accordance with the pseudo-steady-state control characteristic (1). Frequency was sampled at 1-s intervals; the droop of the equivalent governor was taken at 0.5%, resulting in a $\lambda = 200$ kW/Hz power-frequency characteristic given the *C-rate* 1C. Offline testing was justified by the negligible influence of the test specimen on the overall Italian contribution to European primary frequency control.

Figure 3a shows the frequency vs. time for a random day (the ± 10 mHz insensitivity/dead-band window is also shown), whereas Figure 3b reports the probability distribution of recorded frequency values during the whole year: the maximum and minimum recorded values were 50.12 Hz and 49.89 Hz, respectively [35].

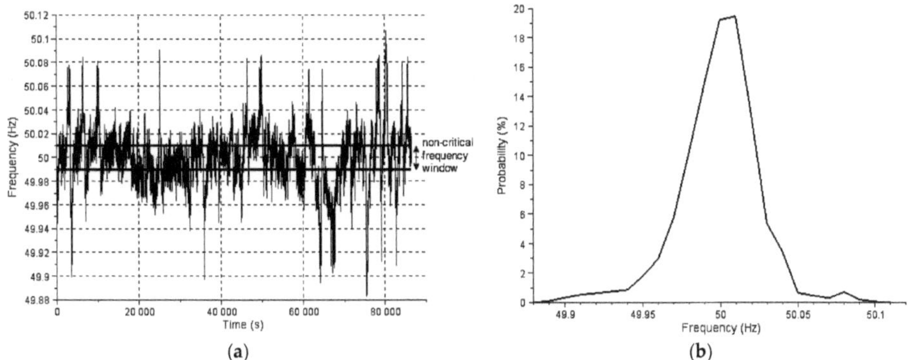

Figure 3. Italian (ENTSO-E) frequency profile measured by Terna and used in PFC tests: (**a**) frequency vs. time during one day; and (**b**) probability distribution of the whole yearly power frequency recording.

4. Battery Energy Storage System Modelling

The model adopted for simulating the $LiFePO_4$ BESS consists of a coupled electrical-thermal model for a battery string, plus a "lifetime" (aging) model to take into account the long-term battery loss of capacity. Validation of the electrical-thermal model, when used in in PFC simulations, against experimental tests by Terna is reported in Section 5.

4.1. The Electrical-Thermal Model

The battery itself is simulated by a coupled electrical-thermal model. The electrical part, shown in Figure 4, is an equivalent Thévenin circuit consisting of a voltage generator, E_m, in series with a single resistance, R_0. The value of the no-load voltage E_m has been taken as a function of SOC, but not of battery temperature T. In fact, during tests performed by Terna with s = 0.5% and C-$rate$ = 1C, battery temperature was almost stable at around 25 °C; moreover, the effect of T on no-load voltage is not significant in the operating range (from 20 to 55 °C), as shown in [36]. Battery resistance R_0 has been assumed to depend on SOC and battery temperature but not on time, i.e., the effect on R_0 of battery ageing due to cycling has been neglected. Note that R_0 takes different values, depending on whether the battery is charging, $R_{0,c}$(SOC,T), or discharging, $R_{0,d}$(SOC,T). The resulting circuit can be regarded as a simplification of the detailed electrical model in [37], which includes an additional shunt-connected voltage generator E_p in series with an impedance Z_p, accounting for parasitic effects, as well as a number of R-C parallel blocks in series with R_0 in order to take into account the dynamic behavior of the battery. Model simplifications (e.g., removal of the shunt parasitic branch) are partly due to uncertainty of battery parameters and lack of data; moreover, the analysis of Terna experimental data suggests that, at least for the PFC-oriented simulations of the paper, the suppression of the R-C parallel blocks does not substantially decrease model performance. The latter remark also applies to the dependence of R_0 on load current (evidenced for instance in [38,39]), which has been disregarded.

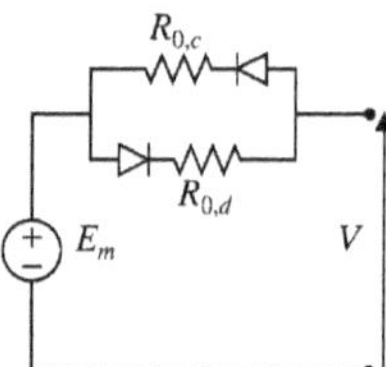

Figure 4. Simplified Thévenin equivalent circuit model of the battery (see text for details).

Figure 5a reports the measured values of the no-load voltage E_m as a function of SOC, yielded by Terna tests on a real $LiFePO_4$ battery (with T constant at 25 °C), as well as the curve used in the model of Figure 4. $R_{0,c}$(SOC,T) and $R_{0,d}$(SOC,T) values, respectively measured during charge and discharge duty, are shown in Figure 5b, which also includes values used in the simulation model. Terna also performed experimental tests to assess the dependence of both $R_{0,c}$ and $R_{0,d}$ on temperature.

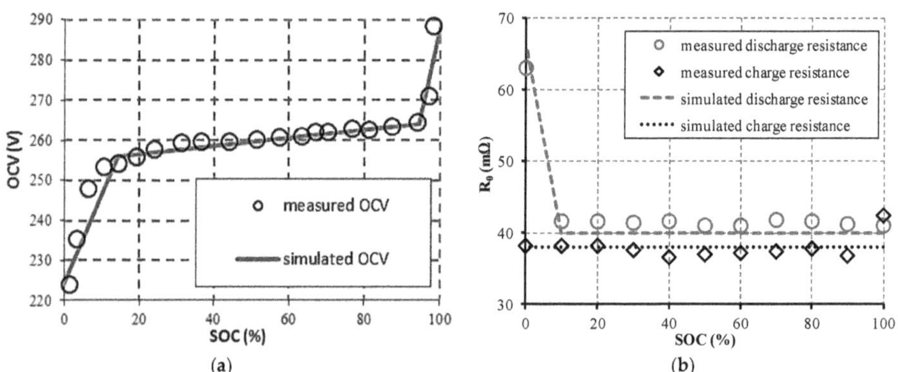

Figure 5. Measured and simulated parameter values for the Figure 4 model, as a function of state-of-charge (SOC). (**a**) No-load voltage E_m; and (**b**) charge and discharge resistances $R_{0,c}$ and $R_{0,d}$.

1-min/23 A current steps (corresponding to a *C-rate* equal to 0.125C) was impressed, measuring $R_{0,c}$ and $R_{0,d}$ for three different temperatures (20 °C, 30 °C, 40 °C) and three different SOC values (10%, 50%, 90%). Tables 1 and 2 report test results for $R_{0,d}$ and $R_{0,c}$, respectively.

Table 1. Battery discharge resistances measured for different SOC values and temperatures.

$R_{0,d}$ at SOC 10% (Ω)	$R_{0,d}$ at SOC 50% (Ω)	$R_{0,d}$ at SOC 90% (Ω)	Temperature (°C)
0.0399	0.0407	0.0374	20
0.0365	0.0348	0.0341	30
0.0307	0.0323	0.0306	40

Table 2. Battery charge resistances measured for different SOC values and temperatures.

$R_{0,c}$ at SOC 10% (Ω)	$R_{0,c}$ at SOC 50% (Ω)	$R_{0,c}$ at SOC 90% (Ω)	Temperature (°C)
0.0377	0.0393	0.0402	20
0.0335	0.0359	0.0352	30
0.0309	0.0310	0.0301	40

Based on such results, the dependence of $R_{0,c}$ and $R_{0,d}$ on temperature has been taken as linear in the operating range (from 20 to 55 °C) with a negative temperature coefficient of about 1%/K. This simplifying assumption seems sufficiently accurate, since temperature coefficient values calculated from Tables 1 and 2 range from 0.852 to 1.25%/K. Moreover, similar values may be inferred from experimental tests reported in [36].

To evaluate battery temperature T and the auxiliary consumptions (due to BMS and to the heating, ventilating, air conditioning, HVAC, system that controls the BESS cabinet temperature), a thermal model was set up and coupled to the equivalent electrical circuit [40]. Battery temperature depends on the balance between battery Joule losses $R_0(t)\cdot i(t)^2$ and thermal power removed by the HVAC:

$$\frac{dT}{dt} = \frac{R_0(t) \cdot i(t)^2 - \Delta T \cdot G}{C_T}, \tag{2}$$

where $\Delta T \cdot G$ is the thermal power removed by the HVAC (G is the thermal conductance, ΔT is the difference between battery temperature, T, and cabinet temperature T_0 set by HVAC) and C_T is the thermal capacitance of the battery. Thermal exchanges with the outside environment are neglected because of the extensive thermal insulation of the cabinet. Steady-state auxiliary consumptions are given by:

$$P_{aux} = \frac{R_0(t) \cdot i(t)^2}{COP} + P_{BMS}, \tag{3}$$

where COP is the HVAC coefficient of performance and P_{BMS} includes the power losses specifically related to the BMS (located outside the cabinet) and the power consumption due to PCS auxiliaries, assumed to be constant.

Values measured by Terna, i.e., G = 60 W/K, C_T = 100 Wh/K, T_0 = 20 °C, COP = 2.5 and P_{BMS} = 400 W, were used in simulations. Figure 6 shows the thermal and electrical power flows considered in the model.

4.2. Ageing Model

The ageing model proposed in [21] was adopted to represent the capacity loss (Q_{loss}) of the battery with charge-discharge cycles. The percentage Q_{loss} is given by:

$$Q_{loss} = B \cdot \exp\left(\frac{-31{,}700 + 370.3 \cdot (C\text{-}rate)}{8.314 \cdot T}\right) \cdot A_h^{0.55}, \tag{4}$$

where *C-rate* is the current charging/discharging rate, A_h is the accumulated charge throughput (Ah), expressed as (cycle number) × (*DOD*) × (full cell capacity), *T* is the absolute temperature (K) and *B* is a numerical factor depending on the *C-rate*. The model described by Equation (4) implicitly takes into account "calendar" (time) aging together with aging due to cycling, as long as the BESS is not idle; this condition is certainly fulfilled in the studied PFC application.

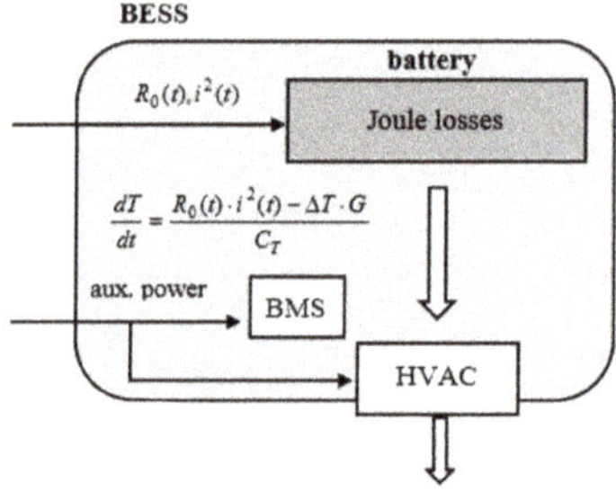

Figure 6. Thermal (white arrows) and electrical (black arrows) power flows associated to the BESS electrical-thermal model.

A power law least square approximation was carried out in order to evaluate *B* values corresponding to *C-rates* in the 0.005–6C range. *B* values reported in [21] for C/2, 2C and 6C, and a *B* value based on 20 years expected calendar life reported by the manufacturer for 0.005C (the latter is the lowest *C-rate* occurred during the simulations reported in this paper), were used. Such values are reported in Table 3. The resulting relationship is:

$$B = 26,222 \cdot (C{-}rate)^{-0.387}. \quad (5)$$

Table 3. *B* values used to calculate the *C-rate* vs. *B* power law approximation and obtained by Equation (5).

C-rate	*B*	*B* from Equation (5)
0.005C	207,000	203,781
C/2	31,630	34,290
2C	21,681	20,052
6C	12,934	13,108

Figure 7 reports the above data and the fitting power curve. End-of-life for the simulated battery is assumed when Q_{loss} equals 20% [22,23]. Figure 8 compares ageing model results with manufacturer's life cycle data, showing a very good agreement.

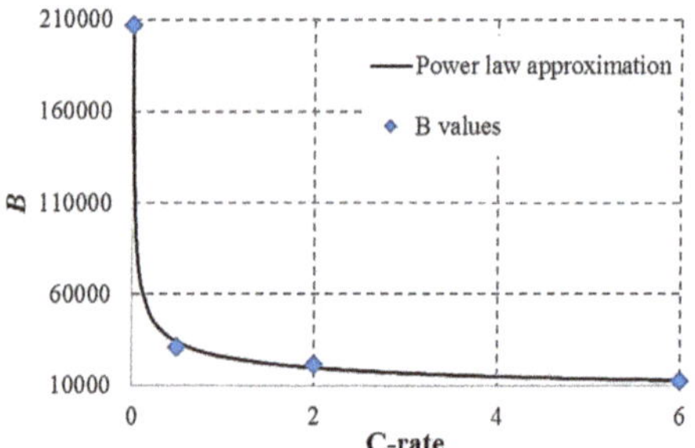

Figure 7. Experimental values [21] and fitting curve of *B* coefficient in the battery model (4), vs. *C-rate*.

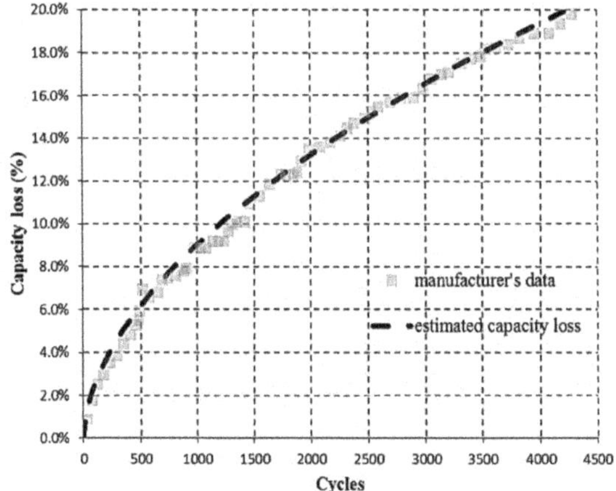

Figure 8. Capacity loss vs. number of cycles for the experimental $LiFePO_4$ battery string: comparison between cycle-life simulation model and experimental data from the manufacturer.

Manufacturer data reported in Figure 8 refer to complete charge-discharge cycles (*DOD* = 100%), with an average *C-rate* of 0.4C; the measured average battery temperature during each test cycle was T = 29 °C. Note that in this context (and throughout the paper) charge and discharge are calculated with reference to the commercial rating of the battery module, i.e., 185 Ah. The estimated capacity loss curve in Figure 8 has been computed by using (4), with T = 302.15 K, *C-rate* = 0.4, and B = 37,382.5 (as yielded by (5) with *C-rate* = 0.4).

5. Results of the Terna Experimental Primary Frequency Control Application

In this section, Terna experimental test results are reported and compared to simulations, in order to validate the electrical-thermal-ageing model presented in Section 4. The experimental test refers to a one-day period (i.e., 86,400 s), during which the battery string has been cycled by using the frequency profile described in Section 3.3, with a *C-rate* 1C and 0.5% droop. During the test, when the battery was completely discharged, PFC service was interrupted and the battery was completely re-charged (recharge time is 4 h), as in the actual operation of the Terna 1 MW/1 MWh BESS. This full recharge phase is mainly necessary in order to recalibrate the SOC estimation (performed by integrating the current flowing through the battery), which otherwise would be increasingly affected by the accumulation of measurement errors.

Figure 9a shows the comparison between measured and calculated battery string voltage during the one-day period, whereas in Figure 9b a zoom of the measured and calculated voltages in the time window between t = 13,000 s and t = 14,000 s is reported. Simulation results agree very well with measured results when the battery string is performing PFC, whereas substantial differences are evidenced during re-charging in Figure 9a, approximately from t = 47,000 s to t = 62,000 s. These differences are due to the lack of capacitances in the electrical model of the battery, which is more suitable for simulating PFC instead of continuative charge or discharge periods. The small differences between measured and calculated voltage shown in Figure 9b depend on small mismatches between actual and calculated SOC of the battery. The average error on voltage, calculated as:

$$V_{error,avg} = \frac{\sum_{i=1}^{N} \left| V_{calc,i} - V_{meas,i} \right|}{N}, \quad (6)$$

equals 6.09 V, i.e., less than 2.4% of rated string voltage.

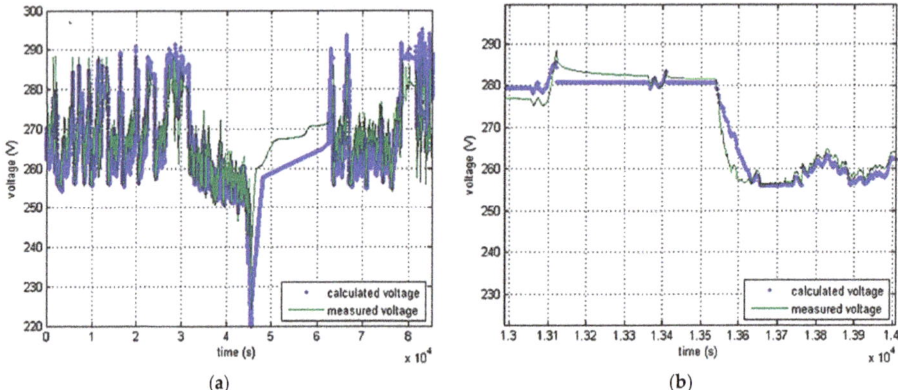

Figure 9. PFC tests, measured and calculated battery voltage vs. time: (**a**) voltage vs. time during 1 day; and (**b**) detail in the time range 13,000–14,000 s.

Figure 10a shows the comparison between measured and calculated battery string current during the one-day test period, whereas Figure 10b details the time window between t = 13,000 s and t = 14,000 s. Both figures show a very good agreement: the average error on current is 3.63 A, i.e., 2% of rated current.

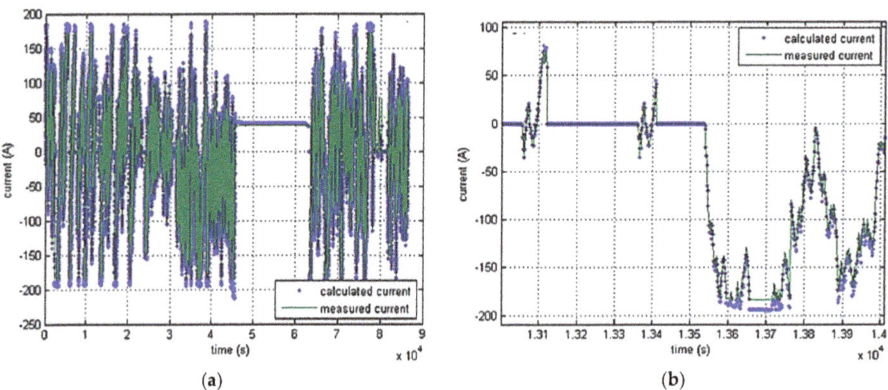

Figure 10. PFC tests, measured and calculated battery current vs. time: (**a**) current vs. time during 1 day; and (**b**) detail in the time range 13,000–14,000 s.

Lastly, Figure 11 reports the measured and calculated battery mean temperature T_m (i.e., the average between the temperatures of each cell in the battery string); the measured "ambient" temperature in the cabinet T_0 (i.e., the temperature imposed by the HVAC) is also shown. Measured and calculated values of T_m are in acceptable agreement, whereas the measured T_0 is always very close to the target value T_0 = 20 °C, thus confirming the approximation made in Equation (2), where ΔT is calculated considering a constant T_0.

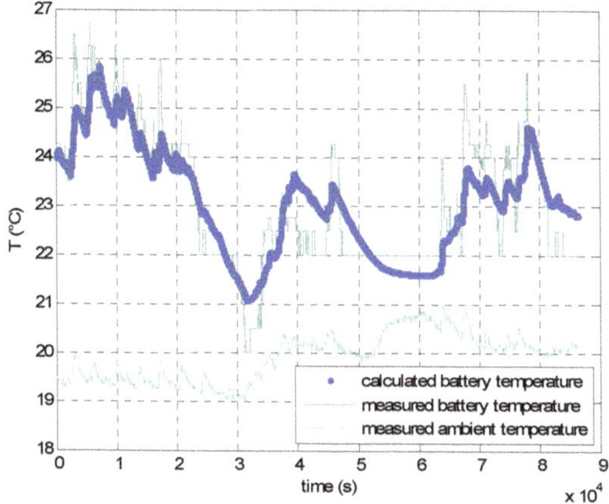

Figure 11. PFC tests, measured and calculated battery mean temperature T_m, and measured ambient (cabinet) temperature T_0, vs. time.

6. Primary Frequency Control Simulation Results

PFC simulations were carried out on a 50 kWh $LiFePO_4$ battery: since the BESSs is modular, results can be easily scaled to other BESS sizes. PFC operation of the 50 kWh battery was simulated for seven different droop values, namely 0.075%, 0.1%, 0.25%, 0.5%, 1%, 2% and 4%, and four different *C-rate* values, i.e., C/2, 1C, 2C and 4C, considering in all cases the frequency profile described in Section 3.3. Battery performance was evaluated in terms of:

- Average number of charge-discharge cycles per day;
- Overall battery efficiency η_{TOT}, including auxiliary consumptions calculated with (3) and assuming 96% PCS efficiency [26];
- Mean temperature, T_m, of the battery in operation (maximum allowed battery temperature T_{max} is 55 °C);
- Mean *C-rate* during the whole operation;
- Battery power-frequency characteristic λ (kW/Hz);
- Expected life;
- Unavailability of the battery rack for the PFC service (due to SOC outside the operating range or T_m exceeding the maximum temperature T_{max} = 55 °C), in percent of the overall operation time.

Results are summarized in Tables 4–7. These also include the equivalent power-frequency characteristic λ (kW/Hz), calculated from rated energy W_{rated} (kWh), *C-rate* and droop *s*:

$$\lambda = \frac{C\text{−}rate \cdot W_{rated}}{f_n} \cdot \frac{100}{s}. \tag{7}$$

Table 4. PFC results: *C-rate* = C/2, T_0 = 20 °C.

s (%)	Cycles per Day	η_{Tot} (%)	T_m (°C)	Mean *C-rate*	λ (kW/Hz)	Life (Years)	Not Operated (%)
0.075	2.51	82.35	21.7	0.21	666.7	5.43	0
0.1	2.03	81.07	21.3	0.17	500	6.03	0
0.25	0.92	72.65	20.4	0.08	200	8.31	0
0.5	0.47	59.67	20.2	0.04	100	10.48	0
1	0.24	41.54	20.1	0.02	50	13.05	0
2	0.12	22.44	20.1	0.01	25	16.07	0
4	0.06	9.31	20.0	0.005	12.5	20	0

Table 5. PFC results: *C-rate* = 1C, T_0 = 20 °C.

s (%)	Cycles per Day	η_{Tot} (%)	T_m (°C)	Mean *C-rate*	λ (kW/Hz)	Life (Years)	Not Operated (%)
0.075	5.09	83.28	27.0	0.42	1333	2.76	4.6
0.1	4.24	82.95	25.5	0.35	1000	3.33	4.6
0.25	1.85	79.31	21.7	0.15	400	6.10	0
0.5	0.94	72.10	20.7	0.08	200	8.11	0
1	0.47	59.55	20.3	0.04	100	10.40	0
2	0.24	41.50	20.1	0.02	50	13.05	0
4	0.12	22.40	20.1	0.01	25	16.07	0

Table 6. PFC results: *C-rate* = 2C, T_0 = 20 °C.

s (%)	Cycles per Day	η_{Tot} (%)	T_m (°C)	Mean *C-rate*	λ (kW/Hz)	Life (Years)	Not Operated (%)
0.075	8.86	79.90	43.9	0.74	2667	0.63	11.1
0.1	7.54	80.41	38.6	0.63	2000	0.99	4.1
0.25	4.14	80.02	28.7	0.34	800	2.64	2.3
0.5	1.88	77.68	22.7	0.16	400	5.51	0
1	0.95	71.04	21.1	0.08	200	7.85	0
2	0.48	59.00	20.4	0.04	100	10.31	0
4	0.24	41.29	20.2	0.02	50	12.94	0

Table 7. PFC results: *C-rate* = 4C, T_0 = 20 °C.

s (%)	Cycles per Day	η_{Tot} (%)	T_m (°C)	Mean *C-rate*	λ (kW/Hz)	Life (Years)	Not Operated (%)
0.075	7.55	72.72	54.5	0.63	5333	0.34	54
0.1	7.73	73.26	54.3	0.64	4000	0.40	48.3
0.25	5.88	75.11	42.2	0.49	1600	0.86	15.5
0.5	3.78	76.73	31.1	0.31	800	2.28	0
1	1.91	74.85	24.5	0.16	400	4.78	0
2	0.96	69.43	21.7	0.08	200	7.49	0
4	0.48	58.30	20.6	0.04	100	10.15	0

Data in Tables 4–7 are re-arranged in graphical form as Figures 12 and 13. Figure 12a plots the overall efficiency vs. expected battery life for different droop values, whereas in Figure 12b curves of efficiency vs. expected life are shown for different *C-rate* values.

Figure 12 shows that PFC operation with low droop values results in a good overall efficiency, even exceeding 80% as shown in Figure 12a, especially with the lower simulated *C-rate* values as shown in Figure 12b. Higher efficiencies, however, are traded with expected life values much shorter than the 20 years conventional BESS calendar life, because very low droops are associated to more sustained cycling. This is the limiting factor on efficiency for the extreme simulated combinations of low droop and high *C-rate*, which are also associated to the onset of operating constraints such as battery overtemperature and SOC limits, which limit battery utilization. Conversely, low *C-rates* combined with higher droop result in much longer battery expected life, at the expense of a sharp decrease in overall efficiency η_{TOT} due to low battery utilization.

Efficiency and expected life values from Tables 4–7 are plotted in Figure 13 as a function of the equivalent power-frequency characteristic λ. Figure 13 shows that combinations of droop and *C-rate* yielding the same λ largely result in similar lifetimes and efficiencies, so that battery power-frequency characteristic could be taken as the actual design parameter for performing PFC with a BESS. As long as the system is linear and time-invariant, the same λ values lead to the same results. However, both the model and the control strategy (which includes dead band, recharge phase, temperature and SOC limits) are not linear, resulting in some differences.

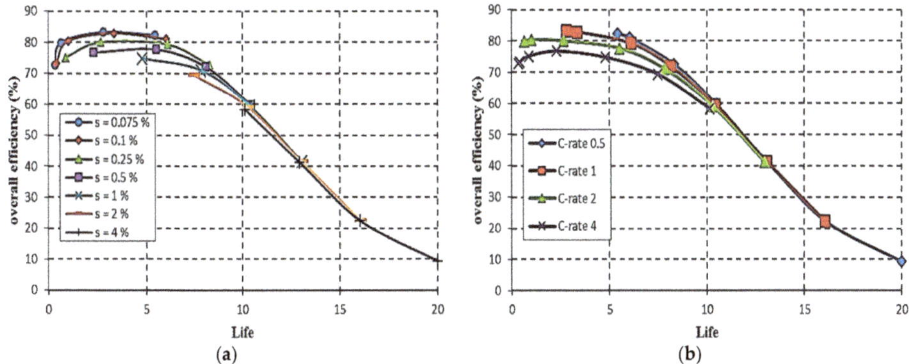

Figure 12. Overall efficiency vs. expected life: (**a**) as a function of droop; (**b**) as a function of *C-rate*.

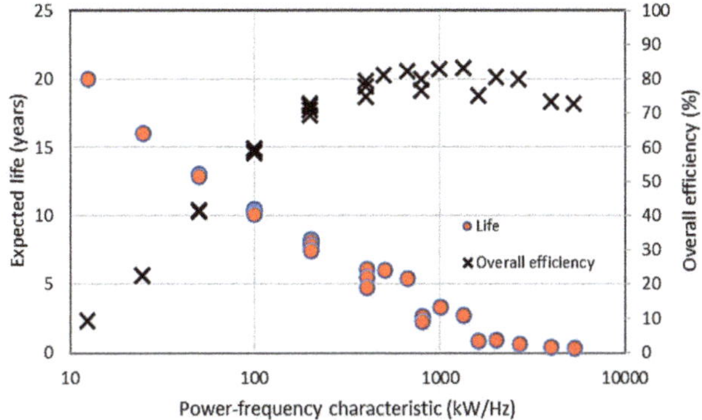

Figure 13. Expected battery life and overall efficiency vs. power-frequency characteristic.

Such differences are smaller when the battery is less stressed (i.e., at low λ values), and increase with increasing λ values. It should be pointed out that the longer BESS lifetimes predicted at low λ values could be offset by components having a shorter life than the battery itself, notably the electronic equipment.

To carry out an economic evaluation of the PFC application, applicable (i.e., national) rules of the electric energy market must be considered, assuming that a PFC market exists. Considering the different national approaches and the relative volatility of the regulatory framework for ancillary services, only general economic remarks can be made here.

Notably, given the short BESS lifetime under intensive cycling, the most favorable scenario would seem to be a capacity-based PFC market (especially in association to high *C-rate* values), whereas operation in an energy-based PFC market, such as in Italy [41], seems much less promising due to need for sustained cycling. Taking the capacity-based German PFC market as the reference market, some rough net present values (*NPV*) calculations may be made for the Terna's 1 MWh BESS. Since such a market remunerates primary control for each MW of reserve deployed when Δf = 200 mHz [42,43], s = 0.4% is the droop value required in order to fully exploit the 1 MWh BESS for PFC (Equation (1)). Table 8 reports PFC results obtained for s = 0.4% and for different *C-rate* values.

Table 8. PFC results for s = 0.4%, T_0 = 20 °C.

C-rate	Cycles per Day	η_{Tot} (%)	T_m (°C)	Mean *C-rate*	λ (kW/Hz)	Life (Years)	Not Operated (%)
0.5	0.58	59.66	20.3	0.05	125	9.8	0
1	1.17	71.85	21.0	0.1	250	7.44	0
2	2.34	77.04	23.8	0.20	500	4.70	0
4 (1)	4.57	75.31	35.1	0.38	1000	1.57	1.5

(1) *C-rate* values higher than 2C could be allowable in the next future if the 30-min criterion (actually adopted in the German PFC market) is relaxed to 15-min criterion.

Figure 14 reports *NPVs*, calculated for s = 0.4%, as a function of *C-rate*. A 3.6% capitalization factor and a 20-year BESS operation period have been considered. With reference to recent Terna BESS projects [24,44], battery cost has been set to 0.4 M€/MWh (if the battery life is shorter than 20 years, replacement cost is accounted as yearly economic losses equal to the ratio between battery cost and estimated battery life); fixed costs (civil works, MV switchgear, control system) have been set to 0.8 M€/MWh; PCS cost has been considered to be 0.2 M€/MW. The cost of losses, which are evaluated by means of the overall efficiency η_{TOT}, has been set at 140 €/MWh, whereas weekly revenues have been set to $R_{w,unitary}$ = 3000 €/MW/week, a typical value for the German market [45]. For the same 1 MWh battery, consideration of different nominal *C-rates* (namely 0.5C, 1C, 2C and 4C) leads to different unitary BESS active power capabilities (0.5 MW, 1 MW, 2 MW and 4 MW, respectively). As a consequence, the weekly revenues (which depends on BESS active power capability) linearly depend on *C-rates*. The yearly incomes $R_{y,tot}$ have thus been evaluated as:

$$R_{y,tot} = \frac{365}{7} \cdot R_{w,unitary} \cdot C{-}rate. \tag{8}$$

Due to the modular design of BESS [24], this implies that the economic evaluation for a larger system could be simply carried out by scaling up the above-described implementation and its attendant costs and revenues.

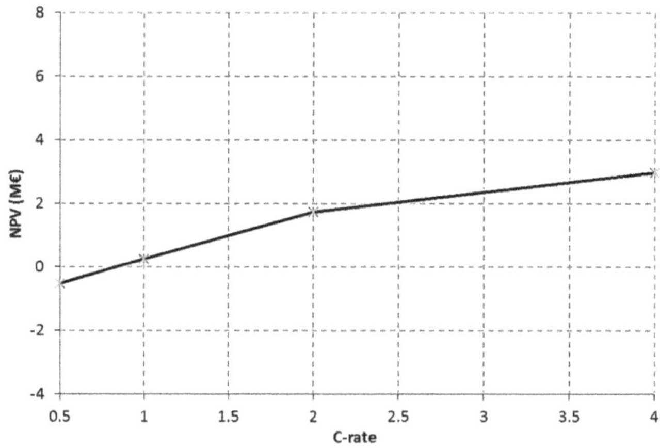

Figure 14. Estimated *NPV* for a 1 MWh BESS, for different *C-rates* (0.4% droop, capacity-based PFC market).

7. Conclusions

The paper studied the application of a $LiFePO_4$ BESS to primary frequency control, in the ENTSO-E Continental Europe grid; technical data from Terna's (the Italian TSO) experimental system has been used for defining BESS characteristics. BESS behavior has been simulated by the combination of an electrical-thermal circuit model with a life-cycle model predicting the capacity loss of the $LiFePO_4$ battery due to charge-discharge cycles. The complete electrical-thermal-ageing model has been subsequently validated by comparisons both with experimental tests carried out by Terna and with manufacturer data. Lastly, numerical PFC simulations have been performed by using actual Italian (ENTSO-E continental system) frequency recordings and considering a conventional, proportional "governor" for the BESS, for a wide range of different droop/*C-rate* combinations.

The main result evidenced by the simulations is that high overall BESS efficiency and expected lifetime are conflicting requirements. High efficiency in PFC service is associated to *C-rate*/droop combinations yielding high values of the power-frequency characteristic λ, which naturally results in sustained cycling of the BESS that drastically shortens expected battery lifetime. As an example, for the simulated and tested 50 kW-50 kWh battery string the choice of λ = 200 kW/Hz (*C-rate* = 1C, 0.5% droop) results in an overall efficiency exceeding 72%, but the expected lifetime is about 8 years. For a given frequency profile, lower λ values (e.g., associated to the usual 4%–5% droop of conventional generators' governors) result in less battery cycling and longer lifetimes, possibly approaching the conventional 20 years value, albeit with much lower efficiencies due to auxiliary losses.

The paper showed the technical feasibility of $LiFePO_4$ BESS use in primary frequency control, evidencing that there is a significant trade-off between expected lifetime and overall efficiency, restricting the choice of operating parameters to a rather narrow band; in the studied system, lifetimes in excess of 10 years are actually associated to efficiencies below 60%, mainly because of BESS underutilization.

Besides the above reported technical issues, an economic evaluation would depend on the specific (national) rules of the PFC market, whether capacity-based or energy-based. Considering a capacity–based market, such as the German one, results show that high *C-rate* values ($\geq$1C) seem to be more profitable.

Author Contributions: Terna S.p.A. (Massimo Rebolini and Francesco Palone) conceived, designed and commissioned the experiments; Francesco Palone and Marco Maccioni analyzed the data; Francesco Palone, Marco Maccioni, Alberto Geri, Fabio Massimo Gatta, Stefano Lauria, Regina Lamedica and Alessandro Ruvio wrote the paper.

Conflicts of Interest: The authors declare no conflict of interest.

References

1. Yang, Y.; Li, H.; Aichhorn, A.; Zheng, J.; Greenleaf, M. Sizing strategy of distributed battery storage system with high penetration of photovoltaic for voltage regulation and peak load shaving. *IEEE Trans. Smart Grid* **2014**, *5*, 982–991. [CrossRef]
2. Choi, J.H.; Kim, J.C. Advanced voltage regulation method of power distribution systems interconnected with dispersed storage and generation systems. *IEEE Trans. Power Deliv.* **2001**, *16*, 329–334. [CrossRef]
3. Cresta, M.; Gatta, F.M.; Geri, A.; Maccioni, M.; Mantineo, A.; Paulucci, M. Optimal operation of a LV distribution network with renewable DG by NaS battery and demand response strategy: A case study in a trial site. *IET Renew. Power Gen.* **2015**, *9*, 549–556. [CrossRef]
4. Borsche, T.; Ulbig, A.; Koller, M.; Andersson, G. Power and energy capacity requirements of storages providing frequency control reserves. In Proceedings of the 2013 IEEE Power and Energy Society General Meeting, Vancouver, BC, Canada, 21–25 July 2013.
5. Mégel, O.; Mathieu, J.L.; Andersson, G. Maximizing the potential of energy storage to provide fast frequency control. In Proceedings of the 4th IEEE Power and Energy Society Innovative Smart Grid Technologies Europe (ISGT Europe 2013), Copenhagen, Denmark, 6–9 October 2013.
6. Oudalov, A.; Chartouni, D.; Ohler, C. Optimizing a battery energy storage system for primary frequency control. *IEEE Trans. Power Syst.* **2007**, *22*, 1259–1266. [CrossRef]
7. Falvo, M.C.; Lamedica, R.; Bartoni, R.; Maranzano, G. Energy saving in metro-transit systems: Impact of braking energy management. In Proceedings of the 2010 IEEE International Symposium on Power Electronics, Electrical Drives, Automation and Motion (SPEEDAM 2010), Pisa, Italy, 14–16 June 2010.

8. Falvo, M.C.; Lamedica, R.; Ruvio, A. An environmental sustainable transport system: A trolley-buses line for Cosenza city. In Proceedings of the 2012 IEEE International Symposium on Power Electronics, Electrical Drives, Automation and Motion (SPEEDAM 2012), Sorrento, Italy, 20–22 June 2012.
9. Falvo, M.C.; Lamedica, R.; Ruvio, A. Energy storage application in trolley-buses lines for a sustainable urban mobility. In Proceedings of the 2012 IEEE International Symposium on Electrical Systems for Aircraft, Railway and Ship Propulsion (ESARS 2012), Bologna, Italy, 16–18 October 2012.
10. Calderaro, V.; Galdi, V.; Graber, G.; Piccolo, A. Siting and sizing of stationary supercapacitors in a metro network. In Proceedings of the 2013 AEIT Annual Conference, Mondello, Italy, 3–5 October 2013.
11. Barton, J.; Infield, D. Energy storage and its use with intermittent renewable energy. *IEEE Trans. Energy Convers.* **2004**, *19*, 441–448. [CrossRef]
12. Schainker, R. Executive overview: Energy storage options for a sustainable energy future. In Proceedings of the IEEE Power Engineering Society General Meeting, Denver, CO, USA, 6–10 June 2004.
13. Leonhard, W.; Grobe, E. Sustainable electrical energy supply with wind and pumped storage—A realistic long-term strategy or Utopia? In Proceedings of the IEEE Power Engineering Society General Meeting, Denver, CO, USA, 6–10 June 2004.
14. Meneses de Quevedo, P.; Contreras, J. Optimal placement of energy storage and wind power under uncertainty. *Energies* **2016**, *9*, 528. [CrossRef]
15. Grasselli, U.; Lamedica, R.; Prudenzi, A. Time-varying harmonics of single-phase non-linear appliances. In Proceedings of the IEEE Power Engineering Society Winter Meeting, New York, NY, USA, 27–31 January 2002.
16. Lamedica, R.; Maranzano, G.; Marzinotto, M.; Prudenzi, A. Power quality disturbances in power supply system of the subway of Rome. In Proceedings of the IEEE Power Engineering Society General Meeting, Denver, CO, USA, 6–10 June 2004.
17. Lamedica, R.; Marzinotto, M.; Prudenzi, A. Harmonic amplitudes and harmonic phase angles monitored in an electrified subway system during rush-hours traffic. In Proceeding of International Conference on Applied Simulation and Modelling (IASTED 2004), Rhodes, Greece, 28–30 June 2004.
18. Falvo, M.C.; Grasselli, U.; Lamedica, R.; Prudenzi, A. Harmonics monitoring survey on office LV appliances. In Proceedings of the 14th International Conference on Harmonics and Quality of Power (ICHQP 2010), Bergamo, Italy, 26–29 September 2010.
19. BYD Company. Available online: http://www.byd.com/energy/reference_ess.htm (accessed on 27 October 2016).
20. A123 Systems to Supply 20MW of Advanced Energy Storage Solutions to AES Gener for Spinning Reserve Project in Chile. Available online: http://www.a123systems.com/ca93980e-389a-40c6-86f9-b869feabe908/media-room-2011-press-releases-detail.htm (accessed on 27 October 2016).
21. Wang, J.; Liu, P.; Hicks-Garner, J.; Shermana, E.; Soukiazia, S.; Verbrugge, M.; Tataria, H.; Musser, J.; Finamore, P. Cycle-life model for graphite-$LiFePO_4$ cells. *J. Power Sources* **2011**, *196*, 3942–3948. [CrossRef]
22. Swierczynski, M.; Stroe, D.I.; Stan, A.I.; Teodorescu, R. Primary frequency regulation with Li-ion battery energy storage system: A case study for Denmark. In Proceedings of the 5th IEEE Annual International Energy Conversion Congress and Exhibition, Melbourne, Australia, 3–6 June 2013; pp. 487–492.
23. Swierczynski, M.; Stroe, D.I.; Stan, A.I.; Teodorescu, R.; Sauer, D.U. Selection and performance-degradation modeling of $LiMO_2/Li_4Ti_5O_{12}$ and $LiFePO_4/C$ battery cells as suitable energy storage systems for grid integration with wind power plants: An example for the primary frequency regulation service. *IEEE Trans. Sustain. Energy* **2014**, *5*, 90–101. [CrossRef]
24. Alì, A.; Gionco, S.; Palone, F.; Rebolini, M.; Polito, R. Sistemi di automazione e soluzioni impiantistiche per i sistemi di accumulo elettrochimici per la rete di trasmissione nazionale. *L'Energia Elettr.* **2014**, *91*, 71–84. (In Italian)
25. Tortora, A.C.; Senatore, E.; Apicella, L.; Polito, R. Sistemi di accumulo di energia elettrochimici per la gestione efficiente delle fonti rinnovabili non programmabili. *L'Energia Elettr.* **2014**, *91*, 35–46. (In Italian)
26. Andriollo, A.; Benato, R.; Bressan, M.; Dambone Sessa, S.; Palone, F.; Polito, R.M. Review of power conversion and conditioning systems for stationary electrochemical storage. *Energies* **2015**, *8*, 960–975. [CrossRef]
27. *Secondary Cells and Batteries Containing Alkaline or Other Non-Acid Electrolytes—Safety Requirements for Portable Sealed Secondary Cells, and for Batteries Made from Them, for Use in Portable Applications*, 2.0 ed.; IEC 62133; International Electrotechnical Commission (IEC): Geneva, Switzerland, 2012.

28. *Safety of Primary and Secondary Lithium Cells and Batteries during Transport*, 2.0 ed.; IEC 62281; International Electrotechnical Commission (IEC): Geneva, Switzerland, 2012.
29. Gatta, F.M.; Geri, A.; Lauria, S.; Maccioni, M.; Palone, F. Arc-flash in large battery energy storage systems—Hazard calculation and mitigation. In Proceedings of the 16th IEEE International on Environment and Electrical Engineering (EEEIC 2016), Florence, Italy, 7–10 June 2016.
30. Künisch, H.J.; Krämer, K.G.; Dominik, H. Battery energy storage another option for load-frequency-control and instantaneous reserve. *IEEE Trans. Energy Convers.* **1986**, *1*, 41–46. [CrossRef]
31. Koller, M.; Borsche, T.; Ulbig, A.; Andersson, G. Review of grid applications with the Zurich 1 MW battery energystorage system. *Electr. Power Syst. Res.* **2015**, *120*, 128–135. [CrossRef]
32. Department of Energy, DoE, Global Energy Storage Database. Available online: http://www.energystorageexchange.org/projects (accessed on 27 October 2016).
33. UCTE Continental Europe Operation Handbook. Available online: https://www.entsoe.eu/publications/system-operations-reports/operation-handbook/Pages/default.aspx (accessed on 27 October 2016).
34. Terna Rete Italia, 'Partecipazione Alla Regolazione di Frequenza e Frequenza-Potenza' (2008). Available online: http://download.terna.it/terna/0000/0105/32.pdf (accessed on 27 October 2016).
35. Terna Rete Italia, 'Qualità del Servizio di Trasmissione Rapporto Annuale per L'anno 2014' (2014). Available online: http://download.terna.it/terna/0000/0108/85.pdf (accessed on 27 October 2016).
36. Ke, M.Y.; Chiu, Y.H.; Wu, C.Y. Battery modelling and SOC estimation of a $LiFePO_4$ battery. In Proceedings of the 2016 International Symposium on Computer, Consumer and Control (IS3C 2016), Xi'an, China, 4–6 July 2016.
37. Huria, T.; Ceraolo, M.; Gazzarri, J.; Jackey, R. High fidelity electrical model with thermal dependence for characterization and simulation of high power lithium battery cells. In Proceeding of the 2012 IEEE International Electric Vehicle Conference (IEVC 2012), Greenville, SC, USA, 4–8 March 2012.
38. Waag, W.; Kabitz, S.; Sauer, D.U. Experimental investigation of the lithium-ion battery impedance characteristic at various conditions and aging states and its influence on the application. *Appl. Energy* **2013**, *102*, 885–897. [CrossRef]
39. Stroe, D.; Swierczynski, M.; Stan, A.I.; Teodorescu, R.; Andreasen, S.J. Experimental investigation on the internal resistance of Lithium iron phosphate battery cells during calendar ageing. In Proceedings of the 39th Annual Conference of the IEEE Industrial Electronics Society (IECON 2013), Vienna, Austria, 10–14 November 2013.
40. Gatta, F.M.; Geri, A.; Lauria, S.; Maccioni, M.; Palone, F. Battery energy storage efficiency calculation including auxiliary losses: Technology comparison and operating strategies. In Proceedings of the 2015 IEEE Powertech, Eindhoven, The Netherlands, 29 June–2 July 2015.
41. Specifiche tecniche per la verifica e valorizzazione del servizio di regolazione primaria di frequenza (2014). Available online: http://download.terna.it/terna/0000/0105/89.pdf (accessed on 27 October 2016). (In Italian)
42. Eckpunkte und Freiheitsgrade bei Erbringung von Primärregelleistung—Leitfaden für Anbieter von Primärregelleistung. Available online: https://www.regelleistung.net/ext/download/eckpunktePRL (accessed on 27 October 2016).
43. Zeh, A.; Müller, M.; Naumann, M.; Hesse, H.C.; Jossen, A.; Witzmann, R. Fundamentals of using battery energy storage systems to provide primary control reserves in Germany. *Batteries* **2016**, *2*, 29. [CrossRef]
44. Terna Open Construction Sites. Available online: http://www.terna.it/en-gb/cantieriapertitrasparenti.aspx (accessed on 27 October 2016).
45. Primary Control Reserve Tender Results. Available online: https://www.regelleistung.net (accessed on 27 October 2016).

Article

Optimal Scheduling and Real-Time State-of-Charge Management of Energy Storage System for Frequency Regulation

Jin-Sun Yang [1], Jin-Young Choi [1], Geon-Ho An [2], Young-Jun Choi [2], Myoung-Hoe Kim [2] and Dong-Jun Won [1,*]

[1] Department of Electrical Engineering, Inha University, Incheon 22212, Korea; sealoveyjs@gmail.com (J.-S.Y.); jy128308@gmail.com (J.-Y.C.)

[2] Department of Power Grid Integration of Research and Development (R&D) Center, Hyosung Corporation, Anyang 14080, Korea; geonhoan@hyosung.com (G.-H.A.); swot87@hyosung.com (Y.-J.C.); runner@hyosung.com (M.-H.K.)

* Correspondence: djwon@inha.ac.kr or djwon777@gmail.com; Tel.: +82-32-860-7404

Academic Editor: William Holderbaum

Received: 9 August 2016; Accepted: 22 November 2016; Published: 30 November 2016

Abstract: An energy storage system (ESS) in a power system facilitates tasks such as renewable integration, peak shaving, and the use of ancillary services. Among the various functions of an ESS, this study focused on frequency regulation (or secondary reserve). This paper presents an optimal scheduling algorithm for frequency regulation by an ESS. This algorithm determines the bidding capacity and base point of an ESS in each operational period to achieve the maximum profit within a stable state-of-charge (*SOC*) range. However, the charging/discharging efficiency of an ESS causes *SOC* errors whenever the ESS performs frequency regulation. With an increase in *SOC* errors, the ESS cannot respond to an automatic generation control (AGC) signal. This situation results in low ESS performance scores, and finally, the ESS is disqualified from performing frequency regulation. This paper also presents a real-time *SOC* management algorithm aimed at solving the *SOC* error problem in real-time operations. This algorithm compensates for *SOC* errors by changing the base point of the ESS. The optimal scheduling algorithm is implemented in MATLAB by using the particle swarm optimization (PSO) method. In addition, changes in the *SOC* when the ESS performs frequency regulation in a real-time operation are confirmed using the PSCAD/EMTDC tool. The simulation results show that the optimal scheduling algorithm manages the *SOC* more efficiently than a commonly employed planning method. In addition, the proposed real-time *SOC* management algorithm is confirmed to be capable of performing *SOC* recovery.

Keywords: energy storage system (ESS); frequency regulation (FR); optimal scheduling; state-of-charge (*SOC*); energy management

---

1. Introduction

Recent power systems are focusing on energy storage systems (ESSs) because of their ability to store energy. Many studies have examined ESSs to utilize renewable integration, peak shaving, ancillary services, and microgrids [1–10]. In [11], the authors presented the feasibility of an ESS for the operation of an island AC microgrid with photovoltaic generation. In [12], the authors proposed an ESS management algorithm for full renewable energy source (RES) exploitation. This algorithm was shown to minimize the curtailment of RES energy generation through energy buffering and forecasting error compensation by an ESS. A one-day-ahead scheduling procedure was combined with a real-time control strategy to improve RES generation and reliability. This study shows that excellent forecasting error compensation can be achieved even with ESSs of moderate size. In recent years, many studies have focused on the ability of ESSs to enable frequency regulation, given that its ramp rate is higher than that of conventional resources. In [13], the authors proposed an alternative market structure with the ability to efficiently demand response (DR) and automatic generation control (AGC) and accommodate the intermittency and uncertainty that are concomitant in renewable generation, thereby leading to the efficient integration of renewables at the market level and reduced regulation requirements at the AGC level. This paper proposed an integrated dynamic market mechanism (DMM) that combined a real-time market clearing procedure with AGC. DMM implementation enabled more frequent economic dispatch than the optimal power flow (OPF). Our proposed algorithm also reduces the root mean square error of the area control error by using an aggregate frequency error. In [14], the authors proposed a hybrid operation strategy for a wind energy conversion system with a battery energy storage system (BESS) to support frequency control. In their study, the output power command of the BESS was determined according to three factors: the state-of-charge (*SOC*), frequency deviation, and load variation. Their proposed operation strategy significantly improved the initial low-frequency response and provided a superior contribution to short-term frequency regulation. The research of [15] proposed an *SOC* feedback control scheme and investigated the performance of the grid frequency deviation response. The control was integrated with the widely used wind turbine blade pitch control and the speed governor control of a local synchronous generator. It was shown that the ESS helped supply power to wind farms to support frequency regulation and effectively regulated the battery *SOC*. An outcome of the increasing number of studies conducted on ESS frequency control is the proposal of a novel approach associated with the actual performance measurement of an ESS conducting frequency regulation [16]. This paper evaluated the current methods for procuring, dispatching, and compensating resources for frequency regulation. The authors also calculated the performance payment of the proposed novel approach by employing a sigmoid function. This approach influenced revenue by adjusting the calculation of the performance score. This approach will provide further insight into administrative price adjustment. The operator would be able to estimate the total mileage expected on an annual cost, and they would be able to calculate an administrative price that would reflect a fair compensation for the resources based on their actual frequency-regulation contribution and performance. In [17], the authors proposed a model that decided the optimal joint bidding strategy of battery storage in joint day-ahead energy, reserve, and regulation markets. Their novel algorithm considering the battery life cycle significantly improved a storage battery's overall economics in performance-based regulation.

At present, an ESS participates in the frequency regulation market by bidding its maximum capacity. However, an ESS is unable to manage its *SOC* because the power system operator sends an AGC signal without considering the *SOC* of an ESS. If an ESS is unable to manage its *SOC*, it does not respond to the AGC signal; therefore, an ESS should be able to manage its *SOC*. Many studies have focused on *SOC* management via day-ahead scheduling or via compensation by renewable generation; however, management by these methods in a real-time operation is difficult.

This study proposes an optimal scheduling algorithm and a real-time *SOC* management algorithm for frequency regulation by an ESS. The optimal scheduling algorithm determines the bidding capacity and base point through the particle swarm optimization (PSO) method for achieving the maximum profit through *SOC* management. Further, the real-time *SOC* management algorithm recovers the *SOC* in a real-time operation via participation in the energy market. This algorithm manages the *SOC* of an ESS by rebidding the scheduled base point in the energy market. Furthermore, a simulation is performed using actual PJM (Monroe, MN, USA) operation data. This paper is organized as follows. Section 2 explains the frequency regulation service and frequency regulation market. Section 3 presents the optimal scheduling algorithm and compares it with a commonly employed planning method. Section 4 presents the proposed real-time *SOC* management algorithm. Section 5 presents simulation cases and discusses the simulation results. Section 6 summarizes the conclusions and suggests future work.

2. Frequency Regulation Market

In a power system, the frequency changes continuously because of an imbalance of supply and demand. The system operators conduct frequency regulation or secondary reserve to reduce the fluctuations and thereby provide stable and reliable system operation. Frequency regulation involves the injection or withdrawal of as much active power as the assigned regulation capacity of the resource [18]. Two types of frequency regulation market exist. The first frequency regulation market type is separated into regulation up and regulation down. Examples include New York Independent System Operator (NYISO, Rensselaer, NY, USA) and California Independent System Operator (CAISO, Folsom, CA, USA), and ISO of Europe. The second frequency regulation market type does not separate regulation up and regulation down and examples for this type include PJM and Midcontinent Independent System Operator (MISO, Carmel, CA, USA). The second frequency regulation market type was assumed in this study because PJM data were used; the frequency regulation market in PJM was selected as the test bed, and simulation was performed according to PJM frequency regulation market rules. To provide frequency regulation, resources bid on the frequency regulation market according to the requisite capacity in each operation time [19]. The resources provide as much active power as the AGC signal whenever the resources receive an AGC signal; this output point is called the "set point". If the resources participate in the energy market, the resources provide as much active power as the bidding quantity every time; this output point is called the "base point". The resources participating in both frequency regulation and the energy market provide as much active power as the base point and perform frequency regulation whenever the resources receive an AGC signal by additionally providing as much active power as the set point [20]. Figure 1 illustrates the concept of the set point and base point of an ESS.

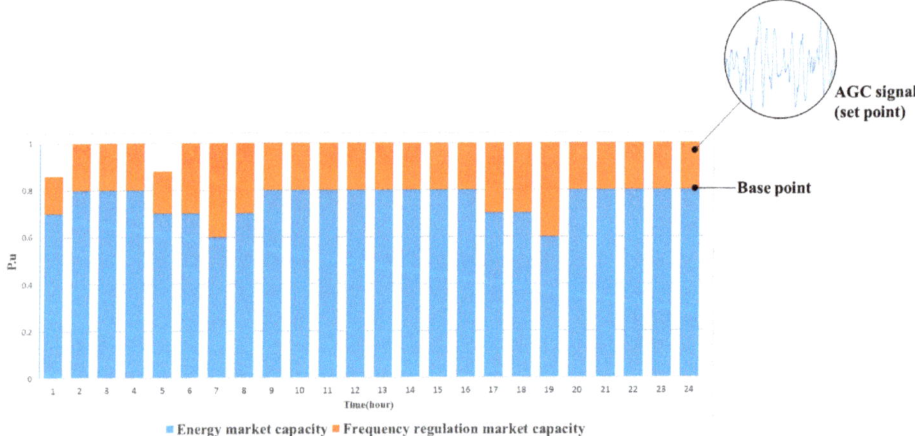

Figure 1. The concept of set point and base point. AGC: automatic generation control.

An ESS always provides active power according to the base point with the assigned energy and provides active power according to the set point when the ESS receives an AGC signal. Therefore, an ESS provides as much active power as the output point. The data used in this paper consist of the actual operation data of the PJM region. The hourly averaged AGC signal, regulation market capacity clearing price, regulation market performance clearing price, and locational marginal price were used to plan the optimal scheduling [21–23]. A real-time operation simulation was performed using an actual AGC signal. The data were collected on 5 September 2013. Figure 2 shows the hourly market price data, hourly averaged AGC signal, and actual AGC signal.

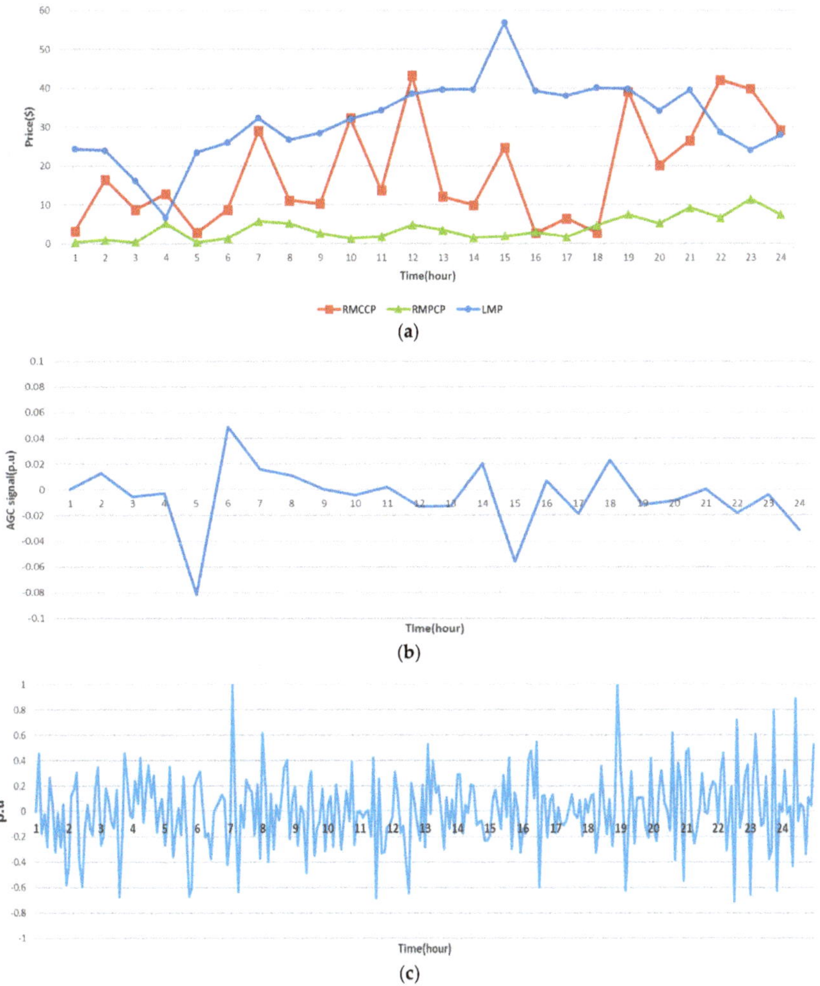

Figure 2. Actual market data in PJM (**a**) hourly market price; (**b**) hourly averaged AGC signal; and (**c**) AGC signal. *RMCCP*: regulation market capability clearing price; *RMPCP*: regulation market performance clearing price; and *LMP*: locational marginal pricing.

3. Optimal Scheduling Algorithm

To provide frequency regulation, ESS owners bid the maximum capacity in the day-ahead market, but they do not manage the *SOC* in the same manner. The optimal scheduling algorithm schedules the bidding plan of an ESS to earn the maximum profit within a stable *SOC* range. When a schedule is determined, ESS owners bid on the frequency regulation market in day-ahead market according to the schedule. For a scheduling plan, this study chose 4 MW/2 MWh Li-ion batteries as the ESS and set the charging/discharging efficiency to 91%. The initial *SOC* was set at 60%. This section compares the *SOC*, frequency regulation profit, and scheduling results of the maximum-capacity bidding plan and the optimal scheduling algorithm.

3.1. Maximum-Capacity Bidding Plan

The maximum-capacity bidding plan is a conventional bidding method in the frequency regulation market. In this plan, the base point is set to zero. Therefore, the bidding capacity is 4 MW, and the base point is set to 0 p.u. The *SOC* changes are calculated as follows Equation (1):

$$\begin{aligned} &\text{If } \overline{AGC}(i) \leq 0, SOC(i+1) = SOC(i) - \frac{c_{bid} \times \overline{AGC}(i)}{C_{rated}} \times \eta_c \times \Delta t; \\ &\text{If } \overline{AGC}(i) > 0, SOC(i+1) = SOC(i) - \frac{c_{bid} \times \overline{AGC}(i)}{C_{rated}} \times \frac{1}{\eta_d} \times \Delta t \end{aligned} \quad (1)$$

where i is the index of time (h), $SOC(i)$ is the *SOC* of the ESS at time i, c_{bid} is the bidding capacity of ESS, C_{rated} is the rated capacity of ESS, $\overline{AGC}(i)$ is the hourly averaged AGC signal, and η_c and η_d are the charging and discharging efficiencies of the ESS, respectively. Figure 3 presents the bidding capacity, base point, and *SOC* when the ESS is bid by the maximum-capacity bidding plan. In addition, Table 1 lists the frequency regulation profit.

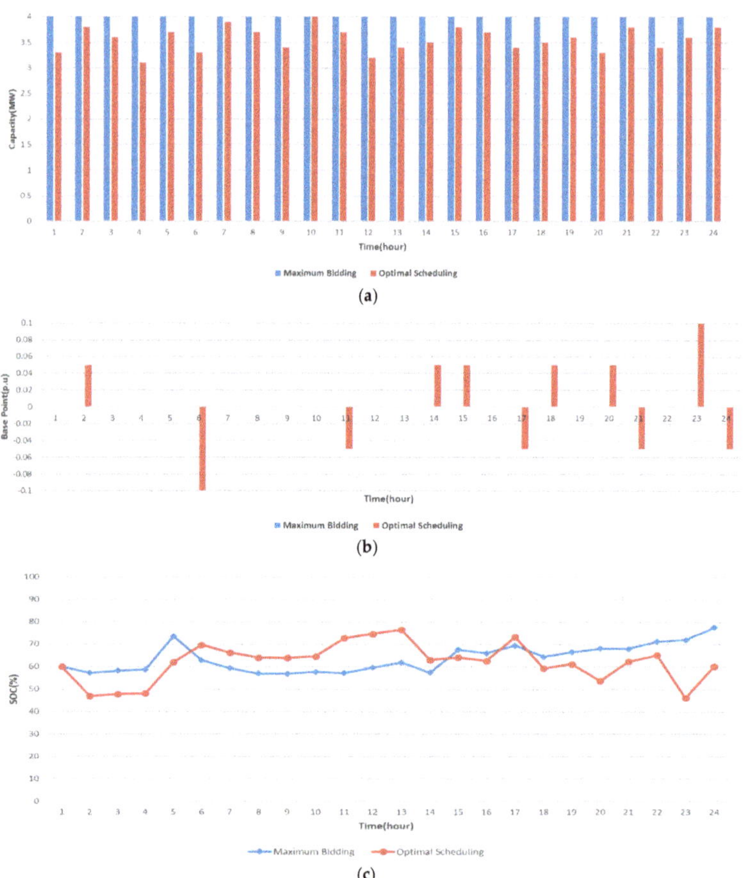

Figure 3. Maximum capacity bidding plan and optimal scheduling algorithm results (**a**) bidding capacity; (**b**) base point; and (**c**) *SOC*.

Table 1. Frequency regulation scheduling profits.

Method	Regulation Profit	Energy Profit	Total Profit
Maximum capacity bidding plan	$2773.4	$0	$2773.4
Optimal scheduling algorithm	$2483.1	$9.4	$2492.5

3.2. Optimal Scheduling Algorithm

The optimal-scheduling algorithm is a method used to earn the maximum profit within a stable *SOC* using the PSO method. PSO is a widely used traditional method for solving problems in power systems [24–27]. PSO solves the optimization problem as follows:

$$x^j_{k+1} = x^j_k + v^j_{k+1} \tag{2}$$

$$v^j_{k+1} = wv^j_k + c_1 r_1 \left(P^j_k - x^j_k\right) + c_2 r_2 \left(P^g_k - x^j_k\right) \tag{3}$$

where P^j_k and P^g_k are the individual best position and the global best position, respectively; k and j represent the iteration number and particle number, respectively; x is the position of the particle; v is the velocity; w is the inertia weight; c_1 and c_2 are acceleration coefficients, and r_1 and r_2 are randomly generated numbers in the range of [0,1]. The PSO method finds a solution to minimize the cost function, $F(x_i)$. To minimize the cost function $F(x_i)$ in particle j, PSO calculates the individual best position of particle j using each x_i and updates each x_i position as much as the velocities to find the individual best position in each iteration k. Finally, PSO is used to compare the individual best position to update the global best position and determine a solution.

To plan the optimal scheduling, the cost function, $F(x_i)$, was formulated as:

$$max \sum_{i=1}^{24} C_o(i) \tag{4}$$

$$C_o(i) = C_c(i) + C_p(i) - C_e(i) \tag{5}$$

$$C_c(i) = c(x_i) \times RMCCP(i) \times PS \tag{6}$$

$$C_p(i) = c(x_i) \times RMPCP(i) \times PS \times MR \tag{7}$$

$$C_e(i) = c(x_i) \times LMP(i) \times bp(x_i) \tag{8}$$

subject to the following constraints:

$$SOC(0) = SOC(24) \tag{9}$$

$$SOC_{\min} \le SOC(i) \le SOC_{\max} \tag{10}$$

$$\begin{aligned} &\text{If } \overline{AGC}(i) \le 0, SOC(i+1) = SOC(i) - \frac{c_{bid} \times \left[\overline{AGC}(i) + bp(x_i)\right]}{C_{rated}} \times \eta_c \times \Delta t; \\ &\text{If } \overline{AGC}(i) > 0, SOC(i+1) = SOC(i) - \frac{c_{bid} \times \left[\overline{AGC}(i) + bp(x_i)\right]}{C_{rated}} \times \frac{1}{\eta_d} \times \Delta t \end{aligned} \tag{11}$$

$$c(x_i) + c(x_i) \times bp(x_i) \le Maximum\ capacity \tag{12}$$

where $C_o(i)$ is the operation profit, $C_c(i)$ and $C_p(i)$ are the regulation capability credit and regulation performance credit, respectively; $C_e(i)$ is the cost by bidding in the energy market; $c(x_i)$ and $bp(x_i)$ are the scheduled bidding capacity and base point in the energy market by PSO, respectively; *RMCCP* (*i*), *RMPCP* (*i*), and *LMP* (*i*) are the hourly regulation market capability clearing price, the hourly regulation market performance clearing price, and the hourly locational marginal pricing, respectively. *PS* and *MR* are the performance score and mileage ratio, respectively. Note that the initial *SOC* and final *SOC* are considered in constraint Equation (9). For operation within a stable

SOC range, the SOC management constraint is set as in Equation (10). The SOC was calculated using constraint Equation (11). Constraint Equation (12) manages the bidding capacity up to the maximum capacity. For the optimal scheduling algorithm, the upper and lower limits of the stable SOC range, S_{min} and S_{max}, are set to 40% and 80%, respectively. Typically, ESSs achieve a performance score of 0.95; thus, PS was assumed to be 0.95. In addition, the MR on 5 September 2013 was 3. To perform optimal scheduling, the data described in Section 3 were used, and Equations (4)–(12) were solved in MATLAB/Simulink (MathWorks, Natick, MA, USA). Figure 3 shows the changes in the bidding capacity, base point, and SOC. Table 1 lists the frequency regulation profits. Regulation profit is the benefit participating frequency regulation market, and energy profit is the cost participating energy market. As shown in Table 1, the maximum-capacity bidding plan earned higher frequency regulation profit than that of the optimal scheduling algorithm. On the other hand, the optimal scheduling algorithm was advantageous in terms of SOC management. The maximum-capacity bidding plan did not manage the SOC, which can cause problems in the operation the next day; in contrast, the optimal scheduling algorithm did manage the SOC by using constraints Equations (9) and (10), which ensured stable operation every day.

4. Real-Time State-of-Charge Management Algorithm

In a real-time operation, the ESS performs frequency regulation in response to an AGC signal every two seconds. The charging/discharging efficiency of the ESS causes an SOC error whenever the ESS provides frequency regulation. Because the SOC error would prevent the ESS from providing frequency regulation, a novel SOC management algorithm is proposed. In a real-time operation, the SOC is calculated as follows:

$$\begin{aligned} &\text{If } AGC\,(z) \leq 0, SOC\,(z+1) = SOC\,(z) - \frac{c\,(x_i) \times [AGC\,(z) + bp\,(x_i)]}{C_{rated}} \times \eta_c \times \Delta t; \\ &\text{If } AGC\,(z) > 0, SOC\,(z+1) = SOC\,(z) - \frac{c\,(x_i) \times [AGC\,(z) + bp\,(x_i)]}{C_{rated}} \times \frac{1}{\eta_d} \times \Delta t \end{aligned} \quad (13)$$

where $AGC\,(z)$ is the actual AGC signal, and z is the frequency regulation interval. To solve the SOC error, this study presents a real-time SOC management algorithm. This algorithm changes the base point through hysteresis loop according to the SOC range. Figure 4 illustrates the concept of a real-time SOC management algorithm. When the SOC is between $SOC_{\text{min\_limit}}$ and $SOC_{\text{max\_limit}}$, the base point is set to $bp\,(x_i)$. The ESS then operates according to the optimal scheduling algorithm. When the SOC is above $SOC_{\text{max\_limit}}$ because of the accumulation of SOC errors, $bp\,(x_i+2)$ is set to BP_{dis} by rebidding the base point after two hours' operation according to PJM market rules. Owing to the changing base point, the ESS provides as much additional active power in the energy market as BP_{dis} until the SOC crosses $SOC_{\text{max\_end}}$. When the SOC crosses $SOC_{\text{max\_end}}$, the ESS bids the base point as $bp\,(x_i+2)$ in the optimal scheduling algorithm results. When the SOC is below $SOC_{\text{min\_limit}}$, $bp\,(x_i+2)$ is set to BP_{ch} by rebidding the base point, and the ESS provides as much additional active power in the energy market as BP_{ch} until the SOC crosses $SOC_{\text{min\_end}}$. When the SOC crosses $SOC_{\text{min\_end}}$, the ESS bids the base point as $bp\,(x_i+2)$ in the optimal scheduling algorithm results. Table 2 lists the parameters selected to maintain a stable SOC range for real-time SOC management. $SOC_{\text{min\_limit}}$ and $SOC_{\text{max\_limit}}$ was set to 45% and 75%, respectively. $SOC_{\text{min\_end}}$ and $SOC_{\text{max\_end}}$ was set to 50% and 70% respectively. The stable SOC range was set from 40% to 80%.

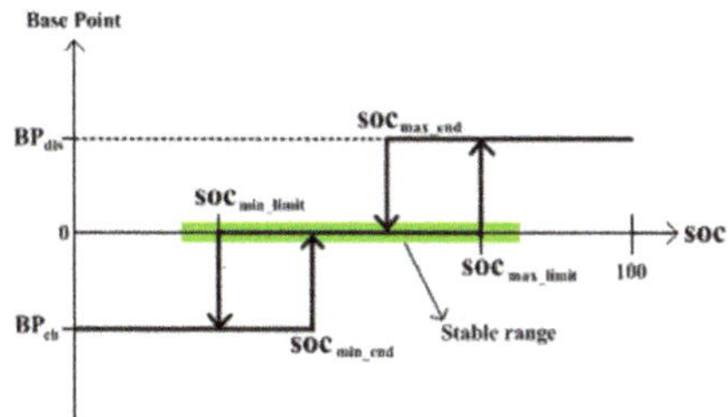

Figure 4. The concept of real-time *SOC* management algorithm.

Table 2. Simulation case results.

Case	Regulation Profit	Energy Profit	Total Profit	*SOC*
Case 1	$911.8	$0	$911.8	10%
Case 2	$1116.2	$−10.4	$1105.8	10%
Case 3 (0.05 p.u)	$2630.3	$−147.4	$2482.9	39.9%
Case 3 (0.1 p.u)	$2555.5	$−159.7	$2395.8	48.9%
Case 3 (0.15 p.u)	$2506.2	$−157.2	$2349	37.4%

5. Simulation Results

This section reports the simulation results to evaluate the performance of the proposed algorithm using a real operation parameter. In the first case, the maximum-capacity bidding plan was performed, the second case used the optimal scheduling algorithm, and the third case applied the real-time *SOC* management algorithm to the optimal scheduling algorithm. The PSCAD/EMTDC (Manitoba HVDC Research Centre, Winnipeg, MB, Canada) was used to confirm the *SOC* changes using Equation (12) when the ESS performed frequency regulation in a real-time operation. The actual AGC signal data in the PJM on 5 September 2013 were used to calculate the *SOC* in real-time. Because the AGC cycle of the PJM is 2 s, the *SOC* was calculated every 2 s. For the simulation, the scheduling was set as described in Section 4. To protect the ESS, the operational *SOC* range was set between 10% and 90%. Therefore, the ESS will shut down to protect itself when the *SOC* exceeds the operational *SOC* range. Case 1 is result by maximum capacity bidding plan, case 2 is result by optimal scheduling algorithm and case 3 is result by optimal scheduling algorithm with real-time *SOC* management by various base point. Cases 1 and 2 were compared to evaluate the optimal scheduling algorithm, which was confirmed by the frequency regulation profit change and increased operation time. In addition, Case 2 was compared with Case 3 to evaluate the real-time *SOC* management algorithm by changing the base point. Table 2 shows total case results.

5.1. Case 1

Figure 5 shows the *SOC* changes when the ESS performed frequency regulation through the maximum-capacity bidding plan. In a real-time operation, the ESS performs frequency regulation by charging and discharging in response to an AGC signal. At 1 a.m., the *SOC* difference between the real-time operation *SOC* (55.8%) and maximum-capacity bidding plan (59.8%) was 4% because of the charging/discharging efficiency. This *SOC* difference increased consistently when the ESS performed the frequency regulation in a real-time operation. At 10 a.m., the *SOC* reached 10%, and the ESS was shut down. When the ESS performed frequency regulation according to the maximum-capacity bidding plan, the performance score decreased because the ESS does not perform frequency regulation when the *SOC* reaches 10%. According to the decreasing performance score, the frequency regulation profit decreased from $2773.40 to $911.80.

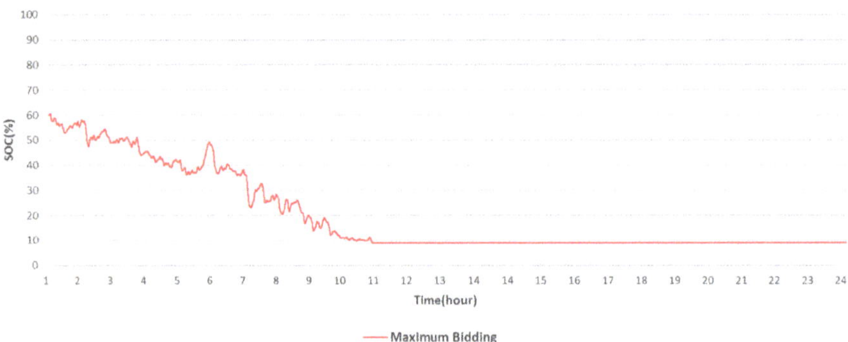

Figure 5. *SOC* of ESS in real-time operation simulation (Case 1—maximum bidding capacity plan).

5.2. Case 2

Figure 6 shows the *SOC* changes when the ESS performed frequency regulation through the optimal scheduling algorithm of Section 4. The *SOC* decreased consistently when the ESS performed frequency regulation. According to the optimal scheduling algorithm, the base point of the ESS was set to −0.1 p.u for charging the *SOC* at 6 a.m. By changing the base point, the ESS recovered the *SOC*. Nevertheless, the real-time *SOC* reached 10% at 1 p.m., and the ESS was shut down. On the other hand, the simulation result of the optimal scheduling algorithm operated 3 h longer than the maximum-capacity bidding plan because the optimal scheduling algorithm managed the *SOC* by constraint in Equation (10). In the planning step, the maximum-capacity bidding plan earned more frequency regulation profit than the optimal scheduling algorithm (more than $280.90); however, in the real-time operation, the optimal scheduling algorithm earned more frequency regulation profit than the maximum-capacity bidding plan (as much as $194.00) through its scheduling capacity and base point. This result means that the optimal scheduling algorithm can increase the operation time of the ESS and earn more frequency regulation profit.

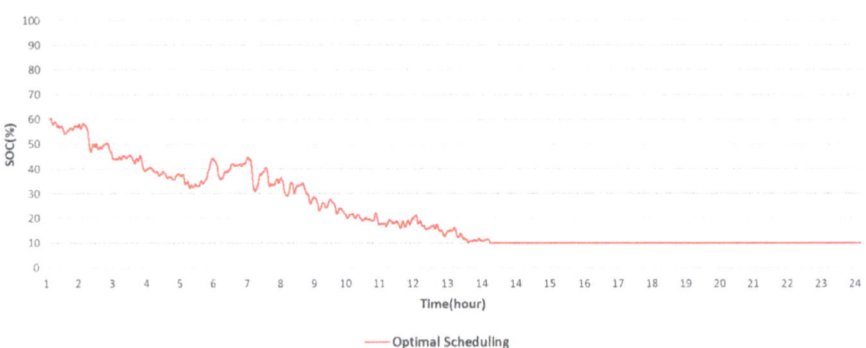

Figure 6. *SOC* of ESS in real-time operation simulation (Case 2—optimal scheduling algorithm).

5.3. Case 3

Figure 7 presents the *SOC* changes when the ESS performed frequency regulation through the optimal scheduling algorithm and real-time *SOC* management algorithm by various base points. In the 0.05-p.u case, at 3 a.m., the *SOC* reached 43.7%. Consequently, the bidding schedule of the base point was changed to −0.05 p.u through the real-time *SOC* management algorithm after 2 h according to the PJM market rules until 9 a.m., when the *SOC* reaches more than 50%. The bidding schedule can change 1 h before the operation time according to PJM market rules, thus, the bidding schedule was changed at 4 a.m. because the bidding schedule of 3 a.m. had already been submitted. Although the *SOC* reached more than 50% at 3 p.m., the bidding schedule was not changed until 4 p.m. for the same reason. Owing to the changed base point, the bidding capacity was also changed to 3.8 MW by constraint Equation (12). In this case, the ESS maintained recovery time for 16 h because 0.05 p.u was not enough for *SOC* recovery. The *SOC* reached 39.9% when the ESS finished operating. The frequency regulation earned a profit of $2482.90. In the 0.1-p.u case, the ESS maintained the recovery time for 8 h, and the 0.1-p.u case recovery was more rapid and stable than that of the 0.05-p.u case. The *SOC* reached 48.9% in the stable range, and the frequency regulation earned a profit of $2395.80. In the 0.15-p.u case, the ESS recovery was more rapid than that of the 0.1-p.u case; however, the 0.15-p.u case recovered too much because the *SOC* reached 81.7% at 5 p.m. The *SOC* management algorithm changed the base point to 0.15 p.u, causing the *SOC* to escape the stable range again. In this case, the ESS maintained the recovery time for 9 h, and the *SOC* reached 37.4%. The frequency regulation earned a profit of $2349. The ESS of Case 3 used the optimal scheduling algorithm and real-time *SOC* management algorithm to perform frequency regulation for a day. In contrast, Case 2, which used only the optimal scheduling algorithm, did not perform for a day. This result shows that when an ESS performs frequency regulation without an additional *SOC* management algorithm, the ESS does not perform for a day, and the proposed *SOC* management algorithm successfully managed the *SOC* of the ESS. Moreover, the base point is a significant factor in this algorithm because the base point determines the recovery quantity.

Figure 7. *SOC* of ESS in real-time operation simulation (Case 3—optimal scheduling algorithm with real-time *SOC* management).

Table 3 lists simulation results for different day. As shown Table 3, the base point is an important factor in this algorithm because the base point determines the recovery quantity of the *SOC*. Furthermore, for the determination base point, real-time *SOC* management algorithm should use accurate forecasting information.

Table 3. Additional simulation case results.

Optimal Scheduling & Real-Time *SOC* Managment	Base Point	Profit	*SOC*
2013.06.10	0.05	\$1254.2 → \$211.7	10%
	0.1	\$1254.2 → \$948.4	25.61%
	0.15	\$1254.2 → \$953.6	40.60%
2013.12.2	0.05	\$2232.4 → \$1857.1	51.42%
	0.1	\$2232.4 → \$1815.1	37.89%
	0.15	\$2232.4 → \$1778.1	42.32%
2014.03.10	0.05	\$4024.3 → \$3536.8	52.12%
	0.1	\$4024.3 → \$3301.4	65.56%
	0.15	\$4024.3 → \$3475.8	41.21%

6. Conclusions

This study proposed an optimal scheduling algorithm for frequency regulation by an ESS to ensure the realization of maximum profit within a stable *SOC* range. The maximum-capacity bidding plan is the best method for realizing frequency regulation profit, but it does not manage the *SOC*. The optimal scheduling algorithm schedules a bidding plan using the PSO method to realize the maximum frequency regulation profit within a stable *SOC* range. Although the optimal scheduling algorithm provides less frequency regulation profit than the maximum-capacity bidding plan, the former maintains stable operation through *SOC* management. The proposed algorithm was evaluated through simulation of three cases by means of the PSCAD/EMTDC tool in a real-time operation. The simulation results show that the optimal scheduling algorithm increases the duration of real-time operation and results in higher profits in comparison to the conventional bidding plan without *SOC* management. While the conventional maximum-capacity bidding plan provides more profit than the optimal scheduling algorithm in the planning step, the optimal scheduling algorithm provides more profit in real-time operations.

This study also proposed a real-time *SOC* management algorithm. The ESS constantly performs frequency regulation using this algorithm by which the ESS changes the base point to recover the *SOC* in a real-time operation. In the case where the real-time *SOC* management algorithm was applied in the PJM regulation market, it effectively recovered the *SOC* when the base point was changed to 0.1 p.u. The proposed real-time *SOC* management algorithm can be applied to various regulation market environments by changing the base point. This algorithm is expected to promote the participation of ESSs in various regulation markets.

Future work will deal with the forecasting error. The cases in this study were simulated using actual operation data, but the ESS uses forecast data during actual operation. Accordingly, the proposed algorithm should consider ESS potential forecasting error.

Acknowledgments: This research was supported by "Development of Operational criteria and Market trading procedure for Regulation Service using Battery Energy Storage" funded by Ministry of Trade, Industry & Energy of Korea (grant number: 1004101337X1); This work was supported by the Power Generation & Electricity Delivery Core Technology Program of the Korea Institute of Energy Technology Evaluation and Planning (KETEP), granted financial resource from the Ministry of Trade, Industry & Energy, Republic of Korea (No. 20142010103010); This research was supported by Basic Science Research Program through the National Research Foundation of Korea (NRF), funded by the Ministry of Science, ICT & Future Planning (grant number: NRF-2013R1A1A1012667).

Author Contributions: Jin-Sun Yang is lead author. Jin-Young Choi and Dong-Jun Won provided guidance. Geon-Ho An, Young-Jun Choi and Myoung-Hoe Kim provided Industrial research support.

Conflicts of Interest: The authors declare no conflict of interest.

References

1. Wu, D.; Tang, F.; Dragicevic, T.; Juan, C.; Vasquez, J.; Guerrero, J. A control architecture to coordinate renewable energy sources and energy storage systems in islanded microgrids. *IEEE Trans. Smart Grid* **2015**, *6*, 1156–1166. [CrossRef]
2. Karami, H.; Sanjari, M.; Hosseinian, S.; Gharehpetian, G. An optimal dispatch algorithm for managing residential distributed energy resources. *IEEE Trans. Smart Grid* **2014**, *5*, 2360–2367. [CrossRef]
3. Erdinc, O.; Paterakis, N.; Mendes, T.; Bakirtzis, A.; Catalão, J. Smart household operation considering bi-directional EV and ESS utilization by real-time pricing-based DR. *IEEE Trans. Smart Grid* **2015**, *6*, 1281–1291. [CrossRef]
4. Jiang, B.; Fei, Y. Smart Home in Smart Microgrid: A cost-effective energy ecosystem with intelligent hierarchical agents. *IEEE Trans. Smart Grid* **2015**, *6*, 3–13. [CrossRef]
5. Tran, D.; Khambadkone, A. Energy management for lifetime extension of energy storage system in micro-grid applications. *IEEE Trans. Smart Grid* **2013**, *4*, 1289–1296. [CrossRef]
6. Daneshi, H.; Srivastava, A. Security-constrained unit commitment with wind generation and compressed air energy storage. *IET Gener. Transm. Distrib.* **2012**, *6*, 167–175. [CrossRef]
7. Su, H.; Gamal, A. Modeling and analysis of the role of energy storage for renewable integration: Power balancing. *IEEE Trans. Power Syst.* **2013**, *4*, 4109–4117. [CrossRef]
8. Rahbar, K.; Xu, J.; Zhang, R. Real-time energy storage management for renewable integration in microgrid: An off-line optimization approach. *IEEE Trans. Smart Grid* **2015**, *6*, 124–134. [CrossRef]
9. Wang, P.; Gao, Z.; Tjernberg, L. Operational adequacy studies of power systems with wind farms and energy storages. *IEEE Trans. Power Syst.* **2013**, *27*, 2377–2384. [CrossRef]
10. Zhang, L.; Li, Y. Optimal Energy management of wind-battery hybrid power system with two-scale dynamic programming. *IEEE Trans. Sustain. Energy* **2013**, *4*, 765–773. [CrossRef]
11. Wu, D.; Tang, F.; Dragicevic, T.; Vasquez, J.; Guerrero, J. Autonomous active power control for islanded AC microgrids with photovoltaic generation and energy storage system. *IEEE Trans. Energy Convers.* **2014**, *29*, 882–892. [CrossRef]
12. Damiano, A.; Gatto, G.; Marongiu, I.; Porru, M.; Serpi, A. Real-time control strategy of energy storage systems for renewable energy sources exploitation. *IEEE Trans. Sustain. Energy* **2014**, *5*, 567–576. [CrossRef]
13. Shiltz, D.; Cvetkovi'c, M.; Annaswamy, A. An integrated dynamic market mechanism for real-time markets and frequency regulation. *IEEE Trans. Sustain. Energy* **2016**, *7*, 875–885. [CrossRef]
14. Choi, J.; Heo, S.; Kim, M. Hybrid operation strategy of wind energy storage system for power grid frequency regulation. *IET Gener. Transm. Distrib.* **2016**, *10*, 736–749. [CrossRef]
15. Dang, J.; Seuss, J.; Suneja, L.; Harley, R. *SOC* feedback control for wind and ESS hybrid power system frequency regulation. *IEEE J. Emerg. Sel. Top. Power Electron.* **2014**, *2*, 79–86. [CrossRef]
16. Papalexopoulos, A.; Andrianesis, P. Performance-based pricing of frequency regulation in electricity markets. *IEEE Trans. Power Syst.* **2014**, *29*, 441–449. [CrossRef]
17. He, G.; Chen, Q.; Kang, C.; Pinson, P.; Xia, Q. Optimal bidding strategy of battery storage in power markets considering performance-based regulation and battery cycle life. *IEEE Trans. Smart Grid* **2015**, *7*, 2359–2367. [CrossRef]
18. Singh, H.; Papalexopoulos, A. Competitive procurement of ancillary services by an independent system operator. *IEEE Trans. Power Syst.* **1999**, *14*, 498–504. [CrossRef]
19. Papalexopoulos, A.; Singh, H. On the various design options for ancillary services markets. In Proceedings of the 34th Hawaii International Conference on System Sciences (HICSS-34), Maui, HI, USA, 3–5 January 2001.
20. PJM Manual 11: Energy & Ancillary Services Market Operations. Available online: http://www.pjm.com/~/media/documents/manuals/m11.ashx (accessed on 9 April 2015).
21. Preliminary Billing Data. Available online: http://www.pjm.com/pub/account/pjm-regulation-data/201309.csv (accessed on 5 September 2013).
22. Market & Operation, Market-Based Regulation. Available online: http://www.pjm.com/~/media/markets-ops/ancillary/mkt-based-regulation/regulation-data.ashx (accessed on 13 August 2014).
23. Hourly Real-Time & Day-Ahead LMP. Available online: http://www.pjm.com/pub/account/lmpmonthly/201309-rt.csv (accessed on 2 January 2014).

24. Li, S.; Zhang, D.; Roget, A.; O'Neill, Z. Integrating home energy simulation and dynamic electricity price for demand response study. *IEEE Trans. Smart Grid* **2014**, *5*, 779–788. [CrossRef]
25. Faria, P.; Soares, J.; Vale, Z.; Morais, H.; Sousa, T. Modified particle swarm optimization applied to integrated demand response and DG resources scheduling. *IEEE Trans. Smart Grid* **2013**, *4*, 606–616. [CrossRef]
26. Nikmehr, N.; Ravadanegh, S. Optimal power dispatch of multi-microgrids at future smart distribution grids. *IEEE Trans. Smart Grid* **2015**, *6*, 1648–1657. [CrossRef]
27. Basu, A.; Bhattacharya, A.; Chowdhury, S.; Chowdhury, S. Planned scheduling for economic power sharing in a CHP-based micro-grid. *IEEE Trans. Power Syst.* **2012**, *27*, 30–38. [CrossRef]

Article

Optimal Scheduling of a Battery Energy Storage System with Electric Vehicles' Auxiliary for a Distribution Network with Renewable Energy Integration

Yuqing Yang [1,2], Weige Zhang [1,2,*], Jiuchun Jiang [1,2], Mei Huang [1,2] and Liyong Niu [1,2]

[1] National Active Distribution Network Technology Research Center (NANTEC), Beijing Jiaotong University, No. 3 Shang Yuan Cun, Haidian District, Beijing 100044, China; yangyuqing@bjtu.edu.cn (Y.Y.); jcjiang@bjtu.edu.cn (J.J.); mhuang@bjtu.edu.cn (M.H.); lyniu@bjtu.edu.cn (L.N.)

[2] Collaborative Innovation Center of Electric Vehicles in Beijing, No. 3 Shang Yuan Cun, Haidian District, Beijing 100044, China

* Author to whom correspondence should be addressed; wgzhang@bjtu.edu.cn; Tel.: +86-138-0100-6306; Fax: +86-10-5168-3907.

Academic Editor: William Holderbaum
Received: 13 August 2015; Accepted: 11 September 2015; Published: 25 September 2015

Abstract: With global conventional energy depletion, as well as environmental pollution, utilizing renewable energy for power supply is the only way for human beings to survive. Currently, distributed generation incorporated into a distribution network has become the new trend, with the advantages of controllability, flexibility and tremendous potential. However, the fluctuation of distributed energy resources (DERs) is still the main concern for accurate deployment. Thus, a battery energy storage system (BESS) has to be involved to mitigate the bad effects of DERs' integration. In this paper, optimal scheduling strategies for BESS operation have been proposed, to assist with consuming the renewable energy, reduce the active power loss, alleviate the voltage fluctuation and minimize the electricity cost. Besides, the electric vehicles (EVs) considered as the auxiliary technique are also introduced to attenuate the DERs' influence. Moreover, both day-ahead and real-time operation scheduling strategies were presented under the consideration with the constraints of BESS and the EVs' operation, and the optimization was tackled by a fuzzy mathematical method and an improved particle swarm optimization (IPSO) algorithm. Furthermore, the test system for the proposed strategies is a real distribution network with renewable energy integration. After simulation, the proposed scheduling strategies have been verified to be extremely effective for the enhancement of the distribution network characteristics.

Keywords: battery energy storage system (BESS); electric vehicles (EVs); optimal scheduling

1. Introduction

Renewable energy generation, such as photovoltaic (PV), wind, biomass, *etc.*, integrated into distribution power systems, expected to be one of the main solutions for clean power supply, will be considerably developed throughout the world during the next couple of decades. Currently, many countries have implemented or are in the process of implementing policies to promote renewable energy in the distribution network. This is because distributed energy resources (DERs) in the distribution power system could provide a better balance between the increasing electricity demand and traditional power exportation, reduce the power losses occurring in the feeders during energy transmission, as well as enhance the controllability of energy deployment, which would be the main component of the next generation distribution network framework, namely the active distribution network, with intelligent monitoring techniques and advanced management measures [1,2].

However, the fluctuation of DERs is still the main concern for large-scale implementation in low or medium voltage networks. Thus, the energy storage system (ESS) has to be involved to mitigate the bad effects of the DERs' integration. Compared to other types of ESS, a battery energy storage system (BESS) is relatively the most stable, easy to access and control, as an extremely effective way to cooperate with DERs. [3] Therefore, the operation strategies for BESS have become a research hotspot from different perspectives.

Actually, many researchers have proposed some optimal strategies to solve the BESS operation problems, as well as for EVs. In [4], an EV scheduling scheme has been proposed with an uncertain real-time price, taking the battery degradation into account. For another, a real-time scheduling strategy for EVs was presented in [5] to increase the voltage margin and tent to minimize the line loss. The two works above were inclined to solve the EV scheduling problem with different visions, the originality of which could also be applied in BESS scheduling. In [6], a mathematical model for a BESS scheduling procedure was proposed to simulate the charging/discharging process, with the objective of minimizing the line losses; however, only the aspect of the power losses was taken into account. Besides, BESS used for ramp rate control, frequency droop response, power factor correction, solar time-shifting and output leveling have been mentioned in [7], focusing on BESS operation to enable solar energy, which tends to solve BESS scheduling in a transient process. Furthermore, BESS was implemented to deal with power quality disturbances and to compensate reactive power in [8], as well as the optimal power flow in [9]. It is noted that the distribution network operation usually deals with power scheduling problems in the steady-state horizon, so in this paper, the BESS scheduling strategies, following the distribution network operation rules in China, are put forward with intervals of 15 min, which is also apparently a compromise decision between precision and computation quantity. Besides, Most of the strategies proposed from the works above were implemented one day ahead, considerably depending on the accuracy of prediction for renewable energy and power demand. Additionally, the errors from forecasting usually do not account for the evaluation of model validity, which may affect the applicability of the model when launching in practice. Therefore, a real-time strategy is the most effective approach for power scheduling.

In this paper, optimal scheduling strategies for BESS operation in both the day-ahead and real-time scale have been proposed, to minimize the renewable energy curtailment, to reduce active power loss, to mitigate the voltage fluctuation, as well as to lower the electricity cost. In addition, the EVs considered as the auxiliary technique are also introduced to attenuate the DERs' influence. Besides, all of the scheduling produced by the proposed strategies has considered the constraints of BESS and EVs' operation, as well as the power flow. Furthermore, the proposed scheduling strategies were simulated in a real distribution network with renewable energy integration, part of Beijing Jiaotong University power network, obtaining promising results and verifying the effectiveness of the proposed strategies.

The reminder of this paper is organized as follows: Section 2 describes the relationships between load variance and power loss, as well as between power deviation and voltage deviation. Additionally, some simplifications have been utilized. Section 3 formulates the multi-objective optimization model. Section 4 presents the procedure of the day-ahead and real-time strategy to solve the optimization problem, as well as the solution of the multi-objective optimization model. Section 5 introduces our numerical studies and analyzes the results, followed by conclusions in Section 6.

2. Problem Derivation

In [10], it has been noted that minimizing distribution system losses could be equally considered as maximizing the load factor and minimizing load variance if the feeder is a single line from the substation with all loads at the end of the line. Actually, this conclusion could be generalized in a radial distribution system under some assumptions as follows:

Assuming that the mean value of each load in the radial distribution power system could keep constant during normal operation.
The voltage fluctuation of the initial nodes is supposed to be neglected, in consideration of these nodes usually being connected to the substation.
Besides, the reactive power in the distribution network could be ignored since the power factor correction facilities take effect.

2.1. Relationship between Load Shaving and Power Loss

The model utilized in this derivation is shown in Figure 1. The active power loss for this single branch could be formulated as Equation (1).

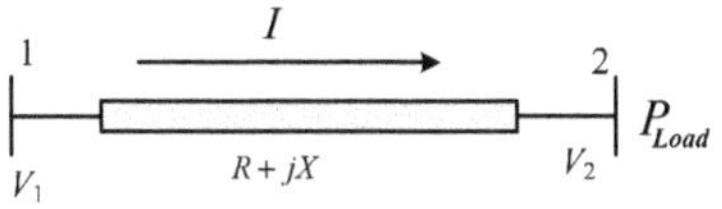

Figure 1. Diagram of a single branch.

$$P_{loss} = \sum_{t}^{T} I_t^2 R = \sum_{t}^{T} \left(\frac{P_t}{U_{rate}}\right)^2 R = \frac{\overline{P^2} R}{U_{rate}^2} \tag{1}$$

where P_{loss} is the active power loss of the single branch; I_t shows the current value running in the corresponding branch at time t; P_t represents the active power of Node 2 at time t; U_{rate} means the rated voltage in the assigned level; and R is the equivalent resistance value [10].

The load variance could be shown as,

$$\sigma_P^2 = \frac{1}{T} \cdot \sum_{t}^{T} (P_t - \overline{P})^2 \tag{2}$$

where σ_P^2 is the load variance and $\overline{P}$ is the mean value of the load; T represents the time duration, which is 24 h in this paper.

The conclusion could be derived in Equations (3) and (4),

$$\sigma_P^2 = \frac{1}{T} \cdot \sum_{t}^{T} (P_t)^2 - 2\overline{P}\frac{1}{T}\sum_{t}^{T} P_t + \frac{1}{T} \cdot \sum_{t}^{T} (\overline{P})^2 \tag{3}$$

$$\overline{P^2} = (\overline{P})^2 + \sigma_P^2 \tag{4}$$

Equation (4) could be plugged into Equation (1); in this way, the relation between P_{loss} and σ_P^2 is expressed in Equation (5),

$$P_{loss} = \frac{R}{U_{rate}^2} \cdot (\overline{P})^2 + \frac{R}{U_{rate}^2} \cdot \sigma_P^2 \tag{5}$$

In Equation (5), the active power loss is linear to the load variance in the single branch, if the deviation of the rated voltage could be neglected.

2.2. Relationship between Load Smoothing and Voltage Deviation

The voltage deviation in Figure 1 could be shown as Equation (6), and the relationship between real power and voltage magnitude could be easily derived based on Kirchhoff's theory in Equation (7).

Then, the combination of Equations (6) and (7) contributed to Equation (8), the relationship between power deviation and voltage deviation.

$$V_{dev} = \sum_{t=0}^{T} \Delta V_2{}^2 = \sum_{t=0}^{T} (V_{2,t+\Delta t} - V_{2,t})^2 \tag{6}$$

$$P_{2,t} = \frac{V_{2,t}(V_{1,t} - V_{2,t})}{R} = \frac{-V_{2,t}^2 + V_{1,t}V_{2,t}}{R} \tag{7}$$

$$P_{dev} = V_{dev} \sum_{t=0}^{T} (\frac{V_{2,t+\Delta t} + V_{2,t} - V_1}{R})^2 \tag{8}$$

In the equations above, $V_{1,t}$ and $V_{2,t}$ represent the voltage magnitude of Nodes 1 and 2 at time t, respectively. Additionally, $P_{2,t}$ is the real power of Node 2 at time t. When the subscript shows $t + \Delta t$ instead of t, it denotes the assigned quantity at time $t + \Delta t$.

In Equation (7), assuming constant $V_{1,t}$, with the consideration that Node 1 is close to the substation, it presents an inverted-U quadratic function relationship between $P_{2,t}$ and $V_{2,t}$. Clearly, $V_{2,t}$ is definitely more than $1/2\ V_{1,t}$ under normal circumstances, which means it would follow the right-half rules of quadratic function; $P_{2,t}$ would be decreasing synchronously with the increasing of $V_{2,t}$. Anyway, the variation absolute values show the same trend. Furthermore, in Equation (8), $(V_{2,t+\Delta t} + V_{2,t} - V_1)^2$ is apparently positive with the quadratic term; this also suggests the positive correlation relationship between $P_{2,t}$ and $V_{2,t}$.

Through the derivation procedure above, the strong positive correlation between load variance and power losses is shown; likewise, the strong positive correlation between power deviation and voltage deviation. For the traditional optimization process with a distribution network, the power flow calculation must be involved with a slow computation speed and week convergence degree. Hence, the tedious iterative work could be avoided, when load variance and power deviation minimizing, to simplify, are utilized for power losses and voltage deviation minimizing, respectively.

3. Model Formulation

In a distribution network with DER integration, BESS and EV are introduced for minimizing the bad side effect from DERs' access. An optimization framework has been proposed for reducing renewable energy curtailment, cutting feeder losses, mitigating voltage fluctuation and lowering electricity expense, considering the constraints of BESS and EV operation. The charging and discharging power of BESS and the charging period of EV are recognized to be control variables, and the optimization would be realized under minimizing one or more objective functions while satisfying the several equality and inequality constraints. Its mathematical model can be established as,

$$\begin{gathered} Min f(x) \\ s.t.\ g(x) = 0, h(x) \leq 0 \end{gathered} \tag{9}$$

where f is the objective function to be optimized; g and h are the equality and inequality constraints, respectively; x is the vector of charging or discharging power or the period selection variable. The detailed description of the objective functions, equality and inequality constraints are stated as follows.

3.1. Minimizing Renewable Energy Curtailment

It is widely accepted that, in power system operation, the power generated is always equal to the power demand in any moment. Assuming that there is no energy storage facility in the regional distribution network, the extra DERs' power has to be injected back to the substation or curtailed, resulting in extra power losses and unfortunate waste of renewable energy, when the DERs' generated power is in excess of the power demand. Therefore, the BESS is integrated into the regional distribution network with renewable energy integration for minimizing DER curtailment [11], as Equation (10) shows,

$$Min \sum_{t=0}^{T} P_{\text{DER}-curtail} \tag{10}$$

where $P_{DER-curtail}$ is the DER power curtailment.

3.2. Minimizing Feeder Losses

Minimizing feeder losses is a crucial indicator for power system economic operation and also a key means for energy conversation [6]. In this paper, reducing feeder losses is also proposed for BESS scheduling and EV coordinated charging. In terms of the theoretical derivation in Section 2, the load variance minimization could be used to substitute feeder loss minimization.

$$Min\ P_{loss} \propto Min\ \sigma_{P}^{2} = \frac{\sum_{t=0}^{T} (P_{load,t} - P_{average})^2}{T} \tag{11}$$

where $P_{load,t}$ is the load power at time t, and $P_{average}$ means the average power for the assigned duration.

3.3. Minimizing Voltage Deviation

With DERs integrated into the distribution network, their intermittent and fluctuation characteristics have aggravated the power and voltage volatility in the distribution power system, serving as the immediate cause of power quality reduction and electric equipment damage. Thus, the objective function of minimizing voltage deviation is proposed to mitigate the fluctuation brought by the DERs' integration, similarly to the last optimization target, which is formulated by the power deviation in Equation (12).

$$Min\ V_{dev} \propto Min\ P_{dev} = \sum_{t=0}^{T} (P_{load,t+1} - P_{load,t})^2 \tag{12}$$

3.4. Minimizing Electricity Cost

Besides the power demand of the regular load, the charging cost of BESS and EV also contribute to electricity bills. On the contrary, the discharging power of BESS and the injected power of DERs facilitate reducing the utility expense. As a consequence, through the optimal scheduling of BESS associated with EV coordinated charging, the electricity cost could be reduced as Equation (13) shows,

$$Min\ C_{bill} = \sum_{t=0}^{T} c_t \cdot P_t \tag{13}$$

where C_{bill} is the overall electricity cost in the distribution network with duration T, and c_t means the time of use (TOU) electricity price at time t.

3.5. Constraints of BESS Operation

For the optimal scheduling strategy of BESS, the constraints of BESS are presented as the charging/discharging power limitation, the energy capacity restriction and the charging/discharging balance requirement.

The constraint of charging/discharging power of BESS refers to the upper/lower power limitation during the BESS charging/discharging process; this means that the charging/discharging process should be within the allowance boundaries, as Equation (14) shows,

$$-P_{\max} \leq P_{storage,t} \leq P_{\max} \tag{14}$$

where $P_{storage,t}$ represents the charging/discharging power of BESS at time t, and $-P_{max}$ and P_{max} signify the upper and lower boundaries of BESS power, respectively.

The constraint of the energy capacity of BESS stands for the upper/lower energy limitation during BESS operation, to guarantee the capability for emergency incidents. In this work, the energy boundaries of BESS are from 10% to 90% of the energy capacity.

$$E_{storage,t} \leq 90\% \cdot E_{storage,\max} \quad (15)$$

$$E_{storage,t}{=}E_{storage,0} + \int_0^t P_{storage,k} dk \quad (16)$$

In Equations (15) and (16), $E_{storage,t}$ and $E_{storage,\max}$ are the state of energy at time t and the designed energy capacity for BESS, respectively. Moreover, $E_{storage,0}$ indicates the initial energy state of BESS.

The constraint of the charging/discharging balance of BESS means that the charging energy is supposed to be equal to the discharging energy during a certain period.

$$\left| \int_{t_a}^{t_b} P_{ch} dt - \int_{t_c}^{t_d} P_{dis} dt \right| \leq \varepsilon \quad (17)$$

In Equation (17), t_a, t_b denote the charging region and, similarly, t_c, t_d the discharging region. Besides, P_{ch}, P_{dis} represent the charging and discharging power, respectively. Additionally, ε stands for the permissible error of the charging/discharging balance.

It is to note that, since the BESS optimal scheduling strategy would be used in the duration of 24 h, the degradation of batteries is not considered in this paper. For long time operation, the boundaries of power and energy limitation should be reset for specified condition.

3.6. Constraints of EVs' Operation

As an auxiliary technique, the coordinated charging strategy of EVs has been used to assist the BESS scheduling, and the constraints of EVs are presented as the constraints of the charging period and the charging pattern.

The constraint of the EV charging period refers to that the charging period selection, which should meet the transport demands of EV users. In this paper, this means the start charging moment should be restricted as Equation (18) shows,

$$T_{arrive,i} \leq T_{start,i} \leq T_{leave,i} - T_{c,i} \quad (18)$$

$$T_{c,i} = (1 - SOC_{arrive,i}) \cdot T_{full} \quad (19)$$

In Equations (18) and (19), $T_{arrive,i}$, $T_{start,i}$ and $T_{leave,i}$ are the arriving, start charging and leaving moment for EV_i, and $T_{c,i}$ represents the charging duration for EV_i. Furthermore, the expression of $T_{c,i}$ in Equation (19) has been given, where $SOC_{arrive,i}$ means the state of charge (SOC) condition of EV_i when arriving at the charging spot, and T_{full} stands for the full charging duration for the assigned EV model.

The constraint of the charging pattern means the constant-current constant-voltage (CC-CV) mode for individual EV charging [12–14].

3.7. Other Constraints

In this paper, the constraints of the balance between power demand and supply and the voltage upper/lower limits are also considered in the optimization procedure.

Besides, the reverse-flow control is also introduced as Equation (20) illustrates, to eliminate the unnecessary power losses.

$$P_{load,t} \geq P_{DG,t} + P_{storage,t} \quad (20)$$

where $P_{DG,t}$ means the sum of the DERs' generation at time t.

4. Solution Technique

To tackle the optimization model proposed in Section 3, day-ahead and real-time scheduling strategies have been presented, respectively. The day-ahead strategy tends to settle the optimization problem one day ahead, which means obtaining the next day BESS operation profile globally and precisely. Alternatively, the real-time scheduling strategy is to solve it with interval updating, to avoid the related effect of regular load and DER forecasting errors.

Both of the strategies are suitable for BESS scheduling and EVs' coordinated charging.

In terms of the positive performance of the optimization solution in [15], both of the multi-objective optimization problems derived from the two strategies above could be tackled by fuzzification multi-objective optimization and an improved particle swarm optimization (IPSO) algorithm. More specifically, the detailed description is illustrated as follows.

4.1. Day-Ahead Strategy

For the day-ahead strategy, DERs, EVs and regular load forecasting should be predicted before this procedure. The BESS scheduling profile of the next day with certain precise time, 15 min in this work, could be optimized, incorporated with minimizing the DERs' curtailment, load variance, load deviation and electricity cost, to be the objective. The detailed procedure flow is demonstrated below and in Figure 2a.

Step 1: According to the forecasting results of the regular load, DERs and EVs, the corresponding 96-point profiles could be set for subsequent optimizing with a 15-min interval.

Step 2: In the day-ahead strategy, the variables for BESS optimizing were $P_{storage} = [P_1, P_2, \ldots, P_{96}]$, and the variables for EVs' coordinated charging were $T_{start} = [T_1, T_2, \ldots, T_{96}]$. For better performance, the initial BESS variables would be set as a zero vector, and the initial EVs' variables would be set as the forecasting vector.

Step 3: The multi-optimization procedure and intelligent algorithm were implemented as in Sections 4.3 and 4.4.

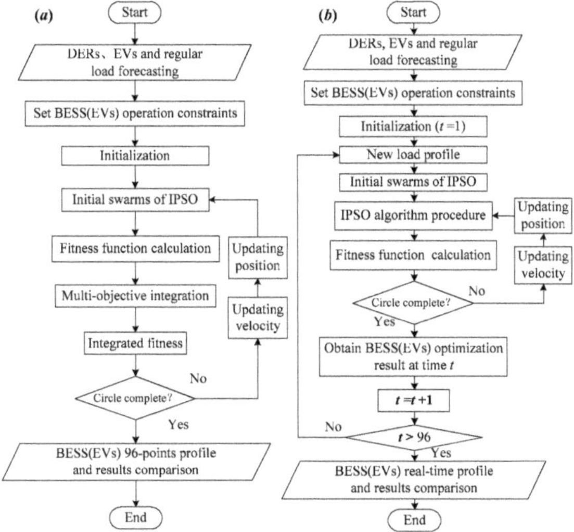

Figure 2. (**a**) Day-ahead strategy flow chart; (**b**) real-time strategy flow chart.

4.2. Real-Time Strategy

In this section, a real-time strategy was proposed to solve the optimization problem, as well as to handle the dependency for forecasting errors. Both the forecasting data and real time capture data have been combined for optimizing globally and meeting the requirements of BESS or EVs' operation.

The real-time optimizing framework is illustrated in Figure 3. In every 15 min, the optimization procedure was performed once to get the optimal operation value for the current point. Additionally, the real-time data have been used for the duration from the beginning to the current point, as the orange lines show in Figure 3, and the forecasting data have been used for the duration from the current point to the end of the day, as the blue lines show in Figure 3. Besides being comparable to the day-ahead strategy, another reason to have optimization in a one-day scale is to ensure the charging and discharging balance for BESS operation. Then, the detailed procedure is shown below and in Figure 2b.

Step 1: According to the forecasting results of the regular load, DERs and EVs, the corresponding 96-point profiles could be prepared for subsequent optimizing with a 15-min interval.

Step 2: The initialization: At time $t = 1$ (t =1, 2, ... , 96), the real-time data of the DERs, EVs and regular load at current time $t = 1$ replaced the forecasting data at time $t = 1$, and the new load profile would be used for optimization. Afterwards, the BESS or EVs' 96-point operation results could be obtained. Only the optimal value at time $t = 1$ was picked to be the operation value for time $t = 1$.

Step Go to the next moment $t = t + 1$;

Step 4: To make the load profile at time t, the real-time data from the beginning to time t have been used to replace the corresponding period of forecasting data. At the same time, the variables for BESS are $P_{storage} = [P_t, P_{t+1}, \ldots, P_{96}]$, 96-$t$ + 1variables in all, and similarly, the variables for EVs are $T_{start} = [T_t, T_{t+1}, \ldots, T_{96}]$.

Step 5: P_t and T_t to obtain the optimal results for the current time t through the following optimization techniques.

Step 6: Go back to Step 3, until completing the circle of 96 points.

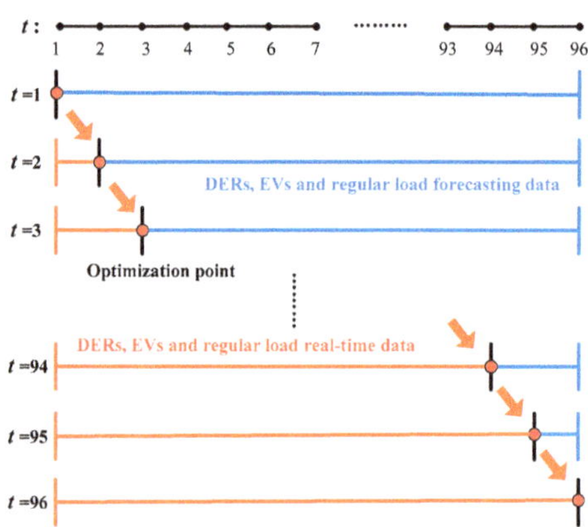

Figure 3. Framework of real-time optimizing.

4.3. Fuzzy Multi-Objective Optimization

In this paper, the optimization problem is actually a multi-objective optimization problem. The fuzzy mathematics method, with the membership function, has been utilized to perform the objective function fuzzification to transfer the multi-objective optimization to a single-objective issue [15,16].

For objectively considering the weight of each optimization objective, the linear membership function is used, which can be described as:

$$\mu_x(x) = \begin{cases} 0 & f_i(x) \le c_{imin} \\ \frac{f_i(x) - c_{imin}}{c_{imax} - c_{imin}} & c_{imin} < f_i(x) < c_{imax} \\ 1 & f_i(x) \ge c_{imax} \end{cases} \quad i = 1,2,3,4 \tag{21}$$

where $f_i(x)$ is the i-th objective function of the fuzzy multi-objective problem; $\mu_i(x)$ is the membership function of $f_i(x)$; m is the number of objective functions; c_{imin}, c_{imax} are the upper and lower limit values of $f_i(x)$, respectively; c_{imin} is the optimal value obtained by the single-objective function; and c_{imax} is initial value of each objective function. Besides, the curves of the membership function by Equation (21) are shown in Figure 4.

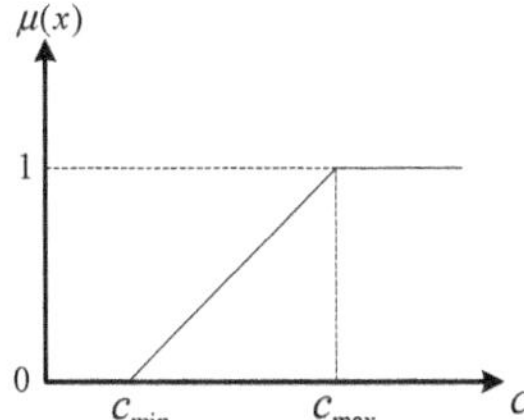

Figure 4. Membership function of the sub-objective.

After the fuzzification, the sub-objectives also need to be integrated. In this work, minimizing the maximum of sub-objectives has been applied as shown in Equation (22),

$$fitness = \max(\mu_1, \mu_2, \mu_3, \mu_4) \tag{22}$$

where *fitness* means the integrated fitness for multi-objective optimization.

4.4. IPSO Algorithm

The optimization model proposed above is considered as a complex multi-constraint, nonlinear optimization problem. Compared to classical algorithms, such as linear programming, quadratic programming, the gradient descending method and other numerical algorithms, the heuristic algorithms are novel algorithms for solving the optimization problem and much easier to implement and extend, such as the genetic algorithms (GA), particle swarm optimization (PSO), differential evolution (DE), artificial immune algorithm and artificial bee colony (ABC) algorithm.

The PSO algorithm possesses superior performance in its implementation and a good trade-off between exploration and exploitation ability, with a simple structure, a simple parameter setting and a fast convergence speed. It has been widely applied in function optimization, mathematical modelling, system control and some other areas [17–19].

In basic PSO algorithms, ω, c_1 and c_2 are fixed values. For the search accuracy and search speed, in this paper, the improved inertia weight is shown in Equation (25). The algorithm may adjust ω dynamically via Equation (25), so that it can optimize dynamically by taking both global search and local search into account during changing The improved PSO is shown as follows:

$$v_{id}^{k+1} = \omega(k)v_{id}^{k} + c_1 r_1 (p_{id} - z_{id}^{k}) + c_2 r_2 (p_{gd} - z_{id}^{k}) \tag{23}$$

$$z_{id}^{k+1} = z_{id}^{k} + v_{id}^{k+1} \tag{24}$$

$$\omega(k) = \omega_{start} - (\omega_{start} - \omega_{end})(\frac{k}{K})^2 \quad (25)$$

where ω_{start} and ω_{end} represent the initial value and the final value of ω, respectively; K is the maximum number of evolutionary generations; k is the current number of evolutionary generations.

5. Case Study

5.1. The Case Setting

For the project requirement, all of the simulation cases in this paper are carried out based on the framework of the campus distribution network and corresponding regular loads, as Figure 5a shows. Furthermore, PV, EVs and BESS are integrated into the network in some cases, of which there are a 300-kWp PV, 100 EVs and a 150-kWh BESS. The details of the case setting are descripted in Table 1. Additionally, the profiles of regular load forecasting, EV load estimation, PV forecasting and the total load prediction of this framework are shown in Figure 5b.

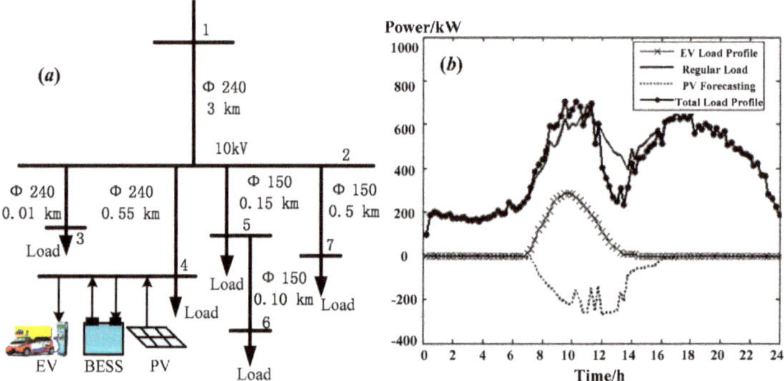

Figure 5. (**a**) The simulation case setting; (**b**) The simulation case data.

Table 1. The simulation cases' settings.

Category	Settings	Optimization scheme
Case 1	Loads	No optimization
Case 2	Loads, PV and uncoordinated charging EVs	No optimization
Case 3	Loads, PV, BESS and uncoordinated charging EVs	Day-ahead optimization
Case 4	Loads, PV, BESS and uncoordinated charging EVs	Real-time optimization
Case 5	Loads, PV, BESS and coordinated charging EVs	Day-ahead optimization
Case 6	Loads, PV, BESS and coordinated charging EVs	Real-time optimization

During the IPSO optimizing, $c_1, c_2 = 1.49$, $\omega_{start} = 0.9$, $\omega_{end} = 0.4$, and the velocity step is 1 kW for BESS optimizing and 15 min for the EVs' coordinated charging.

Besides, the charging/discharging power limitation of BESS is 50 kW; the initial energy state is 75 kWh. Additionally, the arriving time distribution and the arriving SOC distribution of EVs are demonstrated in Figure 6a,b, respectively. Furthermore, the difference between real-time operation data and the forecasting data for the regular load, PV and arriving time of EVs are also given in Figure 6c,d and Figure 7.

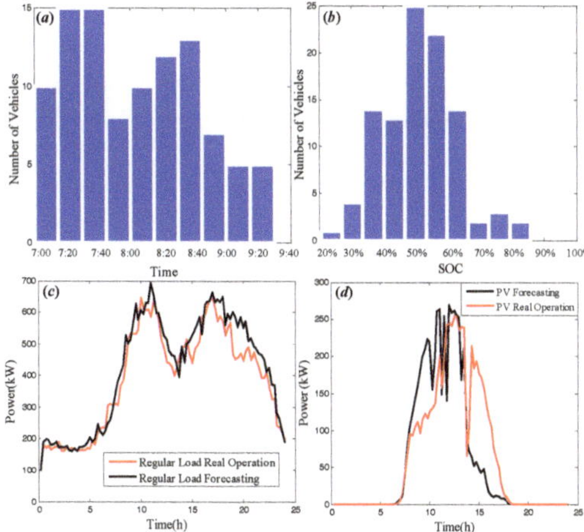

Figure 6. (**a**) EVs' arriving time distribution; (**b**) EVs' arriving SOC distribution; (**c**) real and forecasting data of the regular load; (**d**) real and forecasting data of PV.

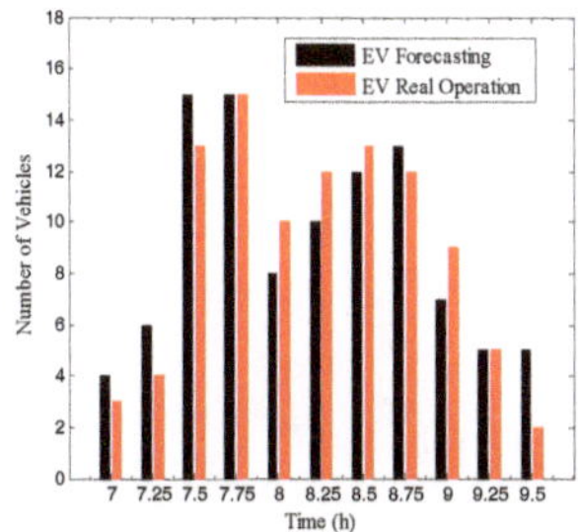

Figure 7. Forecasting and real data of the EVs' arriving time.

In the simulation, there are six cases in all. The simulation case settings and the optimal schemes used are shown in Table 1.

5.2. The Optimizing Results

5.2.1. BESS Optimizing without EVs' Auxiliary

Based on the optimization model and the strategy above, first of all, the BESS day-ahead and real-time optimization without EVs' auxiliary has been performed. The calculation results are shown in Table 2. It can be noticed that, in this work, due to the original case configuration, Objective 1, minimizing the DERs' curtailment, is zero in all cases.

Table 2. Index comparison of BESS optimizing.

	Case 1	**Case 2**	**Case 3**	**Case4**
Objective 1	0.00	0.00	0.00	0.00
Objective 2	30,484	32,478	28,830	27,830
Objective 3	120,300	233,010	112,500	178,290
Objective 4	13,613	12,927	12,767	11,707
P_{loss} (MW)	0.8661	0.8551	0.8526	0.8206
V_{dev} (10^{-5} kV^2)	6.18	11.2	5.59	8.49

In Table 2, it is clear that Objective 2 and 3's values in Case 2 are higher than that in Case 1, which means that the integration of PV and EVs brings about the load variance increase, with the power deviation increasing. Besides, both the day-ahead strategy and real-time strategy are effective for all of the objectives, except for zero DER curtailment, which already achieved the minimum. Comparing between the day-ahead and real-time strategy for BESS optimization without EVs' auxiliary, these two strategies displayed different advantages; the real-time strategy shows better performance in load variance, feeder losses and utility cost control, and the day-ahead strategy shows better effectiveness in the power deviation and voltage deviation control.

The load profiles of before and after BESS day-ahead and real-time optimizing are displayed in Figure 8a,b, respectively. Additionally, the BESS operation power profiles and energy profile for day-ahead and real-time are shown in Figure 8c,d, respectively. It expresses the distinct optimization tracks of day-ahead and real-time optimization.

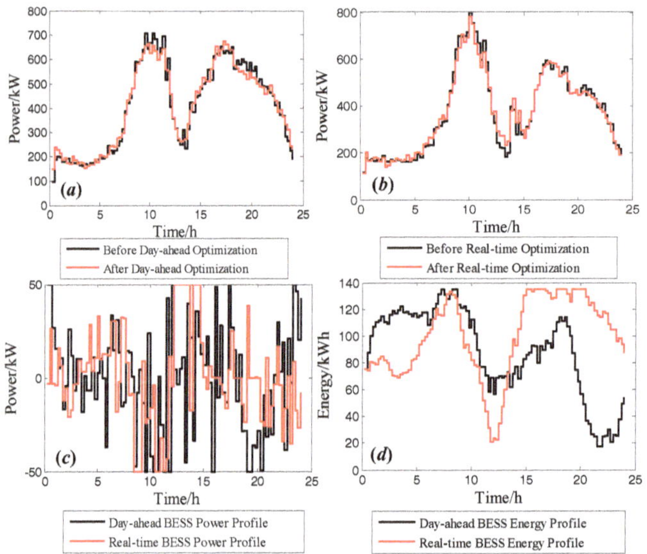

Figure 8. (**a**) Load profiles of before and after BESS day-ahead optimizing (without EVs); (**b**) load profiles of before and after BESS real-time optimizing (without EVs); (**c**) BESS operation power profiles of day-ahead and real-time (without EVs); (**d**) BESS operation energy profiles of day-ahead and real-time (without EVs).

5.2.2. BESS Optimizing with EVs' Auxiliary

Then, the BESS with EVs' auxiliary day-ahead and real-time optimization has been implemented. The calculation results are shown in Table 3.

Table 3. Index comparison of BESS and EV optimizing.

	Case 1	**Case 2**	**Case 5**	**Case6**
Objective 1	0.00	0.00	0.00	0.00
Objective 2	30,484	32,478	29,696	17,465
Objective 3	120,300	233,010	141,050	165,490
Objective 4	13,613	12,927	12,805	10,152
P_{loss} (MW)	0.8661	0.8551	0.8522	0.8156
V_{dev} (10^{-5} kV^2)	6.18	11.2	3.36	7.52

In Table 3, it is obvious that both the day-ahead strategy and real-time strategy for BESS and EV combined optimization are effective for all of the objectives, regardless of Objective 1. Compared between the day-ahead and real-time strategy for combined optimization, these two strategies also show distinct superiority, the same as BESS optimization without EVs' auxiliary; the real-time strategy shows better performance in load variance, feeder losses and utility cost control, and the day-ahead strategy shows better effectiveness in power deviation and voltage deviation control.

The load profiles of before and after combined day-ahead and real-time optimizing are displayed in Figure 9a,b, respectively. Additionally, the BESS operation power profiles and energy profile for day-ahead and real-time are shown in Figure 9c,d, respectively. It expresses the distinct optimization tracks of day-ahead and real-time optimization. Besides, in Figure 10, the EVs' profiles of forecasting, day-ahead and real-time optimization are presented.

In the BESS without and with EVs' auxiliary optimization above, the results illustrated that both of the optimizing scheduling strategies took effect for individual objective and integrated fitness. Moreover, it is also clear that, under the same BESS configuration, strategies with EVs' coordinated charging show significant enhancement for all of the optimization targets. From another perspective, the day-ahead and real-time optimization, no matter if for BESS only or the BESS and EV combination, the optimization routes are totally different, resulting in various profiles of power and energy tendency.

In our view, both strategies show significant effectiveness, and the main distinction is to be applied for different requirements. The day-ahead strategy turns out to be used in the scheduling focusing on the global optimization without bidirectional communication, especially the situation with a high accuracy of profile prediction. Additionally, the real-time strategy is suitable to handle the modest accuracy degree of forecasting and to update the scheduling on the basis of two-way communication.

It should be noted that the BESS did not show the load-shaving effect in the cases above; this is because the BESS power and energy constraints would not allow this and which has been formerly set. Moreover, the main objective of BESS is to alleviate the fluctuation of DERs, which has been considerably verified in this work.

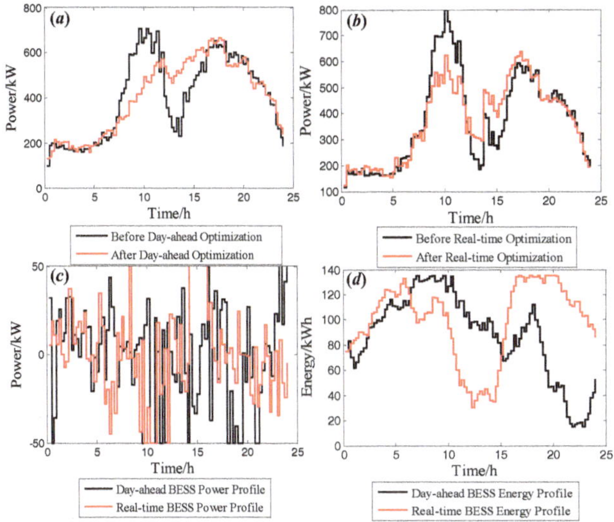

Figure 9. (**a**) Load profiles of before and after combined day-ahead optimizing (with EVs); (**b**) load profiles of before and after combined real-time optimizing (with EVs); (**c**) BESS operation power profiles of day-ahead and real-time (with EVs); (**d**) BESS operation energy profiles of day-ahead and real-time (with EVs).

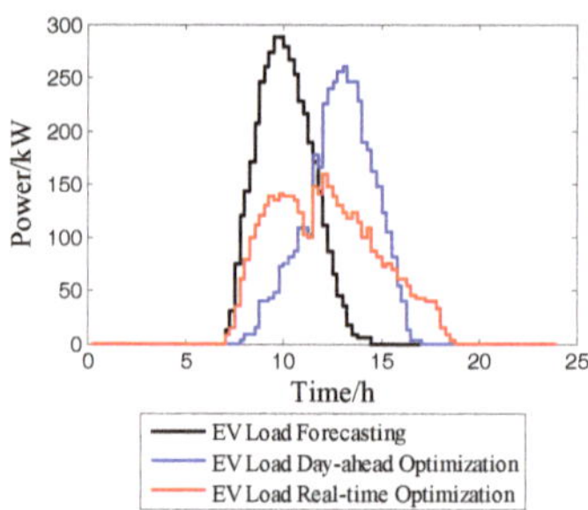

Figure 10. EVs' load profiles of forecasting, day-ahead and real-time optimization.

6. Conclusions

In this paper, the scheduling framework for BESS operation with EVs' auxiliary in a distribution network with renewable energy integration has been proposed, to reduce the renewable energy curtailment, decrease the power losses, mitigate voltage deviation and lower the electricity expense. Moreover, the model also takes into account the BESS operation constraints and EVs' charging limitation. To tackle the scheduling problem, two optimization processes are presented, the day-ahead strategy and the real-time strategy, to be incorporated into the framework, separately. To handle the multi-objective formulation, a fuzzy mathematical method has been launched to turn multi-objective optimization into a single-objective issue; from another aspect, the IPSO algorithm has been implemented to obtain the optimal scheduling results.

Through the simulation results of six cases, it could be concluded that the proposed BESS scheduling and EVs' coordinated charging scheme are effective for the assigned distribution network, and both theday-ahead and real-time strategy procedures show significant performances. Additionally, from our perspective, the mode of the real-time strategy for BESS scheduling with EVs' auxiliary is recommended for the cases with confined load forecasting accuracy, as well as with EVs' access.

Acknowledgments: This work was supported by the National Key Technology R$D Program (2013BAA01B03). In addition, we also want to thank the investigation support from the Office of Energy Saving, Beijing Jiaotong University.

Author Contributions: The initial design and implementation of the optimal BESS scheduling framework and the solution were done by Yuqing Yang. She was also responsible for the simulation implementation. In addition, this work was also performed under the advisement of and regular feedback from Weige Zhang, who also revised the manuscript critically. Moreover, Jiuchun Jiang, Mei Huang and Liyong Niu also gave some useful suggestions for this work and helped revise the manuscript.

Conflicts of Interest: The authors declare no conflict of interest.

References

1. Fan, M.; Zhang, Z.; Su, A.; Su, J. Enabling technologies for active distribution systems. *Proc. CSEE* **2013**, *33*, 12–18.
2. Wang, C.; Li, P. Development and challenges of distributed generation, the micro-grid and smart distribution system. *Autom. Electr. Power Syst.* **2010**, *34*, 10–14.
3. Zhang, W.; Qiu, M.; Lai, X. Application of energy storage technologies in power grids. *Power Syst. Technol.* **2008**, *32*, 1–9.
4. Ortega-Vazquez, M.A. Optimal scheduling of electric vehicle charging and vehicle-to-grid services at household level including battery degradation and price uncertainty. *IET Gener. Transm. Distrib.* **2014**, *8*, 1007–1016. [CrossRef]
5. Luo, X.; Chan, K.W. Real-time scheduling of electric vehicles charging in low-voltage residential distribution systems to minimize power losses and improve voltage profile. *IET Gener. Transm. Distrib.* **2014**, *8*, 516–529. [CrossRef]

6. Teng, J.; Luan, S.; Lee, D.; Huang, Y. Optimal charging/discharging scheduling of battery storage systems for distribution systems interconnected with sizeable PV generation systems. *IEEE Trans. Power Syst.* **2013**, *28*, 1425–1433. [CrossRef]
7. Hill, C.A.; Such, M.C.; Chen, D.; Gonzalez, J.; Grady, W.M. Battery energy storage for enabling integration of distributed solar power generation. *IEEE Trans. Smart Grid* **2012**, *3*, 850–857. [CrossRef]
8. Wasiak, I.; Pawelek, R.; Mienski, R. Energy storage application in low-voltage microgrids for energy management and power quality improvement. *IET Gener. Transm. Distrib.* **2013**, *8*, 463–472. [CrossRef]
9. Gabash, A.; Li, P. Flexible Optimal operation of battery storage systems for energy supply networks. *IEEE Trans. Power Syst.* **2013**, *28*, 2788–2797. [CrossRef]
10. Sortomme, E.; Hindi, M.M.; MacPherson, S.; Venkata, S.S. Coordinated Charging of Plug-In Hybrid Electric Vehicles to Minimize Distribution System Losses. *IEEE Trans. Smart Grid* **2011**, *2*, 198–205.
11. Abed, N.Y.; Teleke, S.; Castaneda, J.J. Planning and operation of dynamic energy storage for improved integration of wind energy. In Proceedings of the 2011 IEEE Power and Energy Society General Meeting, San Diego, CA, USA, 24–29 July 2011.
12. Wen, J. Studies of Lithium-Ion Power Battery Optimization Charging Theory for Pure Electric Vehicle. Ph.D. Thesis, Bejing Jiaotong University, Beijing, China, 2011.
13. Wen, F. Study on Basic Issues of the Li-Ion Battery Pack Management Technology for Pure Electric Vehicles. Ph.D. Thesis, Beijing Jiaotong University, Beijing, China, 2009.
14. Yang, Y.; Jiang, J.; Bao, Y.; Zhang, W.; Huang, M.; Su, S. Dynamic coordinated charging strategy and positive effects in regional power system. In Proceedings of the 11th IEEE International Conference on Control and Automation (ICCA), Taichung, Taiwan, 18–20 June 2014.
15. Yang, Y.; Zhang, W.; Niu, L.; Jiang, J. Coordinated charging strategy for electric taxis in temporal and spatial scale. *Energies* **2015**, *8*, 1256–1272. [CrossRef]
16. He, X.; Wang, W. Fuzzy multi-objective optimal power flow based on modified artificial bee colony algorithm. *Math. Probl. Eng.* **2014**. [CrossRef]
17. Kennedy, J.; Eberhart, R. Particle swarm optimization. In Proceedings of the IEEE International Conference on Neural Networks, Perth, Australia, 27 November–1 December 1995.
18. Shi, F.; Wang, H.; Hu, F.; Yu, L. *MATLAB Intelligent Algorithms 30 Cases Analysis*; Beijing University of Aeronautics and Astronautics Press: Beijing, China, 2011.
19. Zhang, Y.C.; Xiong, X.; Zhang, Q.D. An improved self-adaptive PSO algorithm with detection function for multimodal function optimization problem. *Math. Probl. Eng.* **2013**. [CrossRef]

Article

Distributed Energy Storage Control for Dynamic Load Impact Mitigation

Maximilian J. Zangs †, Peter B. E. Adams †, Timur Yunusov, William Holderbaum * and Ben A. Potter

School of Systems Engineering, University of Reading, Whiteknights Campus, Reading RG6 6AY, UK; m.j.zangs@pgr.reading.ac.uk (M.J.Z.); p.b.e.adams@pgr.reading.ac.uk (P.B.E.A.); t.yunusov@reading.ac.uk (T.Y.); b.a.potter@reading.ac.uk (B.A.P.)

* Correspondence: w.holderbaum@reading.ac.uk; Tel.: +44-118-378-6086; Fax: +44-118-975-1994

† These authors contributed equally to this work.

Academic Editor: Rui Xiong

Received: 31 January 2016; Accepted: 4 August 2016; Published: 17 August 2016

Abstract: The future uptake of electric vehicles (EV) in low-voltage distribution networks can cause increased voltage violations and thermal overloading of network assets, especially in networks with limited headroom at times of high or peak demand. To address this problem, this paper proposes a distributed battery energy storage solution, controlled using an additive increase multiplicative decrease (AIMD) algorithm. The improved algorithm (AIMD+) uses local bus voltage measurements and a reference voltage threshold to determine the additive increase parameter and to control the charging, as well as discharging rate of the battery. The used voltage threshold is dependent on the network topology and is calculated using power flow analysis tools, with peak demand equally allocated amongst all loads. Simulations were performed on the IEEE LV European Test feeder and a number of real U.K. suburban power distribution network models, together with European demand data and a realistic electric vehicle charging model. The performance of the standard AIMD algorithm with a fixed voltage threshold and the proposed AIMD+ algorithm with the reference voltage profile are compared. Results show that, compared to the standard AIMD case, the proposed AIMD+ algorithm further improves the network's voltage profiles, reduces thermal overload occurrences and ensures a more equal battery utilisation.

Keywords: battery storage; distributed control; electric vehicles; additive increase multiplicative decrease (AIMD); voltage control; smart grid

---

1. Introduction

The adoption of electric vehicles (EV) is seen as a potential solution to the decarbonisation of future transport networks, offsetting emissions from conventional internal combustion engine vehicles. The current rate of EV uptake is anticipated to increase with improved driving range, reduced cost of purchase and greater emphasis on leading an environmentally-friendly lifestyle [1]. It is predicted that by 2030, there will be three million plug-in hybrid electric vehicles (PHEV) and EVs sold in Great Britain and Northern Ireland [2], and it is expected that by 2020, every tenth car in the United Kingdom will be electrically powered [3]. It is anticipated that the majority of PHEV/EV will be charged at home, putting additional stress on the existing local low voltage distribution network, which must then cater for the increased demand in energy [4,5]. Uncontrolled charging of multiple PHEV/EV can raise the daily peak power demand, which leads to: increased transmission line losses, higher voltage drops, equipment overload, damage and failure [6–9]. Accommodating the increased demand and mitigation of such failures is a major area of research interest, with the focus mainly placed on the coordinating and support of home charging.

Demand Side Management (DSM) strategies for Distributed Energy Resources (DER), aim to alleviate the impacts of PHEV/EV home-charging and are a favoured solution. Mohsenian-Rad et al. in [10] developed a distributed DSM algorithm that implicitly controls the operation of loads, based on game theory and the network operator's ability to dynamically adjust energy prices. Focusing on financial incentive-driven DSM strategies, in [11], a Time-Of-Use (TOU) tariff and real-time load management strategy was proposed, where disruptive charging is avoided by allocating higher prices to times of peak demand. Financial incentives have also become a drive towards optimising the operation of Battery Energy Storage Solutions (BESS) and Distributed Generation (DG) when including PHEV/EV into the problem formulation [12].

Research focused on grid support has been driven by the need to deliver long-term savings and to avoid the immediate costs and disruption of network reinforcements and upgrades. This area of research proposes the implementation of alternative solutions to support the adoption of low carbon technologies, such as EVs, heat pumps and the electrification of consumer products. To reduce the resulting increased peak demand, Mohsenian-Rad et al. developed an approach of direct interaction between grid and consumer to achieve valley-filling, by means of dynamic game theory [10]. In [13], a Multi-Agent System (MAS) was used to manage flexible loads for the minimisation of cost in a dynamic game. The use of aggregators has been proposed to allow the participation of a number of small providers to participate in network support, such as grid frequency response [14–16]. Yet without the availability of power demand forecasts, real-time control needs to be implemented.

Real-time DSM can either be implemented in a centralised or distributed control approach. In the former, a central controller relays control signals to its aggregated DERs, whereas the latter allows each DER to control itself. A common form of controlling DERs in this mode of operation is set-point control [17]. Using set-point control on multiple identically-configured DERs would yield optimal operation conditions if each DER's control parameters (e.g., bus voltage) were shared. In a system without sharing network information, DER control algorithms have to be improved to prevent, for example, devices located furthest from the substation from being used more frequently than others.

This paper therefore presents an individualised BESS control algorithm that lets distributed batteries respond to fluctuations in real-time local bus voltage readings. The proposed algorithm is based on the robust Additive Increase Multiplicative Decrease (AIMD) type algorithm, yet implements a set-point adjustment based on the location of the controlled BESS. It will be shown how these home-connected batteries can mitigate the impact of additional loads (i.e., EV uptake), whilst assuring that all BESS are cycled equally.

The key contribution of this work can be summarised as a novel distributed battery storage algorithm for mitigating the negative impact of dynamic load uptake on the low-voltage network. This algorithm uses an individualised set-point control to regulate bi-directional battery power flow and, for convergence, extends the traditional AIMD algorithm. As a result, the developed battery control method reduces voltage deviation, over-currents and the inequality of battery usage. Reducing this usage inequality leads to a homogeneous usage of all of the distributed batteries and, hence, prevents unequal degradation rates and unfair device utilisation.

The remainder of this paper is organised as follows: Section 2 gives some background to related work on AIMD algorithms on which this research is based. Section 3 outlines the EV, network and storage models used in the research. Additionally, it explains the assumptions that accommodate and justify these models. Section 4 elaborates on the proposed AIMD control algorithm (AIMD+). Next, Section 5 details the implementation and scenarios used for a set of test cases. For later comparison, this section also outlines a set of comparison metrics. Section 6 presents and discusses the results, followed by the conclusion in Section 7.

2. Related Work

Existing literature addresses the usage of energy storage units in low-voltage distribution networks to assure voltage security [18–22]. An approach used by, e.g., Mokhtari et al. in [21] relies on bus voltage and network load measurements to prevent system overloads. Yet, these kinds of storage control systems do require communication infrastructures to relay the network information and control instructions. This requirement has also been addressed in the comprehensive review on storage allocation and application methods by Hatziargyriou et al. [23]. In the presented work, a control algorithm is proposed that removes the need for such an inter-BESS communication, since it only uses local voltage measurements to infer the network operation. Yet, to prevent conflicting device behaviour, the underlying coordination mechanism is of particular importance. Assuring convergence, the AIMD algorithm is perfectly suited for such coordinated control.

Originally, AIMD algorithms were applied to congestion management in communications networks using the TCP protocol [24], to maximise utilisation while ensuring a fair allocation of data throughput amongst a number of competing users [25]. AIMD-type algorithms have previously been applied to power sharing scenarios in low voltage distribution networks, where the limited resource is the availability of power from the substation's transformer.

For instance, such an algorithm was first proposed for EV charging by Stüdli et al. [26], requiring a one-way communications infrastructure to broadcast a "capacity event" [27,28]. Later, their work was further developed to include vehicle-to-grid applications with reactive power support [29]. The battery control algorithm proposed in this paper builds upon the algorithm used by Mareels et al. [30], where EV charging was organised by including bidirectional power flow and the use of a reference voltage profile derived from network models. Similar to the work by Xia et al. [31], who utilised local voltage measurements to adjust the charging rate, only voltage measurements at the batteries' connection sites were used in this work to control the batteries' operations.

Previous research is therefore extended by the work presented here, as previous work has only utilised common set-point thresholds for controlling each of the DERs. The approach proposed in this paper ensures that unavoidable voltage drops along the feeder do not skew the control decisions, and voltage oscillations caused by demand variation are taken into control considerations. In contrast to previous work, where substation monitoring was used to inform control units of the transformer's present operational capacity, the proposed AIMD+ algorithm does not require this information and, hence, does not require such an extensive communications infrastructure.

3. System Modelling

In this section, the underlying assumptions to validate the research are addressed. Next, a model to describe EV charging behaviour is explained. This is followed by a model of the BESS. Finally, the network models used to simulate the power distribution networks are explained.

3.1. Assumptions

For this work, several underlying assumption were made to obtain the models:

1. The uptake of EVs is assumed to increase and, hence, to have a significant impact on the normal operation of the low voltage distribution network. This assumption is based on a well-established prediction that the majority of EV charging will take place at home [32].
2. The transition from internal combustion engine-powered vehicles to EVs is assumed to not impact the users' driving behaviour. Similar to [33], this assumption allows the utilisation of recent vehicle mobility data [34] to generate leaving, driving and arriving probabilities, from which the EV charging demand can be determined.
3. The transition to low carbon technologies will increase the variability of electricity demand, and therefore, grid-supporting devices, such as BESS, are anticipated to play a more important role [35]. Hence, alongside a high uptake of EVs, an increased adoption of distributed BESS devices is assumed.
4. It is assumed that BESS solutions, or more specifically battery energy storage solutions, start the simulations at 50% SOC and are not 100% efficient at storing and releasing electrical energy, as in [36]. Additionally, its utilisation will degrade the energy storage capability and performance

over time, as shown in [37]. Therefore, the requirements for equal and fair storage usage is of high importance.

5. It is assumed that the load profiles provided by the IEEE Power and Energy Society (PES) are sufficient as base load profiles for all simulations.

3.2. Electric Vehicle Charging Behaviour

From publicly-available car mobility data [33,34] an empirical model was developed to capture the underlying driving behaviour. The raw data, $n_r(t)$, represents the probabilities of starting a trip during a 15-min period of a weekday. Three continuous normal distribution functions, each defined as:

$$\hat{n}_x(t) = \beta_x \frac{1}{\sigma_x\sqrt{2\pi}} \exp\left[-\frac{\left({}^{t}/_{24} - \mu_x\right)^2}{2\sigma_x^2}\right] \text{ where } t = [0, 24] \tag{1}$$

were used to represent vehicles leaving in the morning, $\hat{n}_m(t)$, lunch time, $\hat{n}_l(t)$, and in the evening, $\hat{n}_e(t)$. The aggregate probability of these three functions was optimised using a Generalised Reduced Gradient (GRG) algorithm to fit the original data. In order to represent a symmetric commuting behaviour, i.e., vehicles departing in the morning and returning during the evening, an equality amongst the three probabilities was defined as follows:

$$0 = \int_0^{24} [\hat{n}_m(t) + \hat{n}_l(t) - \hat{n}_e(t)]\, dt \tag{2}$$

The resulting parameters from the GRG fitting of the three distribution functions are tabulated in Table 1. Additionally, the resulting departure probabilities, as well as the reference data $n_r(t)$ are shown in Figure 1.

Table 1. Parameters for normal distributions.

Equation $\hat{n}_x(t)$	μ_x (Mean)	σ_x (SD)	β_x (Weight)
$\hat{n}_m(t)$	0.3049	0.0488	0.00206
$\hat{n}_l(t)$	0.4666	0.0829	0.00314
$\hat{n}_e(t)$	0.7042	0.0970	0.00521

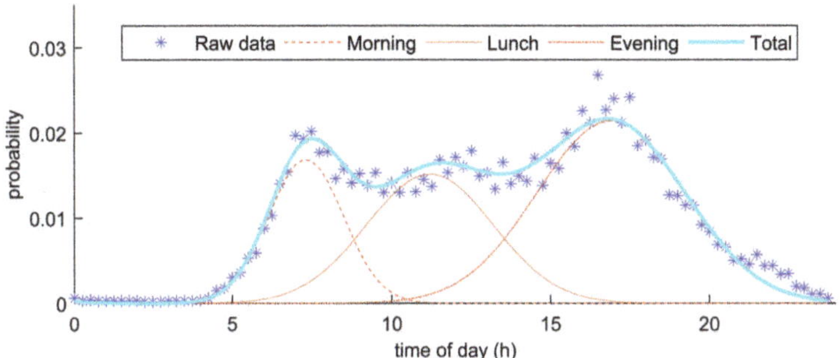

Figure 1. The probability of starting a trip at a particular time during a weekday, extrapolated into three normal distributions (RMS error: 9.482%).

Statistical data capturing the probability distribution of a trip being of a certain distance were also extracted from the dataset. This was done for both the weekdays $w_{wd}(d)$ and weekends $w_{we}(d)$. The Weibull function was chosen to be fitted against the extracted probability distributions and is defined as:

$$\hat{w}_x(d) := \begin{cases} \frac{k_x}{\gamma_x}\left(\frac{d}{\gamma_x}\right)^{k_x-1} \exp\left[-\left(\frac{d}{\gamma_x}\right)^{k_x}\right] & \text{if } d \geq 0 \\ 0 & \text{if } d < 0 \end{cases} \tag{3}$$

Performing the curve fitting using the GRG optimisation algorithm, a weekday trip distance distribution, $\hat{w}_{wd}(d)$, and a weekend trip distribution, $\hat{w}_{we}(d)$, could be estimated. The computed function parameters for these two estimated distribution functions are tabulated in Table 2. Their resulting probability distributions are plotted for comparison against the real data, $w_{wd}(d)$ and $w_{we}(d)$, in Figure 2.

Table 2. Parameters for Weibull distributions.

Equation $\hat{w}_x(d)$	γ_x (Scale)	k_x (Shape)
$\hat{w}_{wd}(t)$	15.462	0.6182
$\hat{w}_{we}(t)$	38.406	0.4653

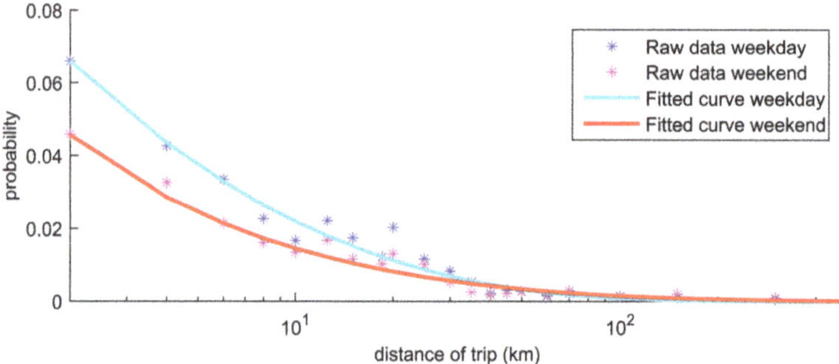

Figure 2. The probability of a trip being of a particular distance during a weekday, extrapolated into a Weibull distribution (RMS error: 3.791%).

In addition to these probabilities, an average driving speed of 56 kmh (35 mph) and an average driving energy efficiency of 0.1305 kWh/kmh (0.21 kWh/mph) are taken from [38]. Using the predicted driving distance and average driving speed with the driving energy efficiency, it is possible to estimate an EV's energy demand upon arrival. Starting to charge from this arrival time until the energy demand has been met allows the generation of an estimated charging profile of a single EV. To do this, a maximum charging power of the U.K.'s average household circuit rating (i.e., 7.4 kW) and an immediate disconnection of the EV upon charge completion were assumed [39].

Generating several of those charging profiles and aggregating them produces an estimated charging demand for an entire fleet of EVs. To provide an example, charge demand profiles for 50 EVs were generated, aggregated and plotted in Figure 3. This plot shows the expected magnitude and variability in energy demand that is required to charge several EVs at consumers' homes based on the vehicles' daily usage.

This model's EV charging behaviour has been implemented to reflect EV demand if applied today without widespread smart charging infrastructure. It does therefore reflect the worst case scenario. Future smart-charging schemes would mitigate the currently present collective EV charging spike, yet the implementation and validation of available smart-charging schemes lies beyond the scope of this paper. This model's data were used to feed additional demand into the power network models, which are outlined in the next section.

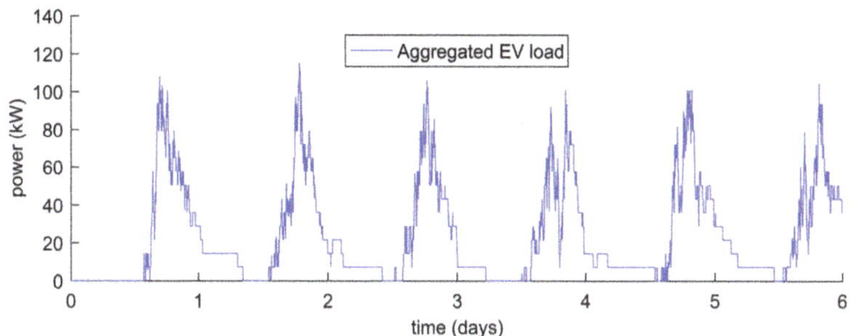

Figure 3. Excerpt from the aggregated 50 EVs; charging powers that were each generated from the empirical models.

3.3. Battery Modelling

For this work, a well-established model that has been used in previous publications by this research group was used [36,40,41]. This model consists of a battery with a self-discharge loss that is dependent on the current battery's State Of Charge (SOC) and an energy conversion loss to represent the energy lost when charging or discharging this battery. A complete list of all notations that are used for this battery model is included in Table 3.

Table 3. Table of the notation used in this section.

Parameter	Description
ine $P_{bat}(t)$	Battery power at time t
$SOC(t)$	Battery state of charge at time t
$\delta_{SOC}(t)$	Change in SOC during time period τ
μ	Self-discharge loss factor
η	Energy conversion efficiency
SOC_{min}	Minimum rated SOC for limited battery operation
SOC_{max}	Maximum rated SOC for limited battery operation
C	Battery capacity
P_{max}	Power rating of battery

When an ideal battery charges or discharges, the change in SOC is related by the battery power, P_{bat}. When sampling battery operation at a regular period, τ, then the energy transferred into the battery can be described as $P_{bat}(t)\tau$. The change in SOC for this ideal battery, δ_{SOC}, is therefore defined as:

$$\delta_{SOC}(t) := \frac{P_{bat}(t)\tau}{C} = \mathrm{SOC}(t) - \mathrm{SOC}(t-\tau) \quad (4)$$

The self-discharge loss is added to this ideal battery model to represent the continual loss of energy in the battery typical of chemical energy storage. This self-discharge loss, $\delta_{SOC,self\text{-}discharge}$, is proportional to the current SOC and is determined using the self-discharge loss factor, μ:

$$\delta_{SOC,self\text{-}discharge}(t) := \mu SOC(t) \quad (5)$$

Additionally, to represent the losses in the power electronics and energy conversion process, an energy conversion loss, $\delta_{SOC,conversion}$, is defined. This loss is proportional to the rate at which the battery's SOC changes, by using the energy conversion efficiency, $\hat{\eta}$ as follows:

$$\delta_{SOC,conversion}(t) := \hat{\eta}\delta_{SOC}(t) \quad (6)$$

Here, the conversion losses in the power electronics are reflected as an asymmetric efficiency, which depends on the direction of the flow of energy. This is done by charging the battery at a lower power when consuming energy and discharging it more quickly when releasing energy. Mathematically, this can be represented as:

$$\hat{\eta} = \begin{cases} \eta & \text{if } \delta_{SOC}(t) \geq 0 \\ \frac{1}{\eta} & \text{if } \delta_{SOC}(t) < 0 \end{cases} \tag{7}$$

When substituting the self-discharge loss and conversion losses, respectively $\delta_{SOC,self\text{-}discharge}$ and $\delta_{SOC,conversion}$, into the SOC evolution equation, the full battery model can be summarised as follows:

$$\begin{aligned} \mathrm{SOC}(t) :&= \delta_{SOC}(t-\tau) - \delta_{SOC,self\text{-}discharge}(t-\tau) - \delta_{SOC,conversion}(t) \\ &= (1-\mu)\delta_{SOC}(t-\tau) - \hat{\eta}\delta_{SOC}(t) \end{aligned} \tag{8}$$

In addition, both the SOC and the P_{bat} are constrained due to the device's maximum and minimum energy storage capabilities, respectively SOC_{max} and SOC_{min}, and maximum charge and discharge rate, P_{max}. These limitations are captured in Equations (9) and (10), respectively.

$$\mathrm{SOC}_{min} \leq \mathrm{SOC}(t) \leq \mathrm{SOC}_{max} \tag{9}$$

$$|P_{bat}(t)| \leq \mathrm{P}_{max} \tag{10}$$

3.4. Network Models

To simulate the low-voltage energy distribution networks, the Open Distribution System Simulator (OpenDSS) developed by the Electronic Power Research Institute (EPRI) was used. It requires element-based network models, including line, load and transformer information, and generates realistic power flow results.

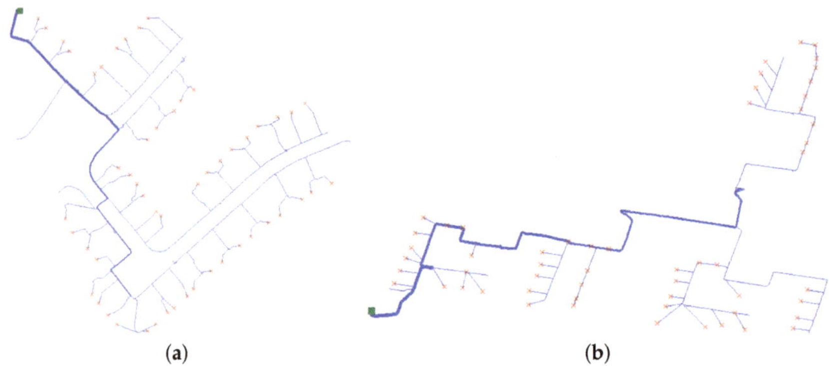

Figure 4. Sample Open Distribution System Simulator (OpenDSS) power flow plots of the used power networks. Consumers are indicated as red crosses and 11/0.416-kV substations are marked with a green square. (**a**) IEEE Power and Energy Society (PES) EU Low Voltage Test Feeder plot; (**b**) Scottish and Southern Energy Power Distribution (SSE-PD) Common Information Model (CIM) (UK) feeder plot.

Simulations were conducted using the IEEE's European Low Voltage Test Feeder [42] and six detailed U.K. feeder models, that are based on real power distribution networks and provided by Scottish and Southern Energy Power Distribution (SSE-PD). The SSE-PD circuit models were provided as Common Information Models (CIM) during the collaboration on the New Thames Valley

Vision Project Project (NTVV) [43]. An example of the IEEE EU LV Test feeder and a U.K. feeder provided by SSE-PD are shown in Figure 4a,b, respectively. A summary of these model's parameters is given in the Table 4.

Table 4. Network model parameters.

Parameter	IEEE EU LV Test Feeder	SSE-PD LV Feeders					
Network number	1 [1]	2 [1]	3	4	5	6	7
Number of customers	55	56	53	91	59	88	37
Median load per customer (VA)	227	227	231	241	224	237	237
Maximum load per customer (kVA)	16.8	16.8	16.8	19.5	16.8	19.5	16.8
Customer connection	Single-phase	Single-phase					
Median substation load (kVA)	24.4	24.9	23.9	41.9	25.6	38.9	16.3
Maximum load per customer (kVA)	72.6	72.7	72.2	92.9	73.5	89.6	60.5
Feeder line model	Three-phase implicit-neutral	Three-phase explicit-neutral					

[1] These networks are shown in Figure 4.

Throughout this paper, all excerpt and time series results were extracted from experiments with the IEEE EU LV Test feeder (i.e., Network No. 1). All concluding results are based on an aggregation of all networks to include network diversity in the analysis.

The model-derived EV data and IEEE EU LV Test feeder consumer demand profiles were used in all simulations. The resultant demand profiles represent the total daily electricity demand of households with EVs. These profiles were sampled at $\tau = 1$ min. The OpenDSS simulation environment was controlled using MATLAB, achieved through OpenDSS's Common Object Model (COM) interface and accessible using Microsoft's ActiveX server bridge.

4. Storage Control

In this section, the control of the energy storage system is explained. Firstly, the additive increase multiplicative decrease algorithm is presented, and its decision mechanism is explained in full. Then, the voltage referencing, used for AIMD+, is outlined.

4.1. *Additive Increase Multiplicative Decrease*

The proposed distributed battery storage control is shown in Algorithm 1. The parameter α denotes the size of the power's additive increase step, and β denotes the size of the multiplicative decrease step. It is worth mentioning that α linearly increases and β exponentially decreases, both charging and discharging powers, where discharging power is represented as a negative power flow, i.e., energy released by the battery. The constants V_{max} and V_{thr} are the maximum historic voltage value and the set-point threshold used to regulate the total demand. In the case when the total demand is too high, the local voltages will fall below V_{thr}, and the batteries reduce their charging power and start discharging. This behaviour reduces total demand on the feeder. At simulation start, V_{max} is set to the nominal voltage of the substation transformer, i.e., 240 V, and V_{thr} is set to a fraction of V_{max}, which was found by solving a balanced power flow analysis. The variable $V(t)$ is the battery's local bus voltage, and P_{max} denotes the maximum charging/discharging power of the battery. The charging and discharging power of the batteries is increased in proportion to the available headroom on the network, which is inferred from the local voltage measurement $V(t)$, to avoid any sudden overloading of the substation transformer.

Algorithm 1 Compute battery power.

1: $R(t) = (V(t) - V_{thr})/(V_{max} - V_{thr})$ ▷ Defines the rate for the current voltage reading
2: **if** $V(t) \geq V_{thr}$ **then** ▷ Given the voltage levels are nominal...
3:     **if** $SOC < SOC_{max}$ **then** ▷ ...and the battery is not fully charged...
4:         $P(t) = P(t-\tau) + \alpha P_{max} R(t)$ ▷ ...increase the charging power
5:     **else** ▷ If the battery has fully charged...
6:         $P(t) = 0$ ▷ ...shut off
7:     **end if**
8:     **if** $P(t) < 0$ **then** ▷ If the battery has been discharging...
9:         $P(t) = \beta P(t-\tau)$ ▷ ...reduce the discharging power by β
10:     **end if**
11: **else** ▷ If voltage levels are not nominal...
12:     **if** $SOC > SOC_{min}$ **then** ▷ ...and battery is charged sufficiently...
13:         $P(t) = P(t-\tau) + \alpha P_{max} R(t)$ ▷ ...increase discharge power
14:     **else** ▷ If the battery is not sufficiently charged...
15:         $P(t) = 0$ ▷ ...shut off
16:     **end if**
17:     **if** $P(t) > 0$ **then** ▷ If the battery has been charging...
18:         $P(t) = \beta P(t-\tau)$ ▷ ...reduce the charging power by β
19:     **end if**
20: **end if**
21: $P(t) = \mathbf{signum}(P(t)) \times \mathbf{min}\{|P(t)|, P_{max}\}$ ▷ Limit the power to battery specifications

The algorithm itself, as shown in Algorithm 1, contains two decision levels. The first determines whether the network is over- or under-loaded by comparing the local bus voltage, $V(t)$, to the battery's set-point threshold, V_{thr}. In the event that the network is not under high load, the battery's SOC is compared to its operation limit to check whether the battery can charge, i.e., $SOC < SOC_{max}$. If there is enough charging capacity left, then the battery's charging power is linearly increased following Line 4. If the battery was previously discharging, the related discharging power is exponentially reduced (Line 9) to reflect the multiplicative decrease.

The second decision level is entered when the network is under load. Here, the discharging power is linearly increased if the battery has enough energy stored, i.e., $SOC > SOC_{min}$ (Line 13). Additionally, if the battery was previously charging, then its charging power is multiplicatively reduced (Line 18). The direction of the charging/discharging power adjustment is determined by the first decision level, as well as the threshold proximity ratio $R(t)$. As the battery's bus voltage, $V(t)$, approaches the threshold voltage, V_{thr}, this ratio tends to zero and, hence, stops the battery operation. Therefore, oscillatory hunting is effectively mitigated. The last step of the algorithm (Line 21) assures that the battery charge/discharge power is within its device rating.

4.2. Reference Voltage Profile

When using a fixed voltage threshold, the difference in the location and load of each customer results in the over-utilisation of batteries located at the feeder end. Similar to Papaioannou et al. [44], yet for the control of BESS instead of EV charging, a reference voltage profile is proposed, which is produced by performing a power flow analysis of the network under maximum demand. An example of a fixed threshold and reference voltage profile is shown in Figure 5.

In the AIMD+, consumers located at the head of the feeder are allocated a higher voltage threshold, while those towards the end of the feeder have similar voltage thresholds to that of the fixed threshold. This replicates the expected voltage drop along the length of the feeder, hence resulting in a more equal utilisation of battery storage units that are located at those distances. The voltage threshold is set in such a way as to limit the maximum voltage drop to 3% at the end of the feeder.

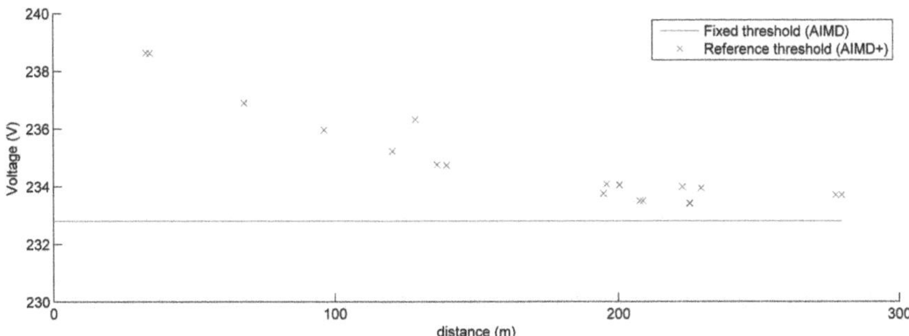

Figure 5. A plot showing the difference between the fixed voltage threshold (AIMD) and the reference voltage profile (AIMD+).

5. Scenarios and Comparison Metrics

In this section, several scenarios are explained that were used to test the performance of the battery control algorithm. Following that is the definition of three comparison metrics. These metrics quantify the improvements caused by the different algorithms in comparison to the worst case scenario.

5.1. Test Cases and Scenarios

In all simulations, the EVs plug-in on arrival and charge at their nominal charging rate until fully charged. The BESS devices were chosen to have a capacity of 7 kWh with a maximum power rating of 2 kW (battery specifications are based on the Tesla Powerwall [45]). Four excerpt cases were defined with different levels of EV and storage uptakes, these are as follows:

- **A** A baseline scenario, where only household demand is used.
- **B** A worst case scenario, in which EV uptake is 100% and no BESS is used.
- **C** An AIMD scenario, in which EV uptake is 100% and each household has a battery energy storage device. Here, each battery was controlled using the AIMD algorithm using a fixed voltage threshold.
- **D** An AIMD+ scenario, in which EV uptake is 100%, and each household has a battery energy storage device. Here, each battery was controlled using the AIMD+ algorithm using the optimised reference voltage profile.

A storage uptake of 100% was adopted to represent the worst case scenario. In addition to the four defined scenarios, a full set of simulations was performed with EV and storage uptake combinations of 0% to 100% in steps of 10%.

5.2. Performance Metric Definition

To obtain comparable performance metrics, three parameters are defined. These parameters capture the improvements in voltage violation mitigation, line overload reduction and the equality of battery usage. All excerpt performance metrics were calculated based on simulations from the IEEE EU LV Test feeder for reproducibility.

5.2.1. Parameter for Voltage Improvement

The first parameters are ζ_C^* and ζ_D^* for, respectively, Cases C and D, and calculate the magnitude of the voltage level improvement by comparing two voltage frequency distributions. More specifically, they find the difference between these probability distributions and compute a weighted sum. Here, the weighting, $\delta^*(v)$, emphasises the voltage level improvements that deviate further from the nominal substation voltage V_{ss}. If the resulting weighted sum is negative, then the obtained voltage frequency distribution was improved in comparison to the associated worst case scenario. In contrast, a positive number would indicate a worse outcome. The performance metric $\zeta^*_{\mathbf{C}}$ is defined as follows.

$$\zeta^*_{\mathbf{C}} := \sum_{v=V_{min}}^{V_{max}} \delta^*(v)\,[P_{\mathbf{B}}(v) - P_{\mathbf{C}}(v)] \tag{11}$$

Here, V_{min} is the lowest recorded voltage, and V_{max} is the highest recorded voltage. $P_{\mathbf{B}}(v)$ is the voltage probability distribution of the worst case scenario (Case B), and $P_{\mathbf{C}}(v)$ is the voltage probability distributions of Case C (i.e., the case with maximum EV and AIMD storage uptake). Similarly, the parameter $\zeta^*_{\mathbf{D}}$ therefore compares Case D, i.e., the AIMD+ case, with Case B.

The aforementioned factor, $\delta^*(v)$, scales down the summation in Equation (11) for voltages within the nominal operating band, where no voltage violations take place. Voltage violations on the other hand are scaled up to increase their impact on the summation. This scaling was produced using a linear function, with its minimum at V_{ss}, that is defined as:

$$\delta^*(v) := \begin{cases} \frac{V_{ss}-v}{V_{ss}-V_{low}} & \text{if } v \leq V_{ss} \\ \frac{v-V_{ss}}{V_{high}-V_{ss}} & \text{otherwise} \end{cases} \tag{12}$$

V_{low} and V_{high} are defined as the lower and upper limits of the nominal operation voltage band, respectively. In general, the proposed voltage comparison parameter, ζ^*, shows an improvement in voltage distribution when it is negative, whereas a positive value implies a voltage distribution with more voltage violations.

5.2.2. Parameter for Line Overload Reduction

Similar to measuring the voltage level improvements, all line utilisation probability distributions between the storage and worst case scenarios were compared. This follows a similar equation to before, but uses a different scaling factor, as described in Equation (11):

$$\zeta^{**}_{\mathbf{C}} := \sum_{c=0}^{C_{max}} \delta^{**}(c)\,[P_{\mathbf{C}}(c) - P_{\mathbf{B}}(c)] \tag{13}$$

Here, C_{max} is the highest line utilisation. $P_{\mathbf{B}}(c)$ and $P_{\mathbf{C}}(c)$ present the line utilisation probability distributions for Cases B and C, respectively, and $\delta^{**}(c)$ is the associated scaling factor. Since the relationship between line current and ohmic losses is quadratic, this scaling factor is defined as an exponential function that amplifies the impact of line currents beyond the line's nominal rating.

$$\delta^{**}(c) = \begin{cases} \left(\frac{c}{1-C_{min}}\right)^2 & \text{if } c \geq C_{min} \\ 0 & \text{otherwise} \end{cases} \tag{14}$$

The capacity scale modifier, C_{min}, defines from where the scaling should start and has been set to 0.5 for this work as only line utilisation above 0.5 p.u. was considered. Therefore, a reduction in line overloads would give a negative ζ^{**}, whereas a positive value implies a higher line utilisation, i.e., worse results.

5.2.3. Parameter for the Improvement of Battery Cycling

The final metric, ζ^{***}, gives an indication of the inequality of battery cycling (one battery cycle is defined as a full discharge and charge of the battery at maximum operating power, i.e., P_{max}) across all battery units. It does this by computing the the ratio between the peak and mean battery cycling. This Peak-to-Average Ratio (PAR) of batteries' cycling is defined in the following equation.

$$\zeta_{\mathrm{C}}^{***} := \frac{\max |C_{\mathrm{C}}|}{B^{-1}\sum_{b=1}^{B} |c_{\mathrm{C}}^{b}|} \tag{15}$$

Here, B is the number of batteries, and c_{C}^{b} is the total cycling of battery b during Scenario C. C_{C} is a vector of $\mathbb{R}_{\geq 0}^{B}$ that contains all batteries' cycling values, i.e., $c_{\mathrm{C}}^{b} \in C_{\mathrm{C}}$. Equally, the battery cycling for Scenario D would be captured by ζ_{D}^{***}. In the unlikely event of an equal cycling of all batteries, ζ^{***} will have a value of one. Yet, as batteries are operated differently, the value of ζ^{***} is likely to be greater than one. Therefore, a resulting PAR closer to one implies a more equal and therefore fairer utilisation of the deployed batteries.

6. Results and Discussion

In this section, the results are outlined that were generated from all simulations. In each of the three subsections, the performances of the AIMD and AIMD+ algorithm are compared against each other. To do so, the performance metrics outlined in Section 5.2 were used. In the following subsections, results from the four test cases defined as A, B, C and D in Section 5.1 are explained first, then the results from the full analysis over the large range of EV and battery storage uptake is presented. In the end, these results are summarised and discussed.

6.1. Voltage Violation Analysis

For the comparison of voltage improvements, results compared the algorithms' performances at reducing bus voltage variation; particularly by increasing the lowest recorded bus voltage. Each load's bus voltage was recorded, from which a sample voltage profile, Figure 6, was extracted, where the bus voltage fluctuation over time becomes apparent. It can be seen that the introduction of EVs has significantly lowered the line-to-neutral voltage. Adding energy BESS devices did raise the voltage levels during times of peak demand, as can be seen between 17:00 and 21:00, where the AIMD+ algorithm has elevated voltages further than the AIMD scenario. To obtain a better understanding of the level of improvement, the voltage frequency distribution of all buses along the feeder was generated and plotted in a histogram in Figure 7.

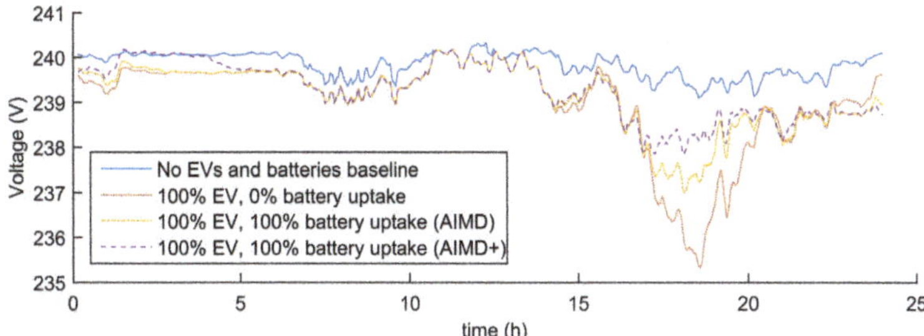

Figure 6. Recorded voltage profile at the bus of the customer closest to the substation over the period of one day with a certain uptake in EV and battery storage devices using a moving average over a window of 5 min. Here, Case A is blue; Case B is red; Case C is yellow; and Case D is violet.

In this histogram, the voltage probability distributions for all four cases were normalised and plotted against each other. Here, the previously seen drop in voltages by introducing EVs is recorded as a shift in the voltage distribution. This voltage drop is mitigated by the introduction of the storage solutions, since the probability distribution is shifted towards higher voltage bands. For the

IEEE EU LV Test feeder, the AIMD+-controlled batteries outperform the AIMD devices as the resulting ζ_{C}^{*} is greater than ζ_{D}^{*}.

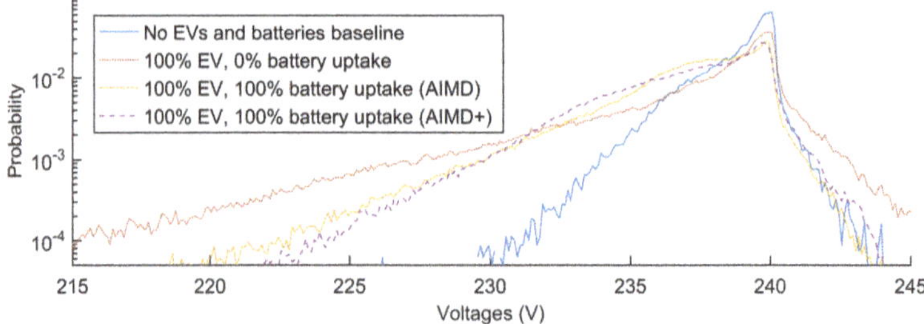

Figure 7. Voltage probability distribution of all loads' buses for certain uptakes of EV and battery storage devices. Here, Case A is blue; Case B is red; Case C is yellow; and Case D is violet; with $\zeta_{\mathrm{C}}^{*} = -0.153$ and $\zeta_{\mathrm{D}}^{*} = -0.135$.

To gain a full understanding of the performance of the AIMD and AIMD+ algorithms, a full sweep of EV and BESS uptake combinations was simulated on all available power distribution networks. The resulting parameters were averaged and plotted in Figure 8.

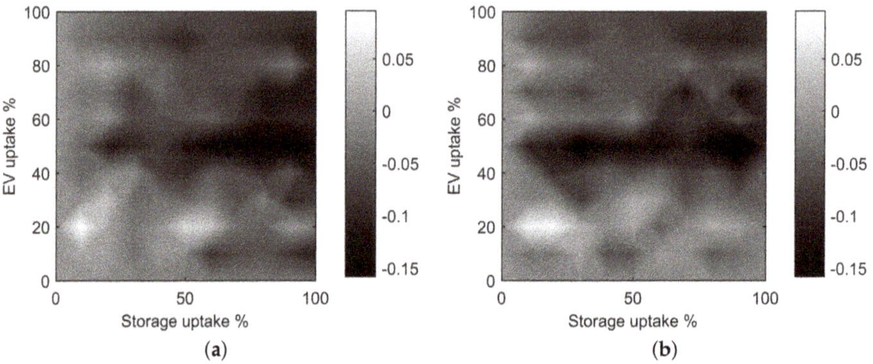

Figure 8. Comparison of voltage improvement indices (i.e., ζ^{*}) for (**a**) AIMD and (**b**) AIMD+. (a) ζ_{C}^{*} indices (AIMD); (b) ζ_{D}^{*} indices (AIMD+).

These figures show that the AIMD+ control algorithm reduces voltage deviation more effectively as the uptake in storage and EVs increases. For low storage uptake, the AIMD algorithm does not perform as strongly since more ζ_{C}^{*} values are positive and larger than their corresponding ζ_{D}^{*} value. This becomes more apparent when averaging all ζ_{C}^{*} and ζ_{D}^{*} values for their common storage uptake and across all EV uptakes. The resulting averaged metrics are plotted in Figure 9.

In this last figure, it can be seen how the sole impact of BESS uptake reflects in a continuing improvement of voltage levels. In fact, both compared algorithms improved the bus voltage, which coincides with the findings in the case studies. On average, this is the case for all BESS uptakes, as $\zeta_{\mathrm{C}}^{*} \approx \zeta_{\mathrm{D}}^{*}$. Nonetheless, it should be noted that the AIMD+ algorithm had reduced the frequency of severe voltage deviations in comparison to the AIMD algorithm and is more effective during scenarios with lower BESS uptake.

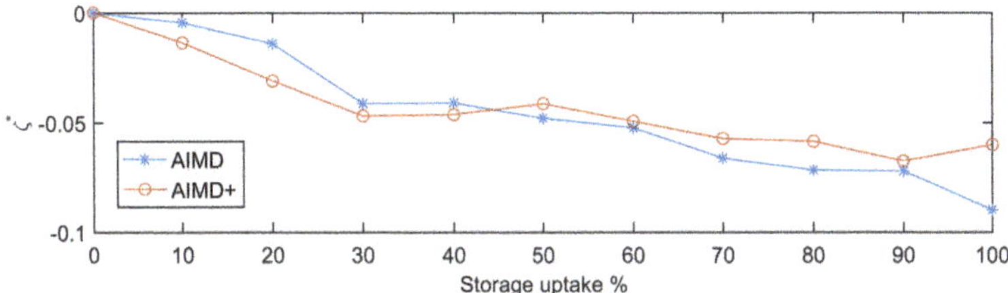

Figure 9. Average ζ^*_{C} (AIMD) and ζ^*_{D} (AIMD+) values recorded against the corresponding storage uptake.

6.2. Line Overload Analysis

Similar to the voltage improvement analysis, a frequency distribution of the line utilisation was generated. Figure 10 shows a probability distribution of the per unit (1 p.u. represents a 100% line usage, i.e., a line current of the same value as the line's nominal current rating) current in all lines, for each of the four scenarios. The corresponding ζ^{**}_{C} and ζ^{**}_{D} values for the AIMD and AIMD+ storage deployment have also been included in the figure's caption. In this figure, the observed high probability of line over-utilisation confirms that the used test network is of insufficient capacity to cater for the chosen EV uptake.

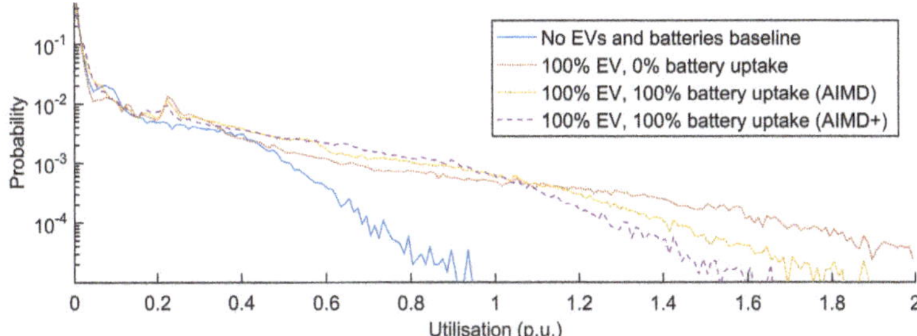

Figure 10. Line utilisation probability distribution of all lines in the simulated feeder for certain uptakes of EV and battery storage devices. Here, Case A is blue; Case B is red; Case C is yellow; and Case D is violet; with $\zeta^{**}_{\mathrm{C}} = -0.360$ and $\zeta^{**}_{\mathrm{D}} = -0.518$.

Here, the AIMD+ controlled storage devices yielded a noticeable reduction in line overloads. This improvement is apparent through the compressed width of the probability distribution and the negative ζ^{**}_{D} value. In contrast, the AIMD controlled storage devices do not fully utilise the line capacity as effectively, which leads to a positive value of ζ^{**}_{C}. To evaluate the line utilisation improvement across all simulations, the full range of EV and storage uptake was evaluated. The resulting plots are shown in Figure 11.

In these figures, it can be seen how the performance metrics change as EV uptake and storage uptake increase. For the AIMD-controlled BESS, the resulting ζ^{**}_{C} values are distributed around zero, whereas the AIMD+ algorithm achieved mostly negative values of ζ^{**}_{D}. These negative values confirm the better usage of available line capacity. This becomes particularly noticeable for scenarios where very low EV uptake is combined with larger BESS uptake. Here, AIMD-controlled storage devices commence their initial charge simultaneously. As they are located closer to the substation, they do not measure a sufficient bus voltage offset to regulate down their charging power. This behaviour causes a number of line overloads at the very beginning of the simulated days. The AIMD+ algorithm on the other hand, with its adjusted thresholds, is more responsive to non-optimal network operation and, therefore, increases the charging rate gradually.

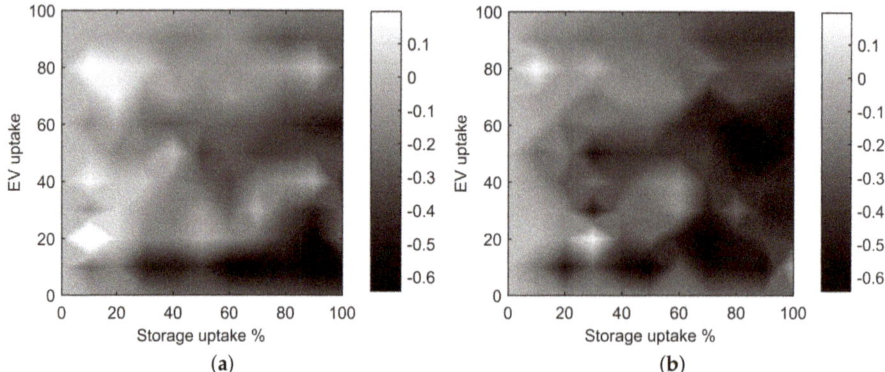

Figure 11. Comparison of line utilisation improvement indices for (**a**) AIMD and (**b**) AIMD+. (a) ζ_{C}^{**} indices (AIMD); (b) ζ_{D}^{**} indices (AIMD+).

This gradual adjustment is based on the fact that the bus voltages in the AIMD+ algorithm are closer to their nominal voltages (i.e., bus voltages found by simulating the feeder with its equally-distributed nominal load) than they are in the conventional AIMD case. A greater voltage disparity, which is the case in AIMD, causes a prolonged additive adjustment to the battery's power. This prolonged adjustment is particularly apparent for batteries situated at the bottom of the feeder, as their voltage measurements deviate the furthest from the substation voltage level. AIMD+ on the other hand prevents this behaviour by setting the voltage threshold based on the network's nominal voltage drop, which is dependent on the distance between the BESS and its feeding substation. As a result, the set-point voltage thresholds at the bottom of the feeder are lower than those closer to the substation. Hence, the additive power adjustment is equalised along the entire feeder. Therefore, by applying these individualised control thresholds, the sensitivity of the algorithm is corrected, whilst successfully mitigating the severity of line overloads.

Averaging the ζ_{C}^{**} and ζ_{D}^{**} values over all EV uptakes gives a clearer indication of performance, as this is now the only variable in the performance analysis. The result is plotted in Figure 12. Here, the hypothesis that AIMD-controlled energy storage devices do not improve line utilisation is confirmed. In contrast, the AIMD+-controlled devices succeed at effectively reducing line overloads. This is also demonstrated by the values of ζ_{C}^{**}, which remain positive yet close to zero, whereas ζ_{D}^{**} decreases with increasing uptake of battery storage devices.

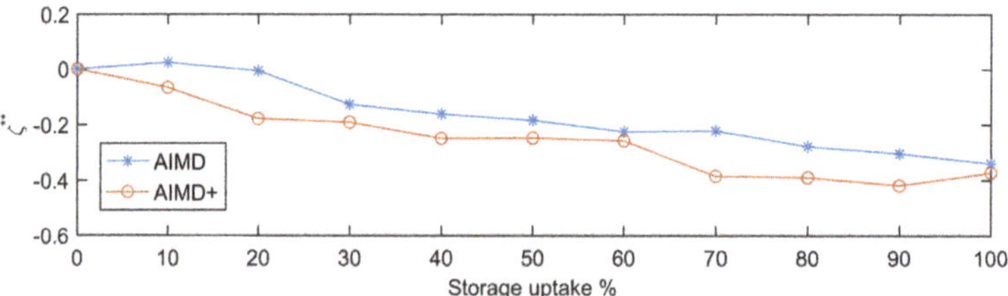

Figure 12. Average ζ_{C}^{**} (AIMD) and ζ_{D}^{**} (AIMD+) values recorded against the corresponding storage uptake.

Whilst the deployment of energy storage has often been seen as a possible solution to defer network reinforcements, the presented results show that this is not always the case. In fact, the importance of choosing an appropriate control algorithm outweighs the availability of the energy storage itself.

This becomes particularly apparent when energy storage devices need to recharge their injected energy for times of peak demand. For the AIMD case, this recharging was not controlled sufficiently, which led to higher line currents. The proposed AIMD+ algorithm was not as susceptible to this kind of behaviour, as it has been designed to take battery location into account. This immunity and well-controlled power flow caused little to no additional strain on the network's equipment, allowing the deployed storage devices to also provide voltage support.

6.3. Battery Utilisation Analysis

In this part of the analysis, the batteries' fairness of usage was evaluated. The battery power profiles were recorded; excerpts are plotted in Figure 13 and are arranged by distance from the substation.

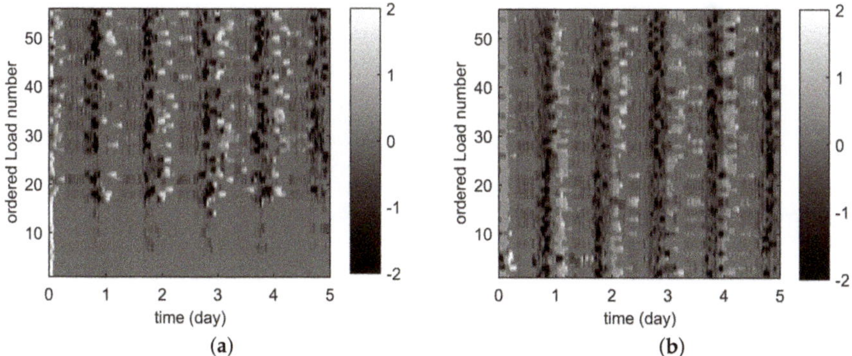

Figure 13. Battery power profiles of each load's battery storage device over four days for (**a**) AIMD and (**b**) AIMD+. (**a**) Case C, 60% EV and 100% AIMD (kW); (**b**) Case D, 60% EV and 100% AIMD+ (kW).

In this figure, it can be seen that only half of the deployed storage devices were active in Case C (AIMD control), whereas all devices are utilised in Case D (AIMD+ control). From the recorded battery SOC profiles, the net cycling of each battery was computed and divided by the duration of the simulation, giving an average daily cycling value. This value is plotted for each load in Figure 14a. The corresponding statistical analysis is presented in Figure 14b.

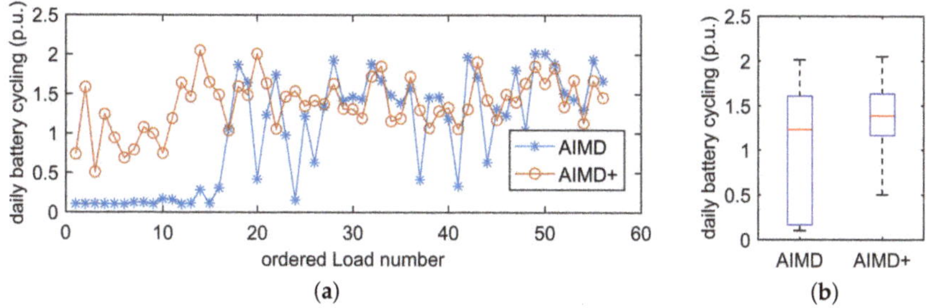

Figure 14. Each load's battery cycling compared for (**a**) 60% EV and 100% AIMD and AIMD+ uptake and (**b**) in a statistical context. Here, $\zeta_{\mathbf{C}}^{***} = 3.89$ and $\zeta_{\mathbf{D}}^{***} = 2.54$. (**a**) Battery cycling for each load; (**b**) statistic.

These two plots show the under-usage of AIMD controlled batteries, as well as the variance in battery usage under AIMD and AIMD+ control. In fact, under AIMD control, 20 out of 55 batteries experienced a cycling of less than 10% per day, whereas the remaining devices were utilised fully.

This discrepancy causes the ζ_C^{***} value to be noticeably larger than ζ_D^{***}. A more detailed comparison is given when plotting the Peak-to-Average Ratios (PAR) of the batteries' daily cycling over the full range of EV and storage uptake scenarios; these plots are shown in Figure 15. Section 5.2.3 gives the detail on the PAR, ζ^{***}.

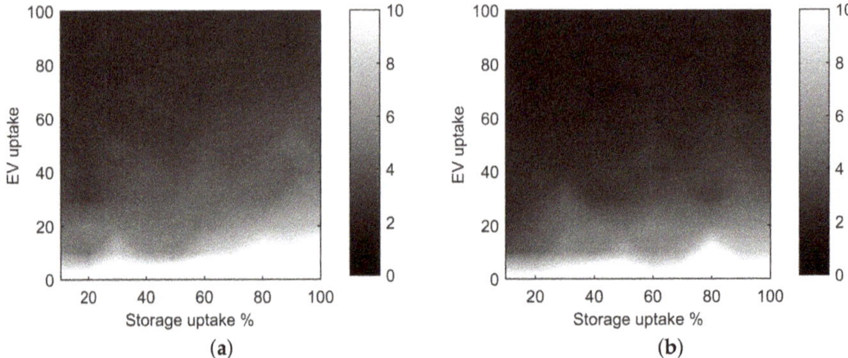

Figure 15. Peak-to-Average Ratios (PAR) of the battery cycling profiles of each load's battery storage device over four days for (**a**) AIMD and (**b**) AIMD+. (**a**) Cycling PAR for AIMD; (**b**) cycling PAR for AIMD+.

The figure shows that for any EV uptake scenario, AIMD-controlled energy storage units were cycled less equally than the AIMD+ controlled devices. Results also show that with a low EV uptake, both the AIMD and AIMD+ algorithm performed worse; yet improved as EV uptake increased.

Averaging the PARs for all batteries' SOC profiles over all EV uptake percentages yields a clear performance difference between AIMD and AIMD+. These resulting PARs, i.e., the ζ_C^{***} and ζ_D^{***} values for their corresponding storage uptake percentages, are presented in Figure 16.

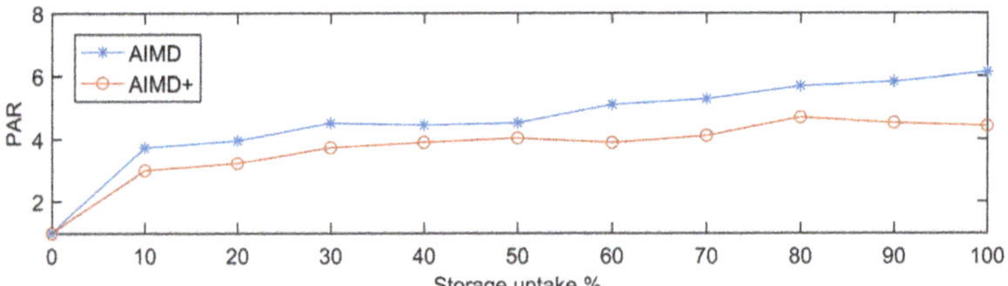

Figure 16. The performance index ζ_C^{***} for AIMD storage and ζ_D^{***} for AIMD+ storage control against storage uptake.

Although the AIMD controlled batteries were, on average, cycled less than the batteries controlled by the proposed AIMD+ algorithm, looking at the average produces a distorted understanding of the performance. In fact, as more than half of the assigned AIMD BESS devices never partook in the network control, a lower average cycling was expected to begin with. The variation in cycling across all batteries, or the cycling PAR, reveals the difference between usage and effective usage. A lower ratio indicates a better usage of the deployed batteries.

7. Conclusions

In this paper, an algorithm is proposed for distributed battery energy storage, in order to mitigate the negative impact of highly variable uncontrolled loads, such as the charging of EVs.

The improved AIMD algorithm uses local bus voltage measurements and implements a reference voltage profile, derived from power flow analysis of the distribution network, for its set-point control. Taking the distance to the feeding substation into account allowed optimising the algorithm's parameters for each BESS. Simulations were performed on the IEEE EU LV Test feeder and a set of real U.K. suburban network models. Comparisons were made of the standard AIMD algorithm with a fixed voltage threshold against the proposed AIMD+ algorithm using a reference voltage threshold. A set of European demand profiles and a realistic EV travel model were used to feed load data into the simulations.

For all conducted simulations, the performance of the energy storage units was improved by using the proposed AIMD+ algorithm instead of traditional AIMD control. The improved algorithm resulted in a reduction of voltage variation and an increased utilisation of available line capacity, which also reduced the frequency of line overloads. Additionally, the same algorithm equalised the cycling and utilisation of battery energy storage, making better use of the deployed battery assets. To take this work further, future work will also consider distributed generation, such as photovoltaic panels, smart-charging EV uptake, as well as decentralised methods for determining voltage reference values, so no prior network knowledge is required.

Acknowledgments: The authors would like to thank SSE-PD for providing their network information for the utilised U.K. feeder models and also Miss Catriona Scrivener for proofreading this manuscript.

Author Contributions: Maximilian J. Zangs and Peter B. E. Adams contributed equally to this piece of work and were supervised by William Holderbaum and Ben A. Potter. Timur Yunusov has provided technical input and feedback throughout.

Conflicts of Interest: The authors declare no conflict of interest.

References

1. Shah, V.; Booream-Phelps, J. *F.I.T.T. for Investors: Crossing the Chasm;* Technical Report; Deutsche Bank Market Research: Frankfurt am Main, Germany, 2015.
2. Department for Business Enterprise and Regulatory (DBER); Department for Transport (DfT). *Investigation into the Scope for the Transport Sector to Switch to Electric Vehicles and Plug-in Hybrid Vehicles;* Technical Report; Department for Business Enterprise and Regulatory (DBER), Department for Transport (DfT): London, UK, 2008.
3. Ecolane, University of Aberdeen. *Pathways to High Penetration of Electric Vehicles;* Technical Report; Ecolane, University of Aberdeen: Bristol, UK, 2013.
4. Clement-Nyns, K.; Haesen, E.; Driesen, J. The impact of Charging plug-in hybrid electric vehicles on a residential distribution grid. *IEEE Trans. Power Syst.* **2010**, *25*, 371–380.
5. Fernández, L.P.; San Román, T.G.; Cossent, R.; Domingo, C.M.; Frías, P. Assessment of the impact of plug-in electric vehicles on distribution networks. *IEEE Trans. Power Syst.* **2011**, *26*, 206–213.
6. Hadley, S.W.; Tsvetkova, A.A. Potential Impacts of Plug-in Hybrid Electric Vehicles on Regional Power Generation. *Electr. J.* **2009**, *22*, 56–68.
7. Putrus, G.; Suwanapingkarl, P.; Johnston, D.; Bentley, E.; Narayana, M. Impact of electric vehicles on power distribution networks. In Proceedings of the IEEE Vehicle Power and Propulsion Conference, Dearborn, MI, USA, 7–10 September 2009; pp. 827–831.
8. Pillai, J.R.; Bak-Jensen, B. Vehicle-to-grid systems for frequency regulation in an islanded Danish distribution network. In Proceedings of the IEEE Vehicle Power and Propulsion Conference (VPPC), Lille, France, 1–3 September 2010.
9. Zhou, K.; Cai, L. Randomized PHEV Charging Under Distribution Grid Constraints. *IEEE Trans. Smart Grid* **2014**, *5*, 879–887.
10. Mohsenian-Rad, A.H.; Wong, V.W.S.; Jatskevich, J.; Schober, R.; Leon-Garcia, A. Autonomous demand-side management based on game-theoretic energy consumption scheduling for the future smart grid. *IEEE Trans. Smart Grid* **2010**, *1*, 320–331.

11. Deilami, S.; Masoum, A.S. Real-time coordination of plug-in electric vehicle charging in smart grids to minimize power losses and improve voltage profile. *IEEE Trans. Smart Grid* **2011**, *2*, 456–467.
12. Masoum, A.S.; Deilami, S.; Member, S.; Masoum, M.A.S. Fuzzy Approach for Online Coordination of Plug-In Electric Vehicle Charging in Smart Grid. *IEEE Trans. Sustain. Energy* **2015**, *6*, 1112–1121.
13. Karfopoulos, E.L.; Hatziargyriou, N.D. A Multi-Agent System for Controlled Charging of a Large Population of Electric Vehicles. *IEEE Trans. Power Syst.* **2013**, *28*, 1196–1204.
14. Wu, C.; Mohsenian-Rad, H.; Huang, J. Vehicle-to-aggregator interaction game. *IEEE Trans. Smart Grid* **2012**, *3*, 434–442.
15. Samadi, P.; Mohsenian-Rad, H.; Schober, R.; Wong, V.W.S. Advanced Demand Side Management for the Future Smart Grid Using Mechanism Design. *IEEE Trans. Smart Grid* **2012**, *3*, 1170–1180.
16. Xu, N.Z.; Chung, C.Y. Challenges in Future Competition of Electric Vehicle Charging Management and Solutions. *IEEE Trans. Smart Grid* **2015**, *6*, 1323–1331.
17. Leadbetter, J.; Swan, L. Battery storage system for residential electricity peak demand shaving. *Energy Build.* **2012**, *55*, 685–692.
18. Sugihara, H.; Yokoyama, K.; Saeki, O.; Tsuji, K.; Funaki, T. Economic and efficient voltage management using customer-owned energy storage systems in a distribution network with high penetration of photovoltaic systems. *IEEE Trans. Power Syst.* **2013**, *28*, 102–111.
19. Toledo, O.M.; Oliveira, D.; Diniz, A.; Martins, J.H.; Vale, M.H.M. Methodology for Evaluation of Grid-Tie Connection of Distributed Energy Resources-Case Study with Photovoltaic and Energy Storage. *IEEE Trans. Power Syst.* **2013**, *28*, 1132–1139.
20. Marra, F.; Yang, G.Y.; Fawzy, Y.T.; Træholt, C.; Larsen, E.; Garcia-Valle, R.; Jensen, M.M. Improvement of local voltage in feeders with photovoltaic using electric vehicles. *IEEE Trans. Power Syst.* **2013**, *28*, 3515–3516.
21. Mokhtari, G.; Nourbakhsh, G.; Ghosh, A. Smart coordination of energy storage units (ESUs) for voltage and loading management in distribution networks. *IEEE Trans. Power Syst.* **2013**, *28*, 4812–4820.
22. Atia, R.; Yamada, N. Sizing and Analysis of Renewable Energy and Battery Systems in Residential Microgrids. *IEEE Trans. Smart Grid* **2016**, *7*, 1204–1213.
23. Hatziargyriou, N.D.; Škrlec, D.; Capuder, T.; Georgilakis, P.S.; Zidar, M. Review of energy storage allocation in power distribution networks: Applications, methods and future research. *IET Gener. Transmi. Distrib.* **2015**, *10*, 1–8.
24. Chiu, D.M.; Rain, R. Analysis of the increase and decrease algorithms for congestion avoidance in computer networks. *Comput. Netw. ISDN Syst.* **1989**, *17*, 1–14.
25. Wirth, F.; Stuedli, S.; Yu, J.Y.; Corless, M.; Shorten, R. *IBM Research Report: Nonhomogeneous Place-Dependent Markov Chains, Unsynchronised AIMD, and Network Utility Maximization*; Technical Report; IBM: New York, NY, USA, 2014.
26. Stüdli, S.; Crisostomi, E.; Middleton, R.; Shorten, R. A flexible distributed framework for realising electric and plug-in hybrid vehicle charging policies. *Int. J. Control* **2012**, *85*, 1130–1145.
27. Studli, S.; Griggs, W.; Crisostomi, E.; Shorten, R. On Optimality Criteria for Reverse Charging of Electric Vehicles. *IEEE Trans. Intell. Transp. Syst.* **2014**, *15*, 451–456.
28. Stüdli, S.; Crisostomi, E.; Middleton, R.; Shorten, R. Optimal real-time distributed V2G and G2V management of electric vehicles. *Int. J. Control* **2014**, *87*, 1153–1162.
29. Stüdli, S.; Crisostomi, E.; Middleton, R.; Braslavsky, J.; Shorten, R. Distributed Load Management Using Additive Increase Multiplicative Decrease Based Techniques. In *Plug in Electric Vehicles in Smart Grids*; Springer: Singapore, Singapore, 2015; pp. 173–202.
30. Mareels, I.; Alpcan, T.; Brazil, M.; de Hoog, J.; Thomas, D.A. A distributed electric vehicle charging management algorithm using only local measurements. In Proceedings of the IEEE PES Innovative Smart Grid Technologies Conference (ISGT), Washington, DC, USA, 19–22 February 2014; pp. 1–5.
31. Xia, L.; Hoog, J.D.; Alpcan, T.; Brazil, M. *Electric Vehicle Charging: A Noncooperative Game Using Local Measurements*; The International Federation of Automatic Control: Cape Town, South Africa, 2014; pp. 5426–5431.
32. Munkhammar, J.; Bishop, J.D.; Sarralde, J.J.; Tian, W.; Choudhary, R. Household electricity use, electric vehicle home-charging and distributed photovoltaic power production in the city of Westminster. *Energy Build.* **2015**, *86*, 439–448.

33. Dallinger, D.; Wietschel, M. Grid integration of intermittent renewable energy sources using price-responsive plug-in electric vehicles. *Renew. Sustain. Energy Rev.* **2012**, *16*, 3370–3382.
34. Institut für angewandte Sozialwissenschaft GmbH; Deutsches Zentrum für Luft-und Raumfahrt e.V. *Mobilität in Deutschland 2008*; Technical Report; Mobilität in Deutschland: Bonn und Berlin, Germany, 2008.
35. National Grid. *Future Energy Scenarios 2015*; Technical Report; National Grid: Warwick, UK, July 2015.
36. Rowe, M.; Member, S.; Yunusov, T.; Member, S.; Haben, S.; Singleton, C.; Holderbaum, W.; Potter, B. A Peak Reduction Scheduling Algorithm for Storage Devices on the Low Voltage Network. *IEEE Trans. Smart Grid* **2014**, *5*, 2115–2124.
37. Laresgoiti, I.; Käbitz, S.; Ecker, M.; Sauer, D.U. Modeling mechanical degradation in lithium ion batteries during cycling: Solid electrolyte interphase fracture. *J. Power Sour.* **2015**, *300*, 112–122.
38. Government Digital Service. *Vehicle Free-Flow Speeds (SPE01)*; Government Digital Service: London, UK, 2013.
39. Office for Low Emission Vehicles. *Electric Vehicle Homecharging Scheme—Guidance for Manufacturers and Installers*; Technical Report; Office for Low Emission Vehicles: London, UK, 2016.
40. Rowe, M.; Holderbaum, W.; Potter, B. Control methodologies: Peak reduction algorithms for DNO owned storage devices on the Low Voltage network. In Proceedings of the 4th IEEE/PES Innovative Smart Grid Technologies Europe (ISGT), Lyngby, Denmark, 6–9 October 2013; pp. 1–5.
41. Rowe, M.; Yunusov, T.; Haben, S.; Holderbaum, W.; Potter, B. The real-time optimisation of DNO owned storage devices on the LV network for peak reduction. *Energies* **2014**, *7*, 3537–3560.
42. Society, I.P. Energy, European Low Voltage Test Feeder. Available online: http://ewh.ieee.org/soc/pes/dsacom/testfeeders/ (accessed on 31 January 2016).
43. Thames Valley Vision—Project Library—Published Documents. Available online: http://thamesvalleyvision.co.uk/project-library/published-documents/ (accessed on 31 January 2016).
44. Papaioannou, I.T.; Purvins, A.; Demoulias, C.S. Reactive power consumption in photovoltaic inverters: A novel configuration for voltage regulation in low-voltage radial feeders with no need for central control. *Prog. Photovolt. Res. Appl.* **2015**, *23*, 611–619.
45. Tesla Motors Inc. *Tesla Powerwall*; Tesla Motors Inc.: Fremont, CA, USA, 2015.

Article

Energy Link Optimization in a Wireless Power Transfer Grid under Energy Autonomy Based on the Improved Genetic Algorithm

Zhihao Zhao [2], Yue Sun [1,2,*], Aiguo Patrick Hu [3], Xin Dai [1,2] and Chunsen Tang [2]

1. State Key Laboratory of Power Transmission Equipment & System Security and New Technology, Chongqing 400044, China; toybear@vip.sina.com
2. College of Automation, Chongqing University, Chongqing 400044, China; zhaozhihao.123@163.com (Z.Z.); cstang@cqu.edu.cn (C.T.)
3. Department of Engineering, The University of Auckland, Auckland 1142, New Zealand; a.hu@auckland.ac.nz

* Correspondence: syue@cqu.edu.cn; Tel.: +86-23-6511-2750

Academic Editor: William Holderbaum
Received: 28 March 2016; Accepted: 18 August 2016; Published: 26 August 2016

Abstract: In this paper, an optimization method is proposed for the energy link in a wireless power transfer grid, which is a regional smart microgrid comprised of distributed devices equipped with wireless power transfer technology in a certain area. The relevant optimization model of the energy link is established by considering the wireless power transfer characteristics and the grid characteristics brought in by the device repeaters. Then, a concentration adaptive genetic algorithm (CAGA) is proposed to optimize the energy link. The algorithm avoided the unification trend by introducing the concentration mechanism and a new crossover method named forward order crossover, as well as the adaptive parameter mechanism, which are utilized together to keep the diversity of the optimization solution groups. The results show that CAGA is feasible and competitive for the energy link optimization in different situations. This proposed algorithm performs better than its counterparts in the global convergence ability and the algorithm robustness.

Keywords: wireless power transfer; wireless power transfer grid; energy link; genetic algorithm

---

1. Introduction

Wireless power transfer (WPT) technology has widely attracted attention from research institutions and companies all over the world. The power could be transferred through magnetic field [1–5], electric field [6–9], Radio Frequency Identification (RFID) [10–13], etc., from the power supply to the load without any electrical connections with the introduction of WPT. It contributes to eliminating the cable constraints in the traditional power transfer pattern and, thus, increases the flexibility of the power supply. Meanwhile, a number of advantages are also brought in as the power transfer process is unaffected by dirt, ice, water and other chemicals and, thereby, is environmentally inert and maintenance free. Consequently, WPT is widely used in numerous applications, including electric vehicles [14], electronics [15], biomedical implants [16], etc.

However, due to the traditional point-point mechanism in the WPT system in which the power is transferred from the power supply to the load directly, the power transfer efficiency (*PTE*) and coverage area are highly restricted. The system operation frequency is usually increased to improve the system performance. However, with the increase of the system operation frequency, the relevant electromagnetic interference (EMI) problem, which is common in high frequency applications, becomes worse in the WPT system. Therefore, another alternative should be proposed to enhance the power transfer performance.

Recent advances in WPT technology have made it possible to transfer sufficient power to the electrical devices over a large air gap with the introduction of repeaters (resonance coils). The *PTE* and coverage area are both increased without sacrificing the operation frequency [17], and thus, the EMI is diminished. However, certain space is occupied by these existing coils, and thus, the space utilization is reduced. Consequently, this pattern is not suitable for the power transfer process within a multi-device system in which devices are randomly distributed, such as a robot soccer match, a rechargeable wireless sensor network and distributed satellites. To achieve the power transfer process within a multi-device system, a regional smart microgrid [18,19] named the wireless power transfer grid (WPTG) [20,21] has been proposed recently. It contributes to ensuring real-time energy supplies and energy load balance during the power transfer process. This energy grid is comprised of battery-powered device nodes. They are equipped with WPT technology and distributed randomly in a certain area. The network routing (multi-hop) theory in the traditional information network is also introduced for the energy link during the power transfer process, as well. As shown in Figure 1, the whole grid could be linked to the external power injection, which could be fulfilled by the power grid. The energy link, which is comprised of sub-links, is the abstract channel of the power flow from the external power injection to the load node. Each sub-link represents the abstract power transfer channel between every two involved device nodes within the energy link. Therefore, with the introduction of device repeaters, the energy could be relayed to enlarge the power transfer area without compromising *PTE*. Any load node within the coverage of the WPTG will be able to be charged. However, the power injection would possibly be cut off in some extreme situations, like earthquakes and power failures. During this broken period of the external power injection, the whole grid is isolated from the external power injection, and then, nodes in this grid are running under the energy autonomy situation. When one member node has power demand, the energy link needs to be established to generate power flow to extend the battery life of this load node. Additionally, this power flow is based on the nodes' own energy storage. Hence, the comprehensive optimization in the energy link needs to be studied. The *PTE* has been taken into consideration to reduce the power dissipation during the power transfer process in previous studies. However, a time delay is also brought in by the introduction of the device repeaters. Furthermore, the energy balance during the power transfer process is also worthy of being noted, as well, in this energy autonomy situation.

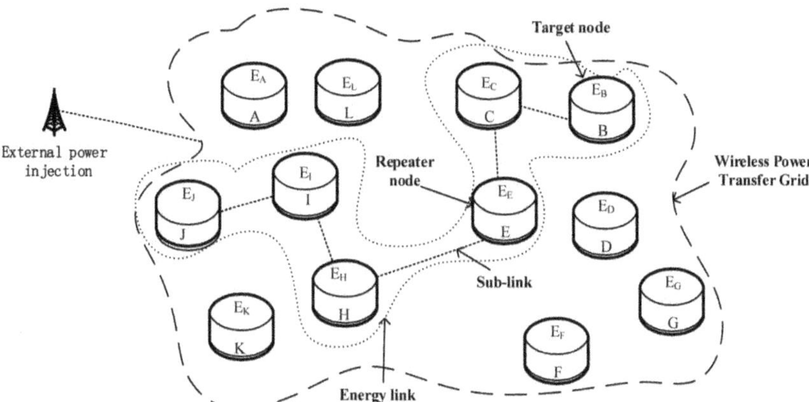

Figure 1. Architecture of a wireless power transfer grid with external power injection.

In this paper, the energy link optimization in WPTG under energy autonomy will be presented with the consideration of power transfer and grid characteristics. The relevant multi-criteria in energy link optimization are analysed. Then, an improved algorithm (concentration adaptive genetic algorithm (CAGA)) is proposed to get the optimal energy link with these multi-criteria. Several key improvements are also introduced to resolve the local optimum problem in the optimization process. They contribute to keeping the diversity of the solution group and, thus, motivate the solution evolution to reach the global optimal result. This paper is organized as follows. Section 2 introduces the WPTG under energy autonomy. Section 3 analyses the energy link in WPTG and presents the optimization model with these multi-criteria in the energy link. The detailed optimization algorithm is provided in Section 4. The simulation results and discussions are demonstrated in Section 5. Section 6 provides conclusions and discussions for future research directions.

2. Wireless Power Transfer Grid under Energy Autonomy

2.1. Grid Introduction

The WPTG consists of battery-powered device nodes. When the external power injection in Figure 1 is cut off, the device repeaters in WPTG are selected to establish the energy links for power transferring to the load node. With the relay function of these repeaters during the transferring process, the energy dissipation is reduced, and the power transfer flexibility is augmented.

2.2. Node Parameters

Device nodes are assumed to be homogeneous as follows.

1. Each node is assumed to operate with the functional load; therefore, the energy storage in each node is declining with time. Each node will reach three energy situations with the variation of its energy storage:
   - Normal situation
   - Energy-poor situation
   - Energy-disabled situation

   Nodes will be removed from WPTG for maintaining its own functional load if they reach the energy-disabled situation. Therefore, the node in the energy-poor situation will call for energy supplies to avoid this worst situation.
2. Each node is assumed to be able to detect the surrounding (neighbouring) nodes' information.
3. Relevant nodes in the energy link act as:
   - Power supply node S
   - Repeater node R
   - Load node T

   Without the external power injection, a certain node will be chosen as the power supply node S. Additionally, the power transfer process could be achieved from the power supply node S to the load node T through the repeater node R. As shown in Figure 2, power supply node A transfers power to load node C through repeater node B. Nevertheless, these roles are not fixed, and they will vary with the change of the load node. Thus, when node A calls for energy supplies and transforms into the load node, power supply node B transfers the power through repeater node C to meet its demand. Meanwhile, the bi-directional wireless power transfer [22,23] technology is introduced, which means the power transfer flow is reversible.

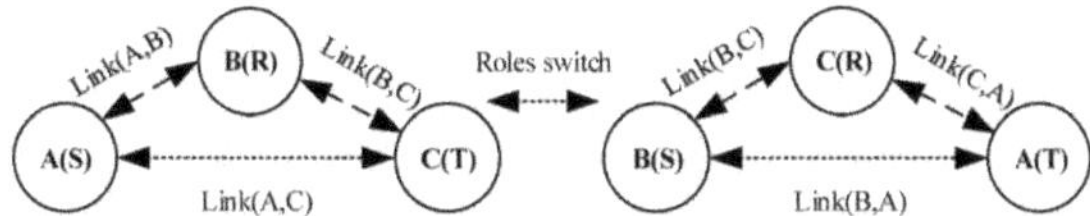

Figure 2. Power transfer roles.

4 During the power transfer process, the involved nodes will suffer from the extra energy load, which is related to the detailed power conditioning performance. Due to the fact that it is beyond the scope of this paper, it will not be explained in detail.

2.3. Grid Operating Mechanism

As shown in Figure 3, the grid operation mechanism could be divided into three phases. During Phase I, the node will broadcast the energy requests and transform itself into load node T if it reaches the energy-poor situation. The surrounding nodes will relay it to their neighbouring nodes after they receive this energy request. Meanwhile, as shown in Figure 4, the nodes compare their own energy storage with the neighbouring nodes. If one node among them has stored more energy than fifty percent of its neighbouring nodes, it will reply with its own energy storage information to node T. Then, nodes that have not replied will not be involved in the consequent operations, and they will be labelled as N in Phase II. Additionally, the involved nodes will be labelled as R_i.

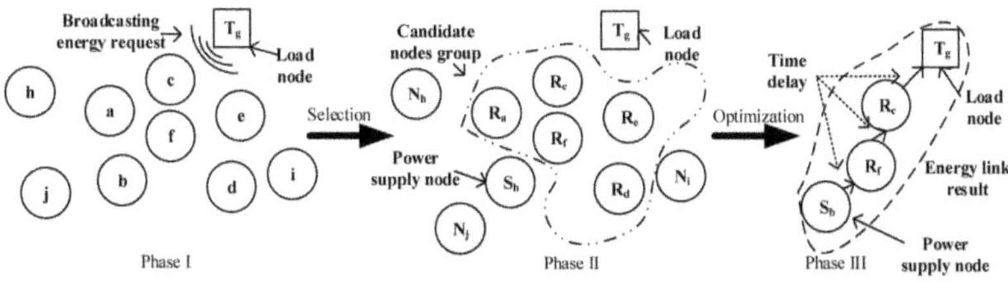

Figure 3. Grid operating mechanism.

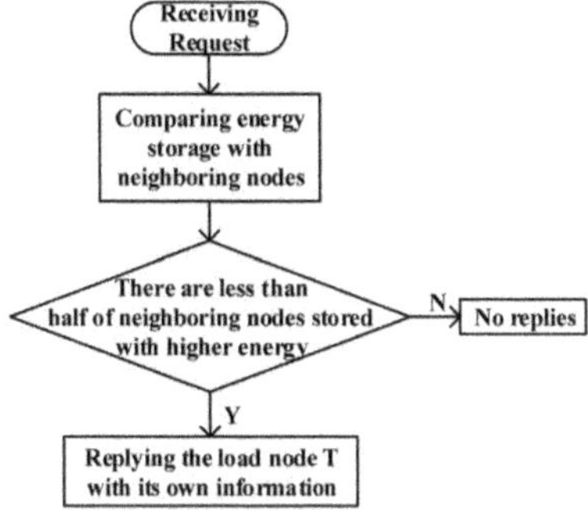

Figure 4. Flowchart of energy request decision.

In Phase II, load node T will compare the received energy storage information, as well. The node with maximum energy storage will be selected to be the power supply node S_i (subscript i denotes that node i is chosen as the power supply node i), and the rest of the communication nodes will be chosen as candidates for repeater nodes. The node with maximum energy storage is selected to achieve the energy balance. This selection mechanism helps to distribute the energy load among the whole grid. With the node information of these candidate nodes shared with the power supply node S_i, node S_i chooses nodes from these candidate nodes to establish the optimal energy link ($\{b, f, c, g\}$ in Figure 3) to transfer the power from node S_i to node T in Phase III. We will focus on this non-linear optimization process in this paper, which could be represented as Equation (1).

$$\begin{cases} \tau_{op} = argmin(J(\tau_i)), \tau_i \in \tau \\ s.t.cons \end{cases} \tag{1}$$

where $\tau = \tau_1, \tau_1, ..., \tau_n (n \in N^*)$ is the solution group, *Cons* represents the constraints, $J(\tau_i)$ is the performance index of the i-th solution and τ_{op} is the optimal energy link with the minimum performance index J. This optimization process will be discussed in detail in subsequent analyses.

3. Energy Link Optimization Model

3.1. Energy Link Analysis

In the traditional point-point WPT system, the *PTE* and power capacity are two main performance indexes. In this paper, the power capacity is assumed to be able to meet the load node's power demand, and thus, the *PTE* could be utilized to measure the power dissipation during the power transfer process. Therefore, during the operating process of WPTG, the power supply node S, as shown in Figure 3, has to select repeater nodes from the candidate nodes' group to establish the optimal energy link with the consideration of the *PTE* index. Furthermore, due to the repeater nodes, the time delay is also introduced including (1) the communication delay and (2) the power conditioning delay. The latter is overwhelmed by the former. Only the communication delay will be taken into consideration in this paper. Meanwhile, the nodes with small energy storage should be chosen to be repeaters with less possibility considering the energy load balance. Therefore, these multi-criteria for this optimization issue are listed as follows:

- Power transfer efficiency *PTE*: During the energy autonomy situation, the power demand of the load node is satisfied by the energy stored in other nodes. The *PTE* should be improved to reduce the power dissipation during the power transfer process.
- Time delay *Del*: The communications are undertaken to inform the repeater nodes to join the energy link; therefore, a time delay is introduced in the power transfer process. In order to improve its real-time performance, the time delay should be reduced.
- Energy load balance S_{mst}: If the nodes with small energy storage are selected to join the energy link, due to the extra energy load during the power transfer process, these will quickly be driven to reach the energy-disabled situation, which should be avoided. Hence, this extra energy load should be distributed among the nodes with higher energy storage to achieve the energy load balance in the WPTG. The minimum energy storage S_{mst} in the selected energy link is utilized to represent this index.

As shown in Figure 5, relevant nodes will be chosen to establish the optimal energy link from the repeater nodes' group.

E_i represents the energy storage in node i. During the comparison, energy link $\tau_j = \{B, C, E, H, I, J\}$ is chosen, in which node D is selected rather than node C due to its higher energy storage; $\{E, H, J\}$ also performs better than its counterpart in the index value of *PTE* and *Del*.

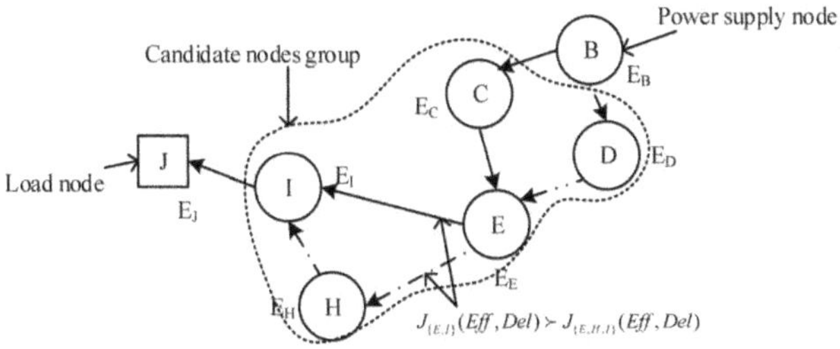

Figure 5. Energy link comparison.

3.2. Optimization Graph

As shown in Figure 5, the potential energy links for power transfer will be compared to get the optimal result. They are the combination of sub-links among relevant nodes in the energy link. Each sub-link is an abstract object of the power transfer channel during the power transfer process between two linked nodes. Meanwhile, the channel parameters are assumed to be constant during the power transfer process. Therefore, the whole optimization grid could be abstracted as an undirected weighted graph $G = < V, E >$ in Figure 6. Node set V represents the device nodes; weighted edge set E means the abstract energy links in which e_{ij} is the abstract energy link between node i and j; weight set $(min_{ij}, pte_{ij}, del_{ij})$ in the edges means the energy link weight, in which min_{ij} stands for the minimum energy storage in two linked nodes; pte_{ij} is the power transfer efficiency during the power transfer process between node i and j; del_{ij} represents the time delay. Meanwhile, the power flow direction is not restrained due to the bi-directional WPT technology mentioned before.

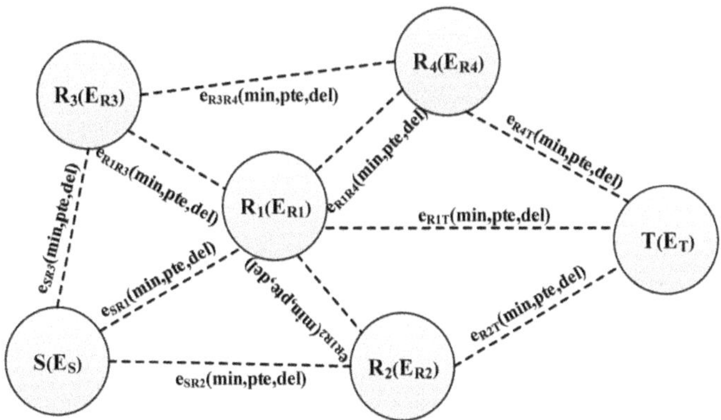

Figure 6. Energy link optimization graph.

Therefore, the multi-criteria in energy link τ_i could be represented as follows:

$$Del(\tau_i) = \sum del_{ij}, \forall e_{ij} \in \tau_i \tag{2}$$

$$PTE(\tau_i) = \prod pte_{ij}, \forall e_{ij} \in \tau_i \tag{3}$$

$$S_{mst}(\tau_i) = Min\{min_{i,j}\}, \forall e_{ij} \in \tau_i \tag{4}$$

The global delay *Del* and *PTE* are both the accumulation of relevant sub-links. The S_{mst} is the minimum value of the energy storage of nodes in the energy link. Then, the global performance index J could be represented as a function of these multi-criteria as below:

$$J(\tau_i) = f(Del(\tau_i), PTE(\tau_i), S_{mst}(\tau_i)) \tag{5}$$

Meanwhile, with the considerations of the performance threshold in the power transfer process, the non-linear optimization model for the energy link could be obtained:

$$\begin{cases} argmin(J(\tau_i)), \tau_i \in \tau \\ s.t. g_1(PTE) = PTE_m - pte_{ij} \leq 0, \forall e_{ij} \in \tau_i \\ \quad g_2(Del) = del_{ij} - D_m \leq 0, \forall e_{ij} \in \tau_i \\ \quad g_3(S_{mst}) = S_m - min_{ij} \leq 0, \forall e_{ij} \in \tau_i \end{cases} \tag{6}$$

where PTE_m, D_m, S_m indicate the threshold value for the power transfer efficiency, the time delay and the energy storage in the energy link solution, respectively. PTE_m is set for the power transfer performance. D_m is set to ensure the real-time energy supplies. S_m is used to manipulate the energy load for the energy balance. This optimization is an NP-hard issue; therefore, the traditional derivation method could not be utilized. Consequently, an intelligent optimization algorithm will be presented in the next section for this optimization issue.

4. Energy Link Optimization Algorithm

The energy link optimization issue could be demonstrated as the routing path problem. Due to its discrete characteristic, a common continuous optimization method, like the particle swarm optimization algorithm [24–26], could not be utilized. The ant colony algorithm [27,28] is easily trapped by local optima due to its open-loop control. In comparison, the genetic algorithm (GA) [29–33] is effective in both discrete and continuous optimization issues. Meanwhile, GA is easily modified to resolve specific issues. As a result, CAGA is proposed to select the optimal energy link. The entire algorithm flowchart is shown in Figure 7.

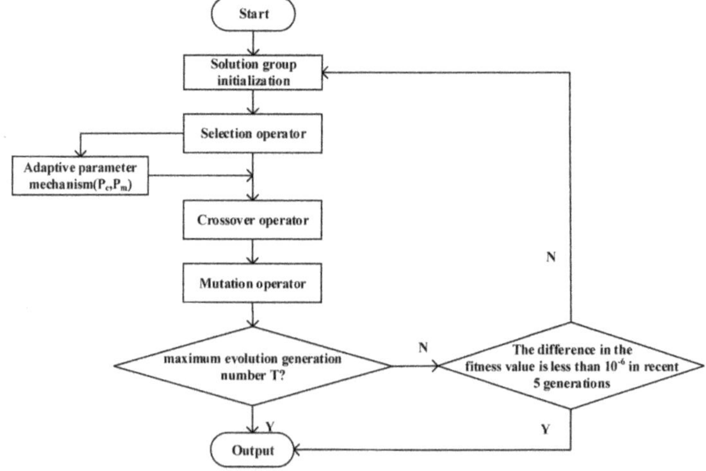

Figure 7. Algorithm flowchart.

4.1. Algorithm Initialization

The variable-length encoding mechanism is adopted for the energy link solution encoding. The position is used to represent the sequence of the nodes in the energy link, and its value is the node ID (gene) within the path. As illustrated in Figure 8, there are N nodes (genes) in the chromosome. Thus, this solution is started from source node S to node T, through repeater nodes $r_1, r_2, ..., r_{N-2}$.

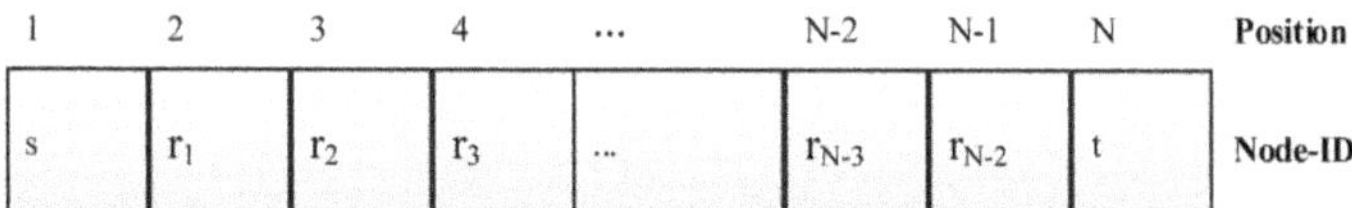

Figure 8. CAGA encoding.

4.2. Fitness Function

The fitness function is used to judge the performance of the solutions during the optimization process. The normalization is undertaken firstly to eliminate the difference in the order of magnitude of these multi-criteria.

$$\begin{cases} PTE'(\tau_i) = PTE(\tau_i)/PTE_{max} \\ Del'(\tau_i) = PTE(\tau_i)/Del_{max} \\ S'_{mst}(\tau_i) = S_{mst}(\tau_i)/E_{max} \end{cases} \tag{7}$$

where PTE_{max} is the maximum power transfer efficiency in WPTG. Del_{max} is the maximum time delay. E_{max} represents the maximum energy storage in the nodes. The fitness function could be expressed with the consideration of the multi-criteria:

$$J = 1/e^{E'(\tau_i)^{\omega_1} \times S'_{mst}(\tau_i)^{\omega_2}/Del'(\tau_i)^{\omega_3}} \tag{8}$$

where ω_1, ω_2, ω_3 represent the performance index weight for the multi-criteria, and they satisfy the following condition:

$$\omega_1 + \omega_2 + \omega_3 = 1 \tag{9}$$

However, with the constraints in the value of the multi-criteria, the penalty function should be introduced firstly to remove the constraints:

$$\psi(g_i, \theta) = \theta \times \Sigma[max\{0, g_i\}]^2 \tag{10}$$

where θ is a large positive number and g_i represents the constraint function in Equation (6). Thus, the energy link can be converted to an optimization issue without any constraints:

$$\Theta = J + \psi(g_i, \theta) \tag{11}$$

Based on Equation (11), the optimal energy link with the best value of the global performance index (with the minimum value of Θ) will be chosen, which means the energy link with high energy storage and power transfer efficiency, as well as a small time delay will be selected.

4.3. Algorithm Operators

The all of the operators of CAGA are comprised of the selection operator, crossover operator and mutation operator. The selection operator is run to select the solutions for the next generation, and the quality of solutions is critical for the optimization performance. To avoid local optima, the concentration mechanism is introduced to compel the unification trend in the solution group. Firstly, the distance factor F is set to measure the diversity of every two solutions. As shown in

Figure 9, the distance factor F of two solutions is set based on the length of their different genes $F(\tau_i - \tau_j) = Length(\tau_{ij})$, and the arrow represents the elimination of the similar genes.

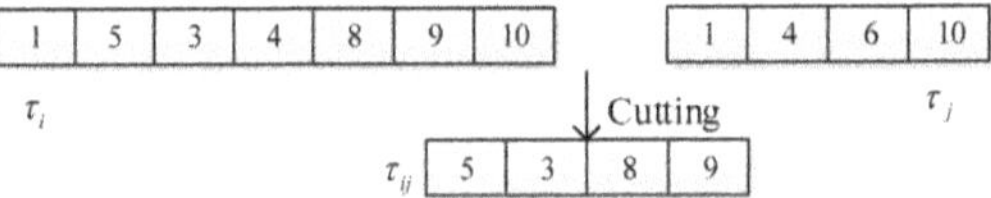

Figure 9. Different genes in two solutions.

Firstly, the concentration value ($C(\tau_i)$) of each solution is set as follows:

$$C(\tau_i) = 1/\Sigma_{j=1}^{G_s} F(\tau_i - \tau_j), i = 1, 2, ..., G_s \tag{12}$$

The concentration value will be high if the solution is similar to others. However, this unification trends should be avoided during the evolution process. Therefore, the selection factor ϕ is introduced to avoid this trend.

$$\phi(\tau_i) = \mid \ln C(\tau_i) \mid / \Sigma_{i=1}^{G_s} \mid \ln C(\tau_i) \mid, i = 1, 2, ..., G_s \tag{13}$$

During the selection process, the elite solution in each generation will be selected out firstly for the next generation, which contributes to compelling the evolution process with better performance. Meanwhile, the rest of the solutions will be chosen based on the selection factor ϕ. This is due to the fact that the normalization trend dominates the evolution process and, thus, contributes to the local optimum. Through the introduction of the concentration factor, the good genes in some solutions with terrible fitness values will be kept to motivate the evolution to reach the global optimum.

Furthermore, to accelerate the evolution speed, a forward order crossover method will be brought in to recombine the genes in the solutions. As shown in Figure 10, firstly, d genes in the candidate solutions A and B will be utilized as the gene segment *core* in Phase I . The solution is divided into three parts, including $(a)forward$, $(b)core$ and $(c)back$; then, these three segments will be combined following the sequence $(back - forward - core)$, and the respective repeating genes in two *core* part will be removed ($A1$ removes the repeating Gene 9, which exists in the *core* of candidate solution B). The new gene segments $A1$ and $B1$ are generated consequently in Phase II. Next, the new gene segments will be put in the respective candidate solution, for example $\{8,4\}$ in $B1$ will replace the original *forward* part in A, and the rest $\{6,9\}$ will replace the *back* part. Finally, the new results A' and B' will be generated. With the operation of the genes close to the source node, the global evolution speed and accuracy could be improved due to the optimization characteristic of the shortest path issue. Meanwhile, the diversity of the solution group is also improved.

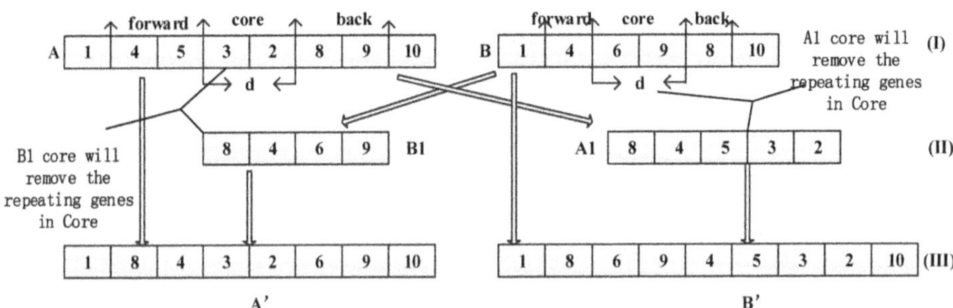

Figure 10. Crossover operator.

As for the mutation operator, the swap mutation method is chosen due to its simpleness and high efficiency. As shown in Figure 11, the arrow means two genes in solution A are randomly picked up to be exchanged in the solution sequence. Finally, a new result solution ($A^{'}$) will be reproduced. This operator contributes to bringing in the variation to the evolution process, and thus, compels the evolution to get rid of the local optimum trap quickly.

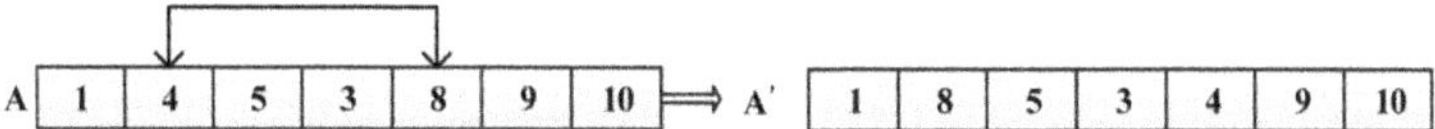

Figure 11. Mutation operator.

4.4. Algorithm Adaptive Parameter Mechanism

The adaptive parameters are usually used to compel the algorithm to ignore the local optimum trap and accelerate the evolution speed. Therefore, the crossover and mutation rate will be modified respectively to compel the evolution to reach the global optimum. The crossover rate P_c is usually set with a higher value, and its counterpart P_m is lower in comparison at the beginning, which contributes to accelerating the evolution speed of the algorithm at the first stage. During the evolution, as mentioned before, the unification trend will appear, in which most of the solutions in the solution group will be similar to the others. It will be difficult to ignore the local optimum due to this trend. Thus, the mutation rate should be increased to improve the solution group diversity, which is measured by the average length of the solutions ($\overline{L(g)}$). This is due to the fact that the solution with a higher length could provide more likely good genes for evolution.

$$\nabla L_{avg}(g) = \overline{L_{off}(g)} - \overline{L_{par}(g)} = \frac{1}{G_s}(\sum_{i=1}^{G_s} L_i(g) - \sum_{i=1}^{G_s} L_i(g-1)) \tag{14}$$

$$\varepsilon(x) = \begin{cases} 1 & ,x \geq 0 \\ -1 & ,otherwise \end{cases} \tag{15}$$

$$\triangle(g) = \sum_{t=0}^{n-1} \varepsilon(\nabla L_{avg}(g-t)), g \geq n \tag{16}$$

$$P_c(g) = P_c(g-1) - \frac{\triangle(g)}{T} \tag{17}$$

$$P_m(g) = P_m(g-1) + \frac{\triangle(g)}{T} \tag{18}$$

where G_s in Equation (14) indicates the size of the solution group, $\overline{L_{off}(g)}$ and $\overline{L_{par}(g)}$ are the average length of the g-th and $(g-1)$-th solution groups, ε in Equation (15) represents the recording of the diversity varying trend, $\triangle$ means the accumulation of the diversity change in the latest n generations, $P_c(g)$ and $P_m(g)$ are the crossover and mutation rate in the g-th generation and T is the threshold value for the maximum evolution generation. Consequently, CAGA is undertaken to adaptively update the rate of crossover and mutation by recording the diversity varying trend during the evolution process.

5. Simulation and Verification

In this section, to verify the proposed method for the energy link optimization, the WPTG with different amounts of nodes (N = 10, 15, 20) will be established with a no weight preferences situation ($\omega_1 = 0.4$, $\omega_2 = 0.3$, $\omega_3 = 0.3$). As shown in Figure 12, each node is assumed to be distributed randomly, and the weighted link between two nodes is used to represent the abstract energy link. Then, random index weight values are set on the energy link to simulate the power transfer and grid characteristics. Consequently, the optimization process will be undertaken based on these grid topologies, respectively. The power supply node is set as Node 1, and the load node is set as node N in these three situations. The rest nodes act as the candidate repeater nodes. Additionally, the optimal energy link is labelled with a bold line in each grid. In comparison with the traditional genetic algorithm [34], CAGA is undertaken to run the optimization by establishing an optimal energy link through selecting the nodes' repeater nodes from the candidates. The detailed algorithm parameters are listed as shown in Table 1. This simulation experiment is performed on a PC with a Intel Xeon(R) 3 GHz processor and 4 Gbytes of runtime memory, running Microsoft Windows 7 professional version. MATLAB is utilized to implement all of the programs.

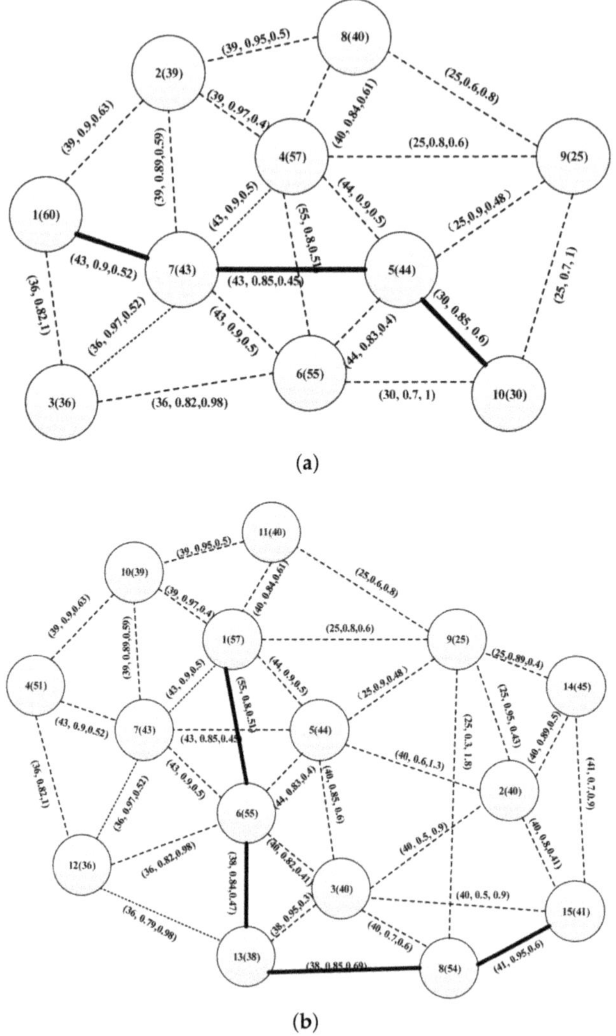

Figure 12. *Cont.*

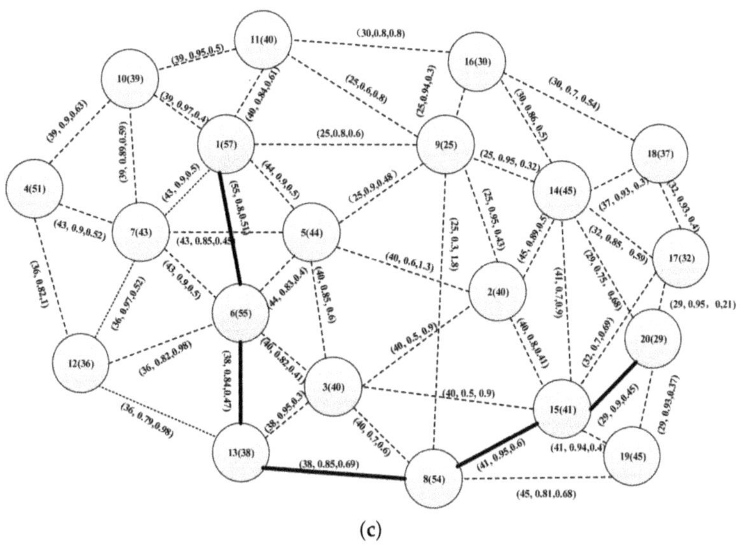

Figure 12. Grid topology (N nodes), $\omega_1 = 0.4$, $\omega_2 = 0.3$, $\omega_3 = 0.3$. (a) $N = 10$; (b) $N = 15$; (c) $N = 20$.

Table 1. Optimization parameters.

Parameter	Value
Solution group size G_s	40
Maximum evolution generations T	50
Initial crossover rate P_c	0.9
Initial mutation rate P_m	0.1
Trend recording number n	3
Constrain value for PTE PTE_m	0.5
Constrain value for time delay D_m/s	1.5
Constrain value for energy storage S_m/kWh	27

The optimization results and their respective time costs of three grids are listed in Table 2. It indicates that the CAGA could get the optimal result in different situations. However, the traditional GA method always gets trapped in the local optimum. This performance difference is due to the improvements in the algorithm operators in CAGA. These improvements contribute to keeping the diversity of the solution groups, which drives the evolution process to the global optimal result. The time cost is also reduced with the improvement in the diversity of the solution group. As for the algorithm robustness, the performance difference is shown in Figure 13. The results show that with the increase in the node number, the failure rate of the proposed algorithm is almost kept steady at a low level. However, the traditional GA changes greatly with the increase in the node number. Thus, the effect of keeping the diversity of the solution group on algorithm robustness is verified, as well.

Table 2. Performance comparison of the convergence.

Optimization Method	N = 10	N = 15	N = 20
Traditional GA method	{1,2,4,5,10}/1.1 s	{1,7,6,13,8,15}/1.97 s	{1,5,3,13,8,15,20}/3.1 s
CAGA	{1,7,5,10}/0.88 s	{1,6,13,8,15}/1.43 s	{1,6,13,8,15,20}/1.91 s

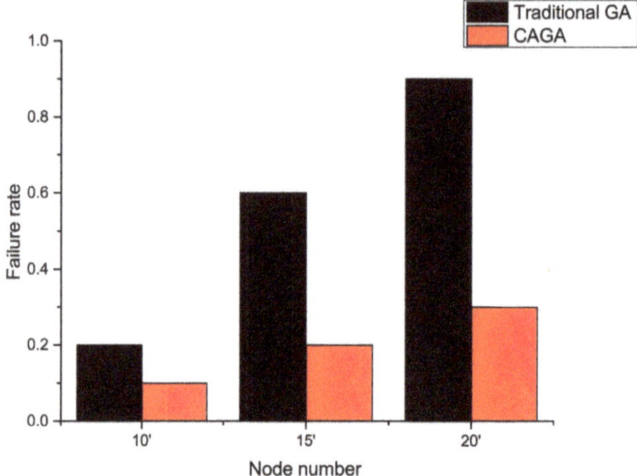

Figure 13. Robustness performance comparison.

To make the algorithm performance difference clearer, the WPTG with 15 nodes is utilized to present it in detail. As shown in Table 3, the convergence results are list to show the two methods' convergence performance in different index weight preferences. The results of the optimized energy link are listed in Table 3, which shows that the energy link ($\{1,6,13,8,15\}$) optimized by CAGA performs better than the traditional genetic algorithm method in the condition of no weight preferences. The time delay is reduced by 17.8%, while the *PTE* is only reduced by 1.24% in the CAGA in comparison with the traditional result. Thus, the comprehensive performance is improved greatly, which is also confirmed in the performance index value (0.3096 $\prec$ 0.3291); During the weight preference situation, the result of CAGA is increased by 8% in comparison with the traditional method in the *PTE* preference situation. The improvement appears in the energy balance and time delay preferences situations as well. Consequently, the feasibility of CAGA proposed in this paper for energy link optimization is verified.

Table 3. Convergence results (N = 15).

Optimization Method	Index Weight	Optimized Energy Link	J	PTE	Del	S_{mst}
Traditional GA method	$\{0.4,0.3,0.3\}$	$\{1,7,6,13,8,15\}$	0.3291	0.5494	2.76	38
CAGA	$\{0.4,0.3,0.3\}$	$\{1,6,13,8,15\}$	0.3096	0.5426	2.27	38
Traditional GA method	$\{0.7,0.2,0.1\}$	$\{1,7,12,13,8,15\}$	0.4011	0.5469	3.29	36
CAGA	$\{0.7,0.2,0.1\}$	$\{1,5,3,13,8,15\}$	0.3763	0.5869	2.69	38
Traditional GA method	$\{0.2,0.7,0.1\}$	$\{1,7,12,13,8,15\}$	0.5244	0.5469	3.29	36
CAGA	$\{0.2,0.7,0.1\}$	$\{1,5,3,8,15\}$	0.493	0.5087	2.3	40
Traditional GA method	$\{0.2,0.1,0.7\}$	$\{1,6,3,8,15\}$	0.0954	0.4362	2.12	40
CAGA	$\{0.2,0.1,0.7\}$	$\{1,5,3,15\}$	0.0922	0.3825	1.82	40

The convergence process and robustness of these two methods are also compared in Figure 14. As shown in Figure 14a, the traditional GA method gets trapped in the 12th evolution generation. In comparison, the proposed CAGA method gets trapped in the 13th generation, but reaches the global optimum in the 15th generation. Meanwhile, the robustness of the algorithm is compared in Figure 14b. The CAGA only fails in the fourth and ninth operation, and thus, the failure rate is 20%. The traditional GA method only succeeds four times in the repeating 10 operations. Therefore, the conclusion can be made that the CAGA method overwhelms the traditional GA method in different weight preference situations, as well as the algorithm robustness. This result is also due to the fact that GA is a kind of metaheuristic algorithm, which tends to get trapped in the local optimal result. The centralization mechanism and a new crossover method named forward order crossover, as well as the adaptive parameter mechanism are introduced to keep the diversity of the solution group in the proposed method. It greatly contributes to driving the evolution process forward to the global optimal result, and thus, it also improves the robustness.

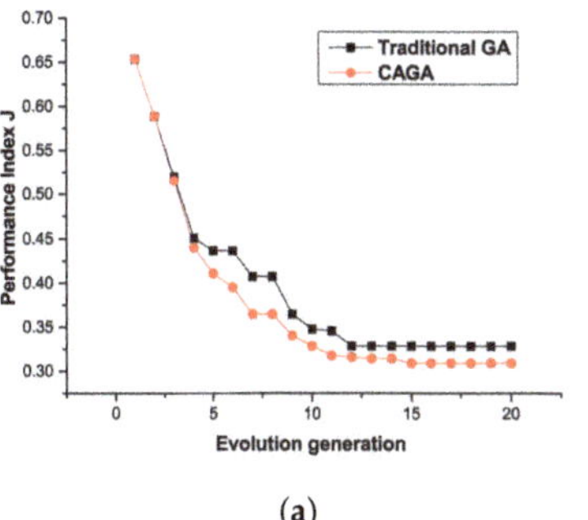

(a)

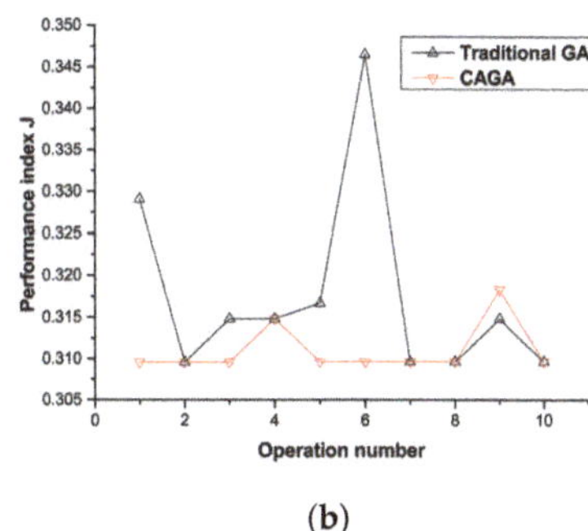

(b)

Figure 14. Algorithm performance comparison, $\omega_1 = 0.4$, $\omega_2 = 0.3$, $\omega_3 = 0.3$. (**a**) Convergence comparison; (**b**) Robustness comparison.

6. Conclusions and Future Directions

WPTG is a novel energy grid for resolving the power transfer issues in device groups. In this paper, an optimization method for the energy link in WPTG is proposed to get the optimal energy link from the power supply node to the load node. The power transfer characteristic (*PTE*) and the grid characteristics (time delay and energy load balance) are taken into consideration to establish the optimization model. An advanced genetic algorithm (CAGA) is proposed to undertake the optimization process. This method is running based on the concentration of the solutions and adaptively varying the algorithm parameters. A forward order crossover operator is also introduced to accelerate the evolution speed. The simulation results confirm the feasibility of the proposed method. In future works, the detailed engineering applications of the wireless power transfer grid will be further studied and developed.

Acknowledgments: This work is financially supported by the National High-tech R&D Program (863 Program, Grant No. 2015AA016201), the National Natural Science Foundation of China (Grant No. 51277192; 51377183; 61573074) and the Fundamental Research Funds of Chongqing (Grant No. cstc2013jcyjA0235).

Author Contributions: Zhihao Zhao designed the optimization method and analysed the results. Zhihao Zhao, Yue Sun and Aiguo Patrick Hu analysed the system characteristics and designed the multi-criteria. Xin Dai and Chunsen Tang provided guidance and key suggestions.

Conflicts of Interest: The authors declare no conflict of interest.

Abbreviations

The following abbreviations are used in this manuscript:

WPT	Wireless power transfer
WPTG	Wireless power transfer grid
PTE	Power transfer efficiency
CAGA	Concentration adaptive genetic algorithm

References

1. Hu, A.P. Selected Resonant Converters for IPT Power Supplies. Ph.D. Thesis, The University of Auckland, Auckland, New Zealand, 2001.
2. Chwei-Sen, W.; Covic, G.A.; Stielau, O.H. Power transfer capability and bifurcation phenomena of loosely coupled inductive power transfer systems. *IEEE Trans. Ind. Electron.* **2004**, *51*, 148–157.
3. Kurs, A.; Karalis, A.; Moffatt, R.; Joannopoulos, J.D.; Fisher, P.; Soljacic, M. Wireless power transfer via strongly coupled magnetic resonances. *Science* **2007**, *317*, 83–86.
4. Wang, Z.-H.; Li, Y.-P.; Sun, Y.; Tang, C.-S.; Lv, X. Load detection model of voltage-fed inductive power transfer system. *IEEE Trans. Power Electron.* **2013**, *28*, 5233–5243.
5. Vijayakumaran Nair, V.; Choi, R.J. An efficiency enhancement technique for a wireless power transmission system based on a multiple coil switching technique. *Energies* **2016**, *9*, 156.
6. Liu, C.; Hu, A.P. Effect of series tuning inductor position on power transfer capability of CCPT system. *Electron. Lett.* **2011**, *47*, 136–137.
7. Dai, J.; Ludois, D.C. Single active switch power electronics for kilowatt scale capacitive power transfer. *IEEE J. Emerg. Sel. Top. Power Electron.* **2015**, *3*, 315–323.
8. Lu, F.; Zhang, H.; Hofmann, H.; Mi, C. A double-sided LCLC-compensated capacitive power transfer system for electric vehicle charging. *IEEE Trans. Power Electron.* **2015**, *30*, 6011–6014.
9. Huang, L.; Hu, A.P. Defining the mutual coupling of capacitive power transfer for wireless power transfer. *Electron. Lett.* **2015**, *51*, 1806–1807.
10. Krikidis, I.; Timotheou, S.; Zheng, G.; Ng, D.W.K.; Schober, R. Simultaneous wireless information and power transfer in modern communication systems. *IEEE Commun. Mag.* **2014**, *52*, 104–110.
11. Lu, P.; Yang, X.-S.; Li, J.-L.; Wang, B.-Z. A Compact frequency reconfigurable rectenna for 5.2- and 5.8-GHz wireless power transmission. *IEEE Trans. Power Electron.* **2015**, *30*, 6006–6010.
12. Yuan, F.; Jin, S.; Huang, Y.; Wong, K.-K.; Zhang, Q.-T.; Zhu, H. Joint wireless information and energy transfer in massive distributed antenna systems. *IEEE Commun. Mag.* **2015**, *53*, 109–116.
13. Ding, Z.-G.; Zhong, C.-J.; Ng, D.W.K.; Peng, M.-G.; Suraweera, H.A.; Schober, R.; Poor, H.V. Application of smart antenna technologies in simultaneous wireless information and power transfer. *IEEE Commun. Mag.* **2015**, *53*, 86–93.
14. Lin, F.-Y.; Covic, G.A.; Boys, J.T. Evaluation of magnetic pad sizes and topologies for electric vehicle charging. *IEEE Trans. Power Electron.* **2015**, *30*, 6391–6407.
15. Raval, P.; Kacprzak, D.; Hu, A.P. Analysis of flux leakage of a 3-D inductive power transfer system. *IEEE J. Emerg. Sel. Top. Power Electron.* **2015**, *3*, 205–214.
16. Sun, T.-J.; Xie, X.; Li, G.-L.; Gu, Y.-K.; Deng, Y.-D.; Wang, Z.-H. A two-hop wireless power transfer system with an efficiency-enhanced power receiver for motion-free capsule endoscopy inspection. *IEEE Trans. Biomed. Eng.* **2012**, *59*, 3247–3254.
17. Lee, C.K.; Zhong, W.-X.; Hui, S.Y.R. Effects of magnetic coupling of nonadjacent resonators on wireless power domino-resonator systems. *IEEE Trans. Power Electron.* **2010**, *27*, 1905–1916.
18. Liang, H.; Zhuang, W. Stochastic modeling and optimization in a microgrid: A survey. *Energies* **2014**, *7*, 2027–2050.
19. Lin, F.Y. Smart distribution: Coupled microgrids. *Proc. IEEE* **2011**, *99*, 1074–1082.
20. Sun, Y.; Yang, F.-X.; Dai, X. Construction of wireless power transfer networks based on improved ant colony optimization. *J. South China Univ. Technol.* **2011**, *39*, 146–164.
21. Xiang, L.-J.; Sun, Y.; Dai, X.; Chen, Y.; Lv, X. Route optimization for wireless power transfer network based on the ce method. In Proceedings of the International Power Electronics and Application Conference and Exposition, Shanghai, China, 5–8 November 2014; pp. 630–634.
22. Dai, X.; Sun, Y.; Su, Y.-G.; Tang, C.-S.; Wang, Z.-H. Study on contactless power bi-directional push mode. *Proc. Chin. Soc. Electr. Eng.* **2010**, *30*, 55–61.
23. Madawala, U.K.; Thrimawithana, D.J. Current sourced bi-directional inductive power transfer system. *IET Power Electron.* **2011**, *4*, 471–480.
24. Clerc, M.; Kennedy, J. The particle swarm-explosion, stability, and convergence in a multidimensional complex space. *IEEE Trans. Evulut. Comput.* **2002**, *6*, 58–73.

25. Zhan, Z.-H.; Zhang, J.; Li, Y.; Chung, H.S.H. Adaptive particle swarm optimization. *IEEE Trans. Syst. Man Cybern. B* **2009**, *39*, 1362–1381.
26. AlRashidi, M.R.; El-Hawary, M.E. A survey of particle swarm optimization applications in electric power systems. *IEEE Trans. Evulut. Comput.* **2009**, *13*, 913–918.
27. Hui, Q.; Zhang, H. Optimal balanced coordinated network resource allocation using swarm optimization. *IEEE Trans. Syst. Man Cybern. Syst.* **2015**, *45*, 770–787.
28. Ho, S.L.; Yang, S. A computationally efficient vector optimizer using ant colony optimizations algorithm for multobjective designs. *IEEE Tans. Magn.* **2008**, *44*, 1034–1037.
29. Gan, R.; Guo, Q.; Chang, H.; Yi, Y. Improved ant colony optimization algorithm for the traveling salesman problems. *J. Syst. Eng. Electron.* **2010**, *21*, 329–333.
30. Ding, C.; Cheng, Y.; He, M. Two-level genetic algorithm for clustered traveling salesman problem with application in large-scale tsps. *Tsinghua Sci. Technol.* **2007**, *12*, 459–465.
31. Ewald, G.; Kurek, W.; Brdys, M.A. Grid implementation of a parallel multiobjective genetic algorithm for optimized allocation of chlorination stations in drinking water distribution systems: Chojnice case study. *IEEE Trans. Syst. Man Cybern. C* **2008**, *38*, 497–509.
32. Liu, F.; Liang, S.; Xian, X. Optimal robot path planning for multiple goals visiting based on tailored genetic algorithm. *Int. J. Comput. Intell. Syst.* **2014**, *7*, 1109–1122.
33. May, G.; Stahl, B.; Taisch, M.; Prabhu, V. Multi-objective genetic algorithm for energy-efficient job shop scheduling. *Int. J. Prod. Res.* **2015**, *53*, 7071–7089.
34. Munetomo, M.; Taka, Y.; Sato, Y. A migration scheme for the genetic adaptive routing algorithm. In Proceedings of the IEEE International Conference on Systems, Man, and Cybernetics, San Diego, CA, USA, 11–14 October 1998; pp. 2774–2779.

Article

A Novel Power-Saving Transmission Scheme for Multiple-Component-Carrier Cellular Systems

Yao-Liang Chung

Department of Communications, Navigation and Control Engineering, National Taiwan Ocean University, Keelung City 20224, Taiwan; ylchung@email.ntou.edu.tw; Tel.: +886-2-2462-2192 (ext. 7224); Fax: +886-2-8369-3022

Academic Editor: William Holderbaum
Received: 23 January 2016; Accepted: 25 March 2016; Published: 2 April 2016

Abstract: As mobile data traffic levels have increased exponentially, resulting in rising energy costs in recent years, the demand for and development of green communication technologies has resulted in various energy-saving designs for cellular systems. At the same time, recent technological advances have allowed multiple component carriers (CCs) to be simultaneously utilized in a base station (BS), a development that has made the energy consumption of BSs a matter of increasing concern. To help address this concern, herein we propose a novel scheme aimed at efficiently minimizing the power consumption of BS transceivers during transmission, while still ensuring good service quality and fairness for users. Specifically, the scheme utilizes the dynamic activation/deactivation of CCs during data transmission to increase power usage efficiency. To test its effectiveness, the proposed scheme was applied to a model consisting of a BS with orthogonal frequency division multiple access-based CCs in a downlink transmission environment. The results indicated that, given periods of relatively light traffic loads, the total power consumption of the proposed scheme is significantly lower than that of schemes in which all the CCs of a BS are constantly activated, suggesting the scheme's potential for reducing both energy costs and carbon dioxide emissions.

Keywords: power-saving; green cellular systems; multiple component carriers

---

1. Introduction

In recent years, 4th generation (4G) cellular systems have been developed and deployed in order to better handle the data demands of ever-increasing numbers of network users, and cellular technologies are even now advancing towards 5th generation (5G) cellular systems and beyond. Importantly, one of the key features of 4G/5G and future cellular systems that allows them to achieve higher capacities than less advanced networks is the ability of base stations (BSs) to utilize multiple component carriers (CCs) together during data transmissions [1,2]. At the same time, the power consumed by such wireless networks, especially by their BSs, has become a matter of increasing concern due to rising energy costs and the environmental impacts of the carbon dioxide (CO_2) emissions that accompany energy production. As a result, the concept of green communications has received increasing attention as a potential means of addressing these concerns [3–7]. The primary goal of green communications is reducing the overall amount of power consumed by the transmission of communications without causing any reduction in the service quality enjoyed by users.

The primary purpose of developing multi-CC BSs was to provide greater capability in handling very large data-based transmissions, while the concept of green communications relies to a large extent on the fact that these multi-CC BSs should be capable of efficiently reducing the amount of energy consumed by their transceivers. That is to say, the fundamental thinking behind green communication efforts is the idea of enhancing the efficiency of multi-CC BSs so that transmission activities will result in energy savings. As such, ongoing research efforts aimed at exploring how the efficiency of data transmissions from multi-CC BSs can be improved, up to and including the point at which such BSs can be seen as "green BSs", are essential.

Moreover, it is worth noting that network operators and network users have different goals and preferences when it comes to the issue of radio resource management. Specifically, network users want radio resource allocation to be both fair and sufficient to guarantee their requirements in terms of service quality, whereas network operators are more concerned, given that radio resources are by nature finite, with maximizing the utilization of those resources as much as possible. Accordingly, certain trade-offs are inevitably required given these competing aims of network users and operators, a subject which has previously been explored by various researchers, including, for example, Rodrigues and Casadevall [8]. However, no past studies have comprehensively examined how the fair scheduling schemes for BSs in multi-CC systems might be refined to yield power savings.

With these points in mind, the goal of the present study was to minimize the amount of power consumed by the operation of BS transceivers with multiple CCs, while still ensuring fairness in resource allocation for various types of users, including the maintenance of sufficient user data rates. To that end, this paper proposes a novel optimization scheme that interprets data transmissions at BSs in a fundamentally different manner than many previously presented resource allocation models. The main contributions of the paper can be summarized as follows:

- A novel and efficient transmission scheme for orthogonal frequency division multiple access (OFDMA)-based multi-CC cellular systems that saves power while concurrently supporting both real-time (RT) (delay-sensitive and high data-rate) and non-real-time (NRT) (non-delay-sensitive) types of downlink traffic and maintaining efficient control of fairness indexes for the two types of users based on their respective data usage needs.
- By adaptively activating and deactivating CCs during periods of relatively light traffic loads, the proposed scheme can yield significant reductions in the power consumed during data transmission. Thus, the proposed scheme has considerable potential in terms of reducing the energy costs and CO_2 emissions associated with cellular networks.

The next sections of this paper are organized as follows: Section 2 gives an overview of existing related literature in the field. Section 3 introduces a system model and power consumption model for multi-CC cellular networks and provides a problem formulation for the model to show the objective function to be optimized and define constraints. Next, in Section 4, a proposed novel transmission scheme for use in the model is detailed, and a time complexity analysis of this scheme is performed. Results in terms of power-saving and fairness performances are subsequently shown and discussed in Section 5. Finally, Section 6 relates the conclusions of this study.

2. Related Work

A substantial amount of the past research regarding designs for radio resource allocation in cellular systems has been concentrated on systems utilizing multi-user single-CC BSs, with several studies having given particular attention to various methods used to improve system performance from the standpoint of individual contributions [9–18]. More specifically, Wong *et al.* [9] proposed a jointly adaptive bit, subcarrier, and power allocation algorithm as a means of improving the performance of a system. In contrast, in the study by Jeong *et al.* [10], the authors proposed a scheme involving high efficiency cross-layer packet scheduling and resource management. In yet another approach, Kivanc *et al.* detailed a set of algorithms relying on greater computational efficiency in determining the allocation of power and subcarriers among system users [11], whereas Madan *et al.* proposed fast algorithms to handle the task of ensuring resource allocation optimization in order to maximize the overall utility of a system [12]. In another study, the subject of how to allocate resources for energy-efficient communication in the context of single-cell downlink environments with numerous transmitting antennas was addressed by Ng *et al.* [13]. Subsequently, the same researchers extended their research framework further to include multi-cell downlink environments with cooperative BSs [14]. Meanwhile, a study by He *et al.* focused on the physical-layer aspect in its investigation of energy-efficient coordinated beamforming and power allocation in multi-cell downlink environments [15]. In a still more recent study, Piunti *et al.* proposed an optimization framework aimed at minimizing the degree of power consumption while also providing the minimum bit rate required for each mobile terminal [16]. Still another approach utilized the development of heterogeneous network deployments with macrocells and small cells in order to enhance energy efficiency [17,18]. Nevertheless, although each of these studies (*i.e.*, [9–18]) made some important contributions, none of them looked explicitly at multi-CC BSs or the important subject of energy savings. As such, their contributions, while valuable, do not have much direct relevance to the growing concerns regarding energy consumption and associated CO_2 emissions discussed in the introduction above.

Nonetheless, as regulatory agencies and people in general have become increasingly aware of the environmental problems linked to energy usage, the subject of green communication technologies, including green cellular systems, has become the focus of a major trend in communications research [3–7,19–28]. Several important studies [3–7], for example, have sought to provide overviews of the various issues related to energy consumption in communication networks. In terms of directly practical research, meanwhile, Niu *et al.* [19] authored a study proposing a cell-zooming-based energy-saving algorithm aimed at providing dynamic adjustments to the transmission power of BSs, while the energy-saving management group of the 3rd Generation Partnership Project (3GPP) presented a number of network architectures aimed at providing energy savings on a system-wide basis [20]. Micallef *et al.* [21], meanwhile, proposed an energy-saving algorithm aimed at exploiting variations in network traffic levels in the context of dual-cell high speed downlink packet access systems. Relatedly, Lorincz *et al.* looked at methods for managing the levels of power used by network devices in the context of realistic traffic patterns as a means of minimizing the overall amount of energy consumed by wireless access networks [22], while Chung and Tsai proposed a model aimed at optimizing both CC activation/deactivation and radio resource allocation in order to save power during BS transmissions [23]. A second study by Lorincz *et al.* proposed an optimization model taking the issue of voice-based transmissions into account in order to better ensure that cellular system resources are allocated in an energy-efficient manner [24]. On the other hand, a hybrid model aimed at providing energy efficiency evaluations for converged wireless/optical access networks was proposed by Aleksic *et al.* [25], while in a study by Enokido and Takizawa, the authors proposed an algorithm dedicated to reducing the total amount of power consumed in distributed models [26]. More recently, Chung proposed a rate-and-power control scheme for the BS transmission to address the problem of energy minimization at BS transceivers while also ensuring required service quality and fairness for all users [27]. After that, Chung further constructed a practical framework for addressing energy-efficient BS transmissions based on the involvement of both radio resource allocation and CC activation/deactivation [28].

Nonetheless, while the aforementioned studies authored by Micallef *et al.* [21], Chung and Tsai [23], and Chung [27,28] did explicitly investigate the issue of energy savings as it relates to BSs with transceivers, the designs utilized in those papers would still not be sufficient to fully address the real-world network environments of the near future. More specifically, Micallef *et al.* [21] did not consider the emerging context of 4G BSs, whereas Chung and Tsai [23] derived their sub-optimal energy-saving transmission algorithm on the assumption of constantly backlogged flows, making the algorithm somewhat less than fully realistic in terms of reflecting system performance given that in real-world situations, traffic is usually dynamic and fluctuating, as opposed to always being backlogged. In addition, the transmission schemes designed by Chung [27,28] were not applicable to the cases of BSs with more than two CCs being utilized. Accordingly, to the best of this author's knowledge, previous studies have yet to fully investigate how to minimize power consumption in relation to multi-CC BS transceivers, especially in the context of the activation/deactivation of multiple CCs. Furthermore, given the aforementioned goal of green communications, the capacity of BSs to avoid wasting power in the operation of their transceivers is only likely to grow in importance.

3. Problem Description

3.1. System Model

For the purposes of the present study, we focused on the downlink data transmissions in an OFDMA-based multi-CC BS system of a single cell. In this scenario, the cell itself is comprised of n user terminals, with each user terminal being indexed as user-k. In this model, the smallest allocation unit for resource scheduling is a sub-channel spanning 12 sub-carriers for which the bandwidth is denoted as b_{sub}. In addition, it is assumed that only a single user can be assigned to a given sub-channel. Moreover, the amount of time, in seconds, spanning a specific number of consecutive OFDMA downlink frames is defined as the smallest transmission unit and denoted as α. The notation s is used to denote the current scheduling round. In this context, α is set equal to 1 in order to avoid re-defining the unit of "power" and in order to allow reasonable required computation time for the scheme employed.

It is further assumed that there are a total of c CCs located in different frequency bands. These CCs, which are indexed by i, can be utilized for transmission. Without loss of generality, for a user with a given specific transmission power, the data rates supported by the CCs in the higher frequency bands are lower than those supported by the CCs in the lower frequency bands. In this study, the CC with $i = 1$ was regarded as the primary CC (PCC), while the others were regarded as the supplementary CCs (SCCs). For data transmissions, the PCC was always used as the main CC and was thus set to always be activated, while the SCCs were only used when the traffic was relatively heavy in order to supplement the PCC. As a result, there are naturally c kinds of combinations for these SCCs and the PCC. More specifically, if Ω is defined as the set of all the combinations of CC configurations involving at least the PCC, then the configuration status vector in round s can be defined as $v^{(z)}(s) = [v_1^{(z)}(s), v_2^{(z)}(s), ..., v_c^{(z)}(s)]$, where z denotes the index for the configuration being utilized in Ω, and $z = 1, 2, \ldots, |\Omega|$. Note, then, that $|\Omega| = c$. We can set $z = x$ to indicate that the CCs with $i = 1$ to x are activated and that no other configurations are permitted. As such, for a specific configuration z in round s, $v_i^{(z)}(s) = 1$ if CC-i is activated; otherwise, it equals 0. Furthermore, we also assume that CC-i consists of ℓ_i sub-channels.

Figure 1 presents a conceptual construction of the system model from a systemic point of view, where it is comprised of a classifier, an RT processing queue, an NRT processing queue, an admission gate, a scheduling queue, a scheduler, and c OFDMA-based CCs. The data of the various users are transmitted at the session level. In the classifier, all the user session requests are categorized as either RT or NRT sessions, and from the classifier, they are then forwarded in sequence to, respectively, either the RT processing queue or the NRT processing queue. If allowed, these session requests then pass, at the start of each scheduling round, through the admission gate, after which they are buffered to the scheduling queue where they then wait for the scheduler to accept or deny their requests for transmissions. For convenience, in some parts of this paper, the NRT users and the RT users are indexed as type-m users, with $m = 1, 2$, respectively. In addition, the term $\Re^{(m)}(s)$ is used to denote the set of type-m users in the cell for round s; $\Re_{\text{w}}(s)$ is used to denote the set of active users with session requests awaiting the scheduler decision in the scheduling queue for round s; and $\Re_{\text{s}}(s)$ is used to denote the set of active users whose sessions are actually being served during round s. Meanwhile, the duration of a delay in addressing a session request from a given user is measured starting from the time the request is forwarded to the RT processing queue or the NRT processing queue to the time his/her data is completely transmitted.

Furthermore, the basic power consumption required by the BS to activate CC-i is denoted with p_i, while $p^{(z)}(s)$ is used to denote the total power consumption of all the active CCs in configuration z during round s. Moreover, T_i denotes the bandwidth (in Hz) of CC-i, while the total system bandwidth is denoted by $T = \sum_{i=1}^{c} T_i$. As such, $p^{(z)}(s)$ can also be expressed as $p^{(z)}(s) = \sum_{i=1}^{c} p_i v_i^{(z)}(s)$ while $\sum_{k \in \Re_{\text{w}}(s) \cup \Re_{\text{s}}(s)} b_k(s) \leqslant T^{(z)}(s)$, where $T^{(z)}(s) = \sum_{i=1}^{c} T_i v_i^{(z)}(s)$ and $b_k(s)$ denotes the bandwidth allocated to the active user-k during round s. It is thus implied that, given a specific configuration z, the total fraction of bandwidths occupied by active users during each round can be distributed arbitrarily over any sub-channels of the activated CCs.

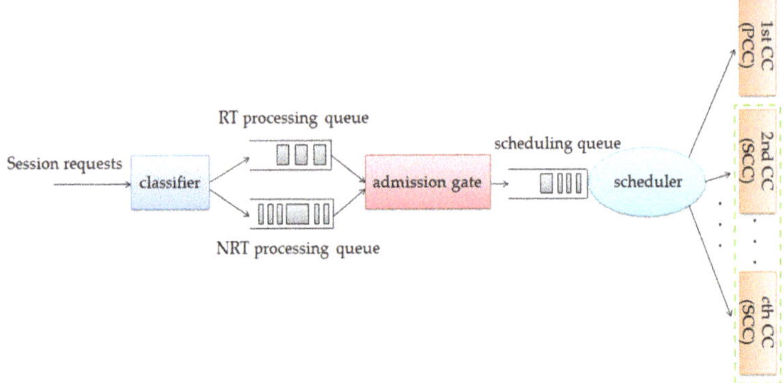

Figure 1. The system model for the present study. The dotted line surrounding the SCCs indicates that those SCCs can be turned off based on the traffic load.

3.2. Power Consumption Model

The power consumption model used in this study refers to the simple, linear model presented by Correia *et al.* [3], as that model allows for the efficient evaluation of the total energy consumption of transceivers at a BS. This model considered both the output power and the input power [3]. The input power is consumed and converted to a certain output power by BS transceivers. This output power is the time-varying radiated power consumed by the BS transceivers that is then used to support user sessions so that their data rate requirements are met; herein, this power is referred to as transmission power. The input power, *i.e.*, the total power required by BS transceivers, comprises the basic power and the transmission power.

Specifically, we let $p_{\text{trans},k}(i,j,s)$ denote the transmission power required to support active user-k, where that active user is from sub-channel-j of CC-i in round s. If we then suppose that the physical-layer coding scheme utilized by future systems will be sufficiently enhanced to achieve the target bit error rate (BER), then according to the analysis of Madan, Boyd, and Lall [12], $p_{\text{trans},k}(i,j,s)$ can be calculated as:

$$p_{\text{trans},k}(i,j,s) = \alpha \frac{N_0 b_{\text{sub}}}{H_k(i,j,s)J} (e^{\frac{r_k(i,j,s)}{b_{\text{sub}}}} - 1) \tag{1}$$

Notice that in Equation (1), for round s, $r_k(i,j,s)$ denotes the physical-layer data rate for user-k from sub-channel-j of CC-i, N_0 denotes the noise power spectral density, $H_k(i,j,s)$ indicates the channel gain betweensub-channel-j of CC-i, and $J = -1.5/\log(5^{\varepsilon})$, where ε denotes the target constant BER. Furthermore, $p_{\text{trans\_max},i}$ is used to denote the maximum transmission power allowed for CC-i.

3.3. Problem Formulation

In this study, our ideal goal, for each scheduling round, was to minimize the overall power consumed of the BS transceivers, $P_{\text{total}}(s)$, subject to the stipulation that the fairness indexes for the different type-m users (*i.e.*, m = 1, 2), $\Phi^{(m)}(s)$, needed to be maintained at their respective desired target values, $\Phi^{(m)}_{\text{target}}$, and, furthermore, that the data transmission rates of the active users, $r_k(s)$, $\forall k \in \Re_w(s) \cup \Re_s(s)$, had to be achieved at their respective minimum required levels, $r_{\text{req},k}$. Furthermore, we let $\beta(s)$ denote the set of available radio resources for round s and let $\varpi_k(s)$ denote the set of the radio resources assigned to active user-k in $\Re_w(s) \cup \Re_s(s)$ during round s. Under those conditions, the optimization problem for each round could then be formulated as follows:

$$\underset{z,\, r_k(s),\, \varpi_k(s)}{\text{minimize}}\ P_{\text{total}}(s) = p^{(z)}(s) + \sum_{k \in \Re_w(s) \cup \Re_s(s)} \sum_{i=1}^{z} \sum_{j=1}^{\ell_i} p_{\text{trans},k}(i,j,s)$$

subject to:

$$z \in \Omega;$$
$$\Phi^{(m)}(s) = \Phi^{(m)}_{\text{target}},\ m = 1, 2;$$
$$r_k(s) \geqslant r_{\text{req},k},\ \forall k \in \Re_w(s) \cup \Re_s(s);$$
$$\sum_{k \in \Re_w(s) \cup \Re_s(s)} \sum_{j=1}^{\ell_i} p_{\text{trans},k}(i,j,s) \leqslant p_{\text{trans\_max},i},\ i = 1, 2, \ldots, z;$$
$$\sum_{k \in \Re_w(s) \cup \Re_s(s)} \varpi_k(s) \subseteq \beta(s),\ \varpi_x(s) \cap \varpi_y(s) = \varnothing,\ x \neq y.$$

Using Jain, Chiu, and Hawe's fairness index formula [29], $\Phi^{(m)}(s)$ can then be expressed as:

$$\Phi^{(m)}(s) = \frac{\left(\sum_{k \in \Re^{(m)}(s)} \phi_k(s)\right)^2}{\left|\Re^{(m)}(s)\right| \sum_{k \in \Re^{(m)}(s)} (\phi_k(s))^2},\ m = 1, 2 \tag{2}$$

where $\phi_k(s)$ denotes the fairness index for the given user-k. Furthermore, according to the aforementioned work by Rodrigues and Casadevall [8], $\phi_k(s)$ is designed to be calculated by:

$$\phi_k(s) = \frac{\frac{\bar{r}_k(s)}{r_{\text{req},k}}}{d_k(s)},\ \forall k \in \Re^{(1)}(s) \cup \Re^{(2)}(s) \tag{3}$$

where $\bar{r}_k(s)$ denotes the average observed data rate for the given user-k until round $s - 1$ (which can be determined using various smoothing methods (for an example of a convergence analysis of smoothing methods, please refer to the study by Liu and Wang [30]), such as exponential filtering methods) and $d_k(s)$ denotes the delay in addressing any session request for user-k till round s.

At this point, one can ascertain that the considered optimization formulation constitutes an integer-nonlinear programming problem. However, an integer-nonlinear optimization problem is highly complex, meaning that in order to find the optimal solution for such a problem, an exhaustive search is generally required; in other words, a problem of this type can be viewed as numerically tractable. Moreover, no optimal solution will be possible when the traffic load is so high that it exceeds what the system can handle. Consequently, in light of the structure of the considered problem, we instead propose, in the following section, a heuristic power-saving transmission scheme that achieves efficient control of fairness for each type of user while also maintaining the data rates for those users at levels above their respective minimum requirements.

4. Proposed Scheme

The scheme proposed and presented herein is composed of five components, namely, the session admission control (SAC), transmission power estimation (TPE), SCC activation/deactivation (SAD), fair bandwidth allocation (FBA), and fair power adjustment (FPA) components.

The significance of each individual component is described as follows. The SAC component is used to periodically determine whether the system should accept or deny new session requests to maintain the data rates of the users being served. The TPE component is utilized to estimate the total transmission power needed to satisfy the active users that have been granted admission by the SAC.

The SAD component is employed in order to activate/deactivate the SCCs as necessary in order to meet the needs of the active users as determined by the TPE. The FBA component is used to ascertain which active users should be assigned to which individual sub-channels in order to maintain the fairness indexes of the different types of users at acceptable levels. The FPA component is then utilized to further adjust the transmission power levels for the individual sub-channels scheduled by the FBA in order to enhance the channel efficiency (which is accomplished by employing the water-filling concept) and maintain the required minimum data rates of the users selected by the FBA.

Moreover, the SAC component is employed in the admission gate, while the TPE, SAD, FBA, and FPA components are employed in the scheduler. A conceptual flow chart of the proposed scheme is presented in Figure 2.

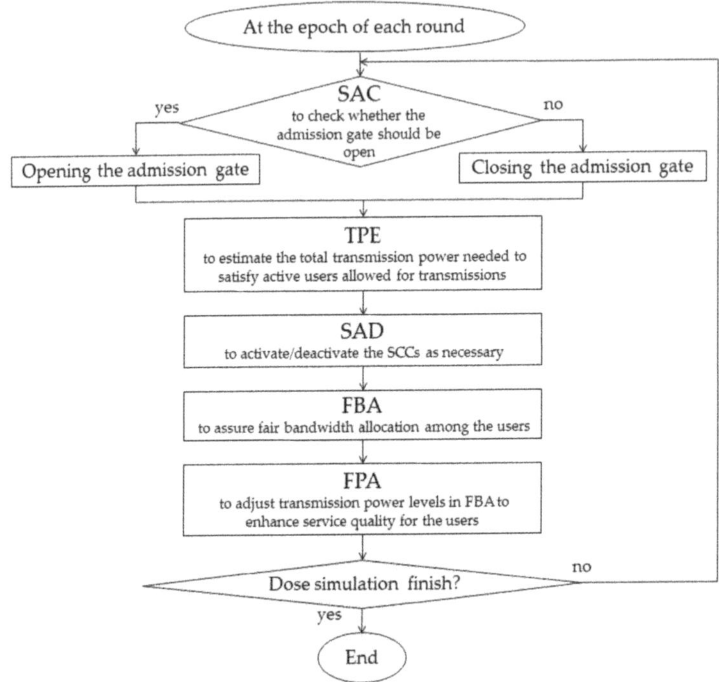

Figure 2. A conceptual flow chart of the proposed scheme.

Notice that the SAC component constitutes the first operation executed at the beginning of every round for the purpose of scheduling session requests. In addition, the average signal-to-noise ratios for those CCs which are activated will be estimated on a periodic basis (*i.e.*, during each round) by the individual users, and those estimates will then be sent back to the BS. Given all the conditions stipulated above, a detailed description of the entire proposed scheme is provided in the following subsections.

4.1. Session Admission Control (SAC)

At the start of each round, the mechanism will check to determine whether or not incoming session requests from the RT or NRT processing queues should be allowed to proceed through the admission gate to join the scheduling queue to access the network. Define $\vartheta(s)$ as the number of non-occupied sub-channels in the system during round s. The key feature of the scheme's design is that if $(\vartheta(s) - |\Re_W(s)|) > 0$, the admission gate is opened to allow some type-m user requests to enter the scheduling queue based on the ratio of the type-m users' required data rate to the required data rate of all the users in the processing queues; otherwise, those session requests in both the RT and NRT processing queues are not allowed to enter the scheduling queue (*i.e.*, the admission gate is kept closed). This design effectively ensures that both the data rate of the users in $\Re_{w}(s)$ and the fairness indexes for the different types of users can be maintained as well as possible at certain levels.

To facilitate description, define $\Re_{rw}^{(m)}(s)$ as the set of type-m users in the processing queue until round s, while also defining $\bar{r}_{rw\_req}^{(m)}(s)$ as the average required data rate of those type-m users in $\Re_{rw}^{(m)}(s)$. In addition, let $n^{(m)}(s)$ denote the number of type-m user requests in the respective processing queues that are permitted to proceed to the scheduling queue in round s. The following pseudo code details the exact operation of this component:

If $((\vartheta(s) - |\Re_{w}(s)|) > 0)$

$$n^{(m)}(s) = \left\lfloor (\vartheta(s) - |\Re_{w}(s)|)\frac{\bar{r}_{rw\_req}^{(m)}(s)\left|\Re_{rw}^{(m)}(s)\right|}{\sum_{n=1}^{2}\bar{r}_{rw\_req}^{(n)}(s)\left|\Re_{rw}^{(m)}(s)\right|} \right\rfloor, \ m = 1,\ 2;$$

Else $n^{(m)}(s) = 0,\ m = 1,\ 2.$

4.2. Transmission Power Estimation (TPE)

After the completion of the SAC component, the total transmission power needed is estimated. This is accomplished by estimating the transmission power required by every session of the respective users currently being served in $\Re_{s}(s)$, as well as by estimating the transmission power required by every session of the respective users in $\Re_{w}(s)$.

First, $p_{trans\_s}(s)$ is defined as the estimated transmission power for the sessions being served in the current round; it is calculated as follows:

$$p_{trans\_s}(s) = \alpha \sum_{k\in\Re_{s}(s)} \sum_{(i,j)\in\aleph_{k}(s)} \frac{N_{0}b_{sub}}{\widetilde{H}_{k}(i,j,s)J}\left(e^{\frac{\widetilde{r}_{k}(i,j,s)}{b_{sub}}} - 1\right) \tag{4}$$

Notice that in Equation (4), $\widetilde{r}_{k}(i,j,s)$ denotes the estimated version of $r_{k}(i,j,s)$, $\widetilde{H}_{k}(i,j,s)$ indicates the estimation of $H_{k}(i,j,s)$, and $\aleph_{k}(s)$ denotes the set of currently assigned radio resources for user-k in $\Re_{s}(s)$.

Next, the transmission power, denoted as $p_{trans\_req}(s)$, needed to meet the data rate demands of the users in $\Re_{w}(s)$ for this round is estimated. To minimize the total amount of power consumed, we simply pre-set the minimum required data rate for each user in $\Re_{w}(s)$ for approximate estimations. Thus, $p_{trans\_req}(s)$ can be calculated as:

$$p_{trans\_req}(s) = \alpha \sum_{k\in\Re_{s}(s)} \frac{N_{0}b_{sub}}{\widetilde{H}_{k}(s)J}\left(e^{\frac{r_{req,k}}{b_{sub}}} - 1\right) \tag{5}$$

under the assumption that only a single sub-channel is assigned to an active user session. Notice also that in Equation (5), $\widetilde{H}_{k}(s)$ denotes the estimation of the average channel gain between user-k and each sub-channel of the active CCs.

If $p_{trans\_tot}(s)$ is used to denote the above estimated total transmission power, then:

$$p_{trans\_tot}(s) = p_{trans\_req}(s) + p_{trans\_s}(s) \tag{6}$$

In order to lessen the impact of any undesired fluctuations over a short-term interval, a smoothing exponential filtering version of $p_{trans\_tot}(s)$, denoted as $\hat{p}_{trans\_tot}(s)$, is utilized. Next, the maximum allowable transmission power of the active CCs in the current configuration z, denoted as $p_{\text{trans}}^{(z)}$, can be calculated as follows:

$$p_{\text{trans}}^{(z)} = \sum_{i=1}^{z} p_{\text{trans\_max},i} \tag{7}$$

with that value being calculated for comparison with $\hat{p}_{\text{trans\_tot}}(s)$ to determine whether $\hat{p}_{\text{trans\_tot}}(s)$ can be supported by the current active CCs.

4.3. SCC Activation/Deactivation (SAD)

In order to actually prevent any unnecessary power consumption, the information of the TPE component is utilized by the SAD component to determine when to turn on (or turn off) any necessary (or unnecessary) SCCs. The following pseudo code expresses the details of this component's operation:

```
//Initialize counting variables y1 = 0 and y2 = 0 at the first execution of the proposed scheme//
If (p_trans^(z-1) < p̂_trans_tot(s) ⩽ p_trans^(z))
    y1 = 0;
    y2 = 0;
    z = z; //set that p_trans^(0) = 0//
Else If (p̂_trans_tot(s) > p_trans^(z))
    y1 = y1 + 1;
    y2 = 0;
    z = z;
    While (y1 = y_th) do
        z = z + 1;
Else (p̂_trans_tot(s) ⩽ p_trans^(z-1))
    y2 = y2 + 1;
    y1 = 0;
    z = z;
    While (y2 ⩾ y_th) do
        CC-z denies new arrivals;
        While (all sessions on the CC-z are completely transmitted) do
            z = z - 1.
```

4.4. Fair Bandwidth Allocation (FBA)

The respective fairness control parameters of the type-m, m = 1, 2, users in round s, denoted as $\upsilon^{(m)}(s)$, must then be obtained to control the fairness index for each type of user around the respective target values. The value of $\upsilon^{(m)}(s)$ can be expressed in the form of a stochastic gradient search:

$$\upsilon^{(m)}(s) = \upsilon^{(m)}(s-1) - \eta^{(m)} \cdot (\Phi^{(m)}(s) - \Phi_{\text{target}}^{(m)}) \tag{8}$$

for m = 1, 2, respectively, where $\eta^{(m)}$ denotes the step-size parameters for type-m users.

Since the inherent differing characteristics of the fairness indexes for the RT users and NRT users are respectively used in the calculations for evaluating those indexes, the radio resources are dynamically divided into two parts based on the ratio of the type-m users' required data rate to the required data rate of all the users in the scheduling queue. In addition, the design includes the stipulation that each type of user has their own region of greater priority than the other type of users. By dint of this approach, certain relative levels of fairness between the RT users and NRT users can be maintained. Specifically, we use $\Re_{\text{W}}^{(m)}(s)$ to denote the subset of type-m users in $\Re_{\text{W}}(s)$ and use $\bar{r}_{\text{req}}^{(m)}(s)$ to denote the average required data rate of type-m users in $\Re_{\text{W}}^{(m)}(s)$. In addition, $b_i(s)$ is defined, for round s, as the number of non-occupied sub-channels of CC-i. Next, $b_{\text{NRT},i}(s)$ is used to indicate the number of sub-channels of CC-i for which the NRT users are given higher priority than the RT users to share in during round s; this term can be calculated as follows:

$$b_{\mathrm{NRT},i}(s) = \left\lfloor b_i(s) \frac{\bar{r}_{\mathrm{req}}^{(1)}(s) \left|\Re_{\mathrm{w}}^{(1)}(s)\right|}{\sum_{m=1}^{2} \bar{r}_{\mathrm{req}}^{(m)}(s) \left|\Re_{\mathrm{w}}^{(1)}(s)\right|} \right\rfloor \tag{9}$$

Meanwhile, $b_{\mathrm{RT},i}(s)$ denotes the number of sub-channels of CC-i for which the RT user sessions are given a higher priority than the NRT user sessions to share in during round s; this term can be calculated as follows:

$$b_{\mathrm{RT},i}(s) = b_i(s) - b_{\mathrm{NRT},i}(s) \tag{10}$$

Next, we schedule waiting user session requests in the scheduling queue by starting from the unassigned sub-channels of the active CCs based on SAD with the index $i = 1$ in the configuration z. Referring to the aforementioned study by Rodrigues and Casadevall [8], the optimal fair assignment vector, which is indicated by tuple(i^*,j^*,k^*), can be searched via:

$$\mathrm{tuple}(i^*, j^*, k^*) = \arg \max_{i,j,k \in \Re_{\mathrm{w}}(s)} \left(w_k(s)\tilde{r}_k(i,j,s)\right) \tag{11}$$

under the assumption that equal power allocation is applied for each sub-channel. Note that in Equation (11), $w_k(s)$ denotes the weighting factor for user-k in $\Re_{\mathrm{w}}(s)$ during round s. With appropriate modification of the weighting factor presented in the aforementioned study by Rodrigues and Casadevall [8] to fit our model, $w_k(s)$ can be calculated as follows:

$$w_k(s) = \left(\frac{d_k(s)}{\bar{r}_k(s)}\right)^{\nu^{(m)}(s)}, \ \forall k \in \Re_{\mathrm{w}}(s) \tag{12}$$

The set of optimal tuple(i^*,j^*,k^*)s is the solution that results in a value of $\Phi^{(m)}(s)$ closest to the $\Phi_{\mathrm{target}}^{(m)}$ value.

4.5. Fair Power Adjustment (FPA)

After the FBA component is completed, the power applied for the sub-channels selected in Equation (11) is further adjusted on the basis of the multi-level water-filling concept, with the aim being to better meet the requirements of the constraints in the considered optimization problem. For convenience, the term $p_{i,j}^*(s)$ is defined as the optimal power allocation for sub-channel-j of CC-i during round s, and can be expressed as:

$$p_{i,j}^*(s) = \left[\sigma_k(s) - \frac{1}{\tilde{\gamma}_k(i,j,s)}\right]^+, \ k \in \Re_{\mathrm{w\_sel}}(s) \tag{13}$$

Notice that in Equation (13), $\Re_{\mathrm{w\_sel}}(s)$ indicates the set of active users whose requests have been selected based on Equation (11), $\sigma_k(s)$ is the water level for user-k in $\Re_{\mathrm{w\_sel}}(s)$ during round s of the water-filling problem, $\tilde{\gamma}_k(i,j,s)$ indicates the estimated channel gain-to-noise ratio between user-k and sub-channel-j of CC-i during round s, and $[x]^+ \equiv \max(0,x)$. Lastly, with regard to those requests from the users in $\Re_{\mathrm{w\_sel}}(s)$, if their allocated power levels are less than those necessary to satisfy their minimum required data rates, they are forced to continuously wait in the scheduling queue.

4.6. Time Complexity Analysis

In this subsection, a study of the worst-case performance of the proposed scheme in terms of the time complexity as a function of $|\Re_{\mathrm{w}}(s)|$, $|\Re_{\mathrm{w\_sel}}(s)|$, and ℓ_i, with $i = 1,2,...,c$, is provided. In the SAC component, the amount of time required to perform comparisons is referred to using the constant a_1, while the amount of time required for statements is referred to using the constant a_2. In the TPE component, the amount of time required for statements is referred to using the constant a_3. In the SAD component, the amount of time required to perform comparisons is referred to using the constant a_4, while the amount of time required for statements is referred to using the constant a_5. In the FBA component, the amount of time required for statements is referred to using the constant a_6. In addition, the FBA component sorts the fair assignment vectors, and the worst-case cost for the FBA component is indicated by $O(|\Re_{\text{w}}(s)|\,(\sum_{i=1}^{c}\ell_i)\log(\sum_{i=1}^{c}\ell_i))$. In the FPA component, the amount of time required for statements is referred to using $a_7\,|\Re_{\text{w\_sel}}(s)|$, where a_7 is a constant. As such, the worst-case time complexity of the proposed scheme can be calculated as $O(|\Re_{\text{w}}(s)|\,(\sum_{i=1}^{c}\ell_i)\log(\sum_{i=1}^{c}\ell_i))$. It should be noted that c typically falls in the range of 2–5 (with an especially high likelihood of being equal to or smaller than 3) in today's environments. As a result, the complexity should generally be low.

5. Results and Discussion

In this section, numerical examples are presented to compare the power consumption of the proposed scheme for transmitting data in a multi-CC BS system with that of a conventional scheme in order to quantify the amount of power that could be saved by using the proposed scheme. In the conventional scheme, important elements of the proposed scheme, namely, the SAC, TPE, and SAD components, are simply excluded; in other words, all the CCs are constantly activated, regardless of any fluctuations in the traffic load.

For our experiments, the example of a medium-sized urban macro-cell in which the BS utilizes three CCs for data transmissions was considered. Of the three CCs, one was assumed to be in the 700 MHz frequency band, while the other two were assumed to be in the 2 GHz frequency band. Recall that the three CCs were indexed with i, with i = 1, 2, and 3, respectively. In addition, the following parameters were also assumed: T_i = 5 MHz; b_{sub} = 180 kHz; p_i = 10 W; $p_{\text{trans\_max},i}$ = 10 W; and one OFDMA downlink frame spans 25 sub-channels. We further assumed that all the users of the cell had a uniform spatial distribution, while the movements of the users were assumed to occur at 3 km/h in random directions. Path loss was assumed as the channel propagation model, with the Hata and COST 231 models being adopted for one CC in the 700 MHz frequency band and the other two CCs in the 2 GHz frequency band, respectively [31,32]. The Hata path loss, denoted as L_{Hata} (in dB), is expressed as:

$$L_{\text{Hata}} = 69.55 + 26.16\log_{10}f_1 - 13.82\log_{10}h_{\text{b}} - \varphi(h_{\text{d}}) + (44.9 - 6.55\log_{10}h_{\text{b}})\log_{10}d \tag{14}$$

Note that, in Equation (14), f_1 is the carrier frequency for the Hata path loss model, which is herein set equal to 700 MHz, h_{b} is the effective BS antenna height (in m), h_{d} is the user-terminal antenna height (in m), $\varphi(h_{\text{d}})$ is a correction factor for the user-terminal antenna height, which is given by:

$$\varphi(h_{\text{d}}) = 3.2\log_{10}(11.75h_{\text{d}})^2 - 4.97 \tag{15}$$

and d is the distance between the BS and the user-terminal (in km). In turn, the COST 231 path loss, denoted as L_{COST231} (in dB), is expressed as:

$$L_{\text{COST231}} = 46.3 + 33.9\log_{10}f_2 - 13.82\log_{10}h_{\text{b}} - \varphi(h_{\text{d}}) + (44.9 - 6.55\log_{10}h_{\text{b}})\log_{10}d \tag{16}$$

where f_2 is the carrier frequency for the COST 231 path loss model, which is herein set equal to 2000 MHz.

In compliance with the application-layer service trend of the 4G and future 5G networks, we looked at RT sessions consisting of YouTube activities and NRT sessions consisting of file transfer protocol, hypertext transfer protocol, and social network activities as a study case. Session times of RT users were set to have a mean of 3 min and an exponential distribution. For NRT users, 20% of the total file size was assumed to be contributed from file sizes with a truncated lognormal distribution and a mean of 2 MB, while the other 80% was assumed to be contributed from files with a fixed size of 100 kB each. For a given user, the inter-arrival time of user session requests, meaning the period of time between the end of a session and the generation of a new request, was assumed to have an exponential distribution and a mean of 1 s. For the sake of convenience, $\Re^{(m)}$ refers to the set of type-m users in the given cell. For the NRT users, the minimum required data rate $r_{\text{req},k}, k \in \Re^{(1)}$ was set to 300 kbps, while for the RT users, the minimum required data rate $r_{\text{req},k}, k \in \Re^{(2)}$ was set to 500 kbps. Moreover, $\Phi^{(m)}_{\text{target}}, m = 1, 2$, were both set at 1; $\eta^{(m)}, m = 1, 2$, were both set at 0.1; $h_b = 100$ m; $h_d = 1$ m; the threshold y_{th} in the SAD component was set at 10; and the simulation time was set to 1 h.

For demonstrations, the number of users was the variable, with the probability of a given user being an RT user set at 30%, meaning that the probability of a given user being an NRT user was set at 70%. The considered variation in the number of users over time is shown in Figure 3. In that figure, an increase in users from one time point to the next indicates the addition of new users, while a decrease indicates users leaving the cell. Accordingly, the numbers of RT and NRT user sessions being transmitted versus time in the 3-CC cellular system are respectively shown in Figure 4. As shown in Figures 3 and 4 during the interval from 35 to 40 min, there is a gap between the total number of users in the cell and the total number of sessions being transmitted. This gap is mainly due to the system capacity limit, such that some users' requests are forced, by the check of SAC, to be buffered in their respective processing queues to avoid causing service quality degradation of other user sessions that are currently being served.

Given the conditions specified above, Figure 5 presents a comparison of the total power consumption levels of the proposed scheme and the conventional scheme. As shown, the proposed scheme resulted in significantly reduced total power consumption at the BS transceivers during periods of relatively light traffic. This effect was achieved because under the proposed scheme, some SCCs were dynamically deactivated when the traffic load decreased. As such, unnecessary power consumption was significantly reduced even as the respective minimum requirements of all the user sessions were satisfied and the fairness indexes for the different types of users were still efficiently controlled at acceptable levels.

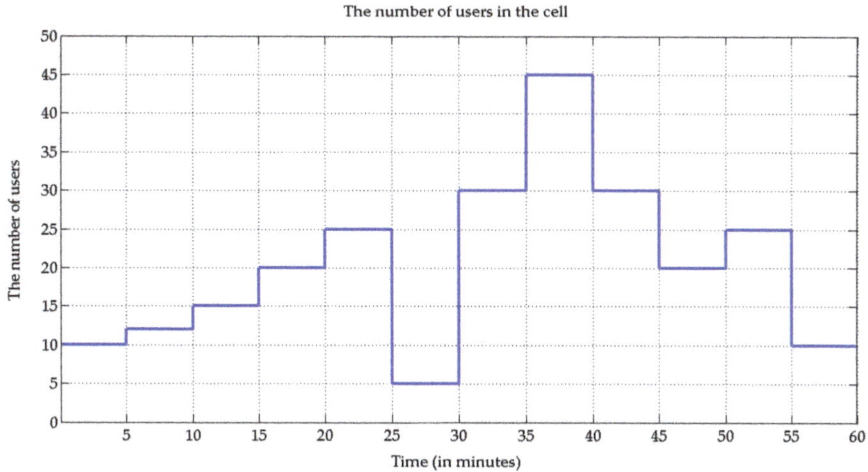

Figure 3. The number of users of the cell over time.

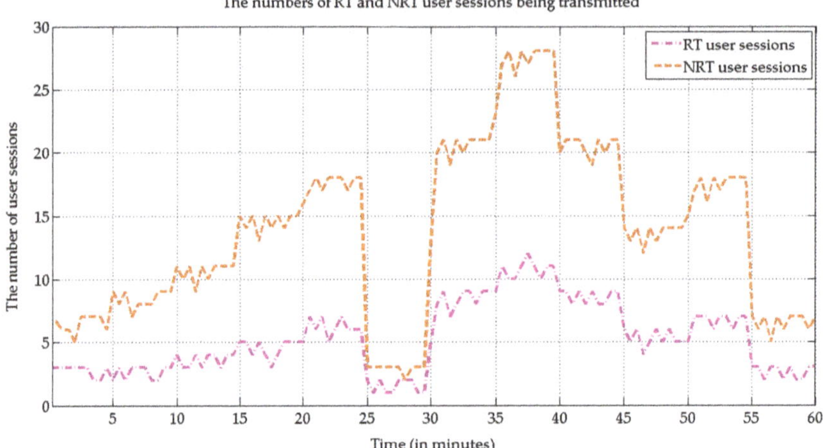

Figure 4. The numbers of RT and NRT user sessions over time.

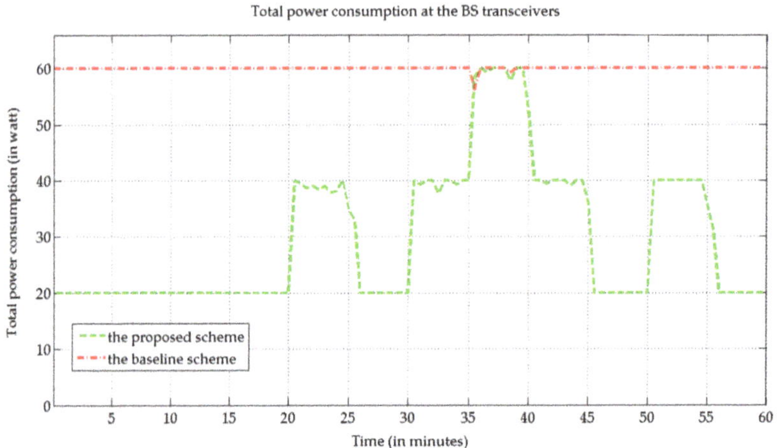

Figure 5. A comparison of the levels of total power consumption for the proposed scheme and the conventional scheme.

Next, as can be seen in Figures 3 and 5 under the proposed scheme, for any periods in which the total number of users was equal to or less than 20, the power consumed by the BS transceivers was consistently reduced to a relatively low level (*i.e.*, around 20 W). In contrast, under the baseline scheme, the power consumed by the BS transceivers remained at a much higher level (*i.e.*, around 60 W) for almost the entire simulation. This was due to the fact that, for a given configuration of CCs, if the water-filling approach based on Equation (13) was applied, the maximum available transmission power was completely utilized for distribution over each sub-channel.

Next, in order to more clearly illustrate the behavior of the two schemes with regard to traffic load, we introduce two graphs (Figures 6 and 7) in which the results for the basic levels of power consumed and for the levels of transmission power consumed over time, respectively, are plotted. By referencing Figures 3 and 6 it can be seen that, under the proposed scheme, the first SCC (CC-2) becomes activated when the number of users is increased to 25, while the second SCC (CC-3) is subsequently activated when the number of users reaches 45. In contrast, when the number of users is relatively low, both of the SCCs are deactivated. These results imply, then, that the proposed scheme can, in comparison to the other scheme, effectively activate/deactivate the SCCs according to uncertain variations in traffic loads over time in order to lessen or prevent unnecessary energy consumption. From Figure 7, meanwhile, the following fact, which was likewise illustrated above with Figure 5, can clearly be seen: due to the water-filling approach, the maximum available transmission power was almost fully utilized for any given configuration of CCs under both schemes. Accordingly, the channel efficiency enhanced. The latter point is illustrated in Figure 8, which shows the fairness indexes for both the RT and NRT users over time.

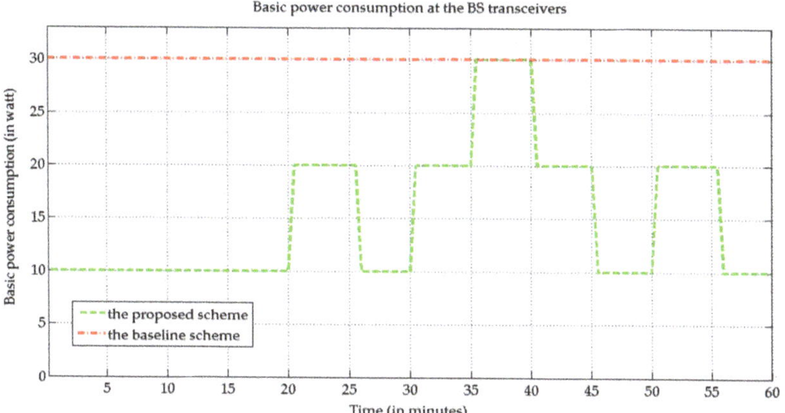

Figure 6. A comparison of the levels of basic power consumption for the proposed scheme and the conventional scheme.

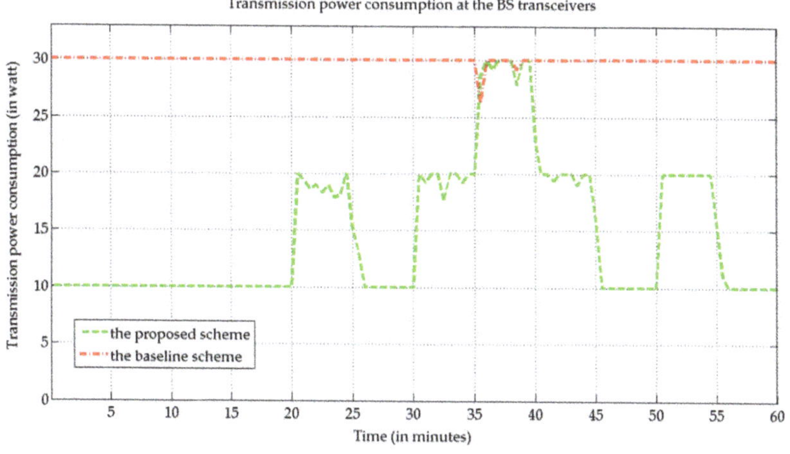

Figure 7. A comparison of the levels of transmission power consumption for the proposed scheme and the conventional scheme.

It was also found, as shown in Figure 8, that for a specific number of users of the cell, the fairness indexes for the different types of users were nearly convergent, consistently reaching a level of less than 2% difference from the target. Moreover, even when the arrival rate of new users increased significantly, the fairness indexes for both the RT and NRT users were nearly maintained at their respective target levels following no more than a few minutes of scheduling rounds (iterations) (the variable results for each round of the 5-min intervals shown in Figure 8, where the intervals were based on the traffic pattern indicated in Figure 3, are unrelated to any of the optimization variables for the power minimization problem being considered. Accordingly, these results do not reflect the stability of the proposed scheme in any way). Relatedly, while the fairness indexes for both types of users were instantly degraded whenever the traffic load was significantly increased, they were still maintained at levels above 0.7 during almost all those brief periods of degradation.

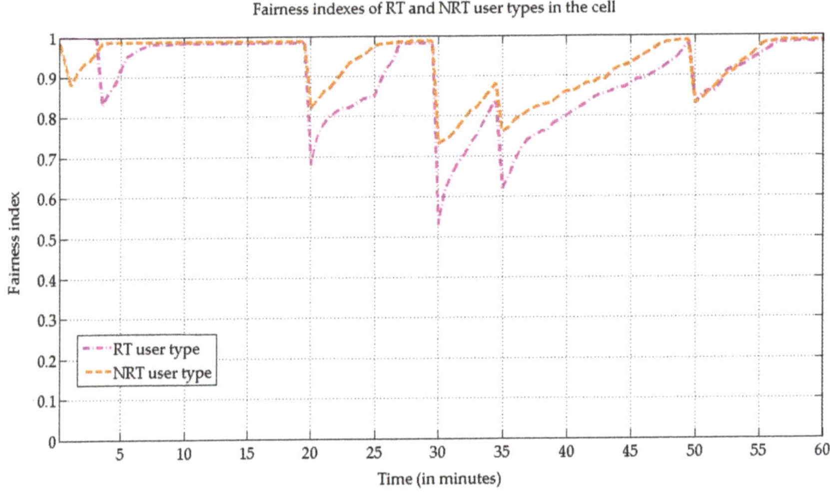

Figure 8. Fairness indexes over time for the RT user type and NRT user type under the proposed scheme.

For the sake of completeness, Table 1 lists the average levels of total power consumption, under the proposed scheme and the conventional scheme, for different minimum required data rates for the NRT users and RT users given the traffic pattern indicated in Figure 3. As indicated by Table 1, the proposed scheme outperforms the conventional scheme in terms of power-saving effectiveness for various minimum required data rates for the NRT users and RT users. This better performance results from the proposed scheme having the substantial advantage of flexibility in terms of activating/deactivating the SCCs according to dynamically fluctuating traffic loads, thus allowing unnecessary power consumption to be avoided. In addition, when the minimum required data rates are set relatively low (*i.e.*, $r_{\text{req},k}$ = 300 kbps for $k \in \Re^{(1)}$ and $r_{\text{req},k}$ = 500 bps for $k \in \Re^{(2)}$), the resulting power-saving ability is particularly significant (with consumption lowered by 50.26%). This is due to the fact that, under a given a traffic pattern, and as indicated by Equation (1), when the minimum required data rate for each user is higher, the accumulated transmission power will also be larger, which, in the event that the accumulated transmission power exceeds the maximum transmission power of the currently active CC(s), will lead to the activation of more SCC(s) to aid in the transmissions. As a result, more power will be consumed. Meanwhile, when the minimum required data rates are set lower, the resulting power consumption is also lower. For the conventional scheme, on the other hand, because the two SCCs were continuously active regardless of the lightness or heaviness of the traffic load, the level of total power consumption remained close to 60 W almost continuously. Furthermore, as the minimum required data rate was increased, the total power consumed slightly decreased, as can be seen from the column for the conventional scheme in Table 1. The reason for this is that the possibility of the higher data rate requirement being satisfied is relatively low under heavy traffic load conditions, such that given the FPA design, some sessions may be forced to continuously queue in the scheduling queue, in which case they will not consume power at the BS transceivers.

Table 1. The average levels of total power consumption, under the proposed scheme and the conventional scheme, when the minimum required data rates for NRT users and RT users are set at different levels given the traffic pattern indicated in Figure 3.

$r_{\mathbf{req},k}$, $k \in \Re^{(1)}/r_{\mathbf{req},k}$, $k \in \Re^{(2)}$	The Proposed Scheme	The Conventional Scheme
300 kbps/500 kbps	29.70 W	59.71 W
500 kbps/1 Mbps	32.15 W	59.06 W
1 Mbps/2 Mbps	39.16 W	58.38 W

Essentially, the advantages of the proposed scheme indicated by the above experimental results spring from the fact that the proposed scheme was based on the radio resource and power allocation approach previously proposed by Rodrigues and Casadevall [8]. The primary aim of their approach was to ensure efficient control of the fairness of resource allocation, and we have simply modified that approach further to fit the purposes of our proposed model. Simply put, the proposed scheme provides powerful power-saving ability, while also satisfying the respective minimum required data rates of all the user sessions and efficiently controlling the fairness of resource allocations to different types of users. More specifically, by monitoring fluctuations in the traffic load and providing dynamic activation/deactivation of the SCCs of the BS accordingly, the proposed scheme significantly reduces any unnecessary consumption of power. Stated differently, the proposed approach allows for efficient use of system resources with regard to power usage, bandwidth efficiency, user fairness, and computational time. Based on the above observations, it can be firmly concluded that the proposed scheme would constitute an effective means of reducing power consumption on the BS side of cellular systems.

6. Conclusions

A novel and efficient power-saving scheme for data transmission in multi-CC cellular systems has been proposed and successfully tested. The proposed scheme is more adaptive and flexible than the conventional scheme with regard to SCC usage, while also maintaining the capacity to satisfy the respective minimum required data rates of all the user sessions and simultaneously manage the fairness indexes for different user types. In terms of limitations, it should be noted that the significant gains of the proposed scheme in terms of power consumption come at the expense of some degradation in fairness indexes for the different types of users whenever the traffic load is significantly increased. However, those degradations are always brief, with the fairness levels for the different types of users being re-established within minutes. It is believed that this novel power-saving scheme is an excellent solution to be employed for use in 4G and future 5G multi-CC cellular systems at the BS side for data transmissions. As such, the present work constitutes a reasonable and feasible means of addressing the problems of rising energy costs and CO_2 emissions associated with cellular systems. In the future, we suggest extending the present system model and scheme to a multi-cell environment. Nevertheless, the detailed design of the spectrum management and the transmission performance may need to be re-evaluated. Furthermore, in addition to providing energy savings/efficiency, another key goal of 5G and future cellular systems will be ensuring spectral efficiency [33]. As such, the question of how these two different objective performance metrics can be efficiently coupled and integrated should be given far greater attention in the coming years [34]. Lastly, it is reasonable to believe, in any event, that continued efforts in this research direction may aid in mitigating the problem of global warming by contributing to the establishment of environmental sustainability. Moreover, by conserving and utilizing energy in a judicious manner, various other types of environmental degradation and resource depletion can be avoided or ameliorated.

Acknowledgments: The author would like to express his gratitude for the financial support provided by the Ministry of Science and Technology (MOST) in Taiwan, R.O.C. under the Contracts MOST 104-2221-E-019-053- and MOST 104-2218-E-019-004-.

Conflicts of Interest: The author declares no conflict of interest. The founding sponsor had no role in the design of the study; in the collection, analyses, or interpretation of data; in the writing of the manuscript, and in the decision to publish the results.

Abbreviations

The following abbreviations are used in this manuscript:

4G	4th Generation
5G	5th Generation
BER	Bit Error Rate
BS	Base Station
CC	Component Carrier
CO_2	Carbon Dioxide
FBA	Fair Bandwidth Allocation
FPA	Fair Power Adjustment
NRT	Non-Real-Time
OFDMA	Orthogonal Frequency Division Multiple Access
PCC	Primary Component Carrier
RT	Real-Time
SAC	Session Admission Control
SAD	Supplementary component carrier Activation/Deactivation
SCC	Supplementary Component Carrier
TPE	Transmission Power Estimation

References

1. Yuan, G.; Zhang, X.; Wang, W.; Yang, Y. Carrier aggregation for LTE-advanced mobile communication systems. *IEEE Commun. Mag.* **2010**, *48*, 88–93. [CrossRef]
2. Hossain, E.; Hasan, M. 5G cellular: Key enabling technologies and research challenges. *IEEE Instrum. Meas. Mag.* **2015**, *18*, 11–21. [CrossRef]
3. Correia, L.M.; Zeller, D.; Blume, O.; Ferling, D.; Jading, Y.; Gódor, I.; Auer, G.; Van der Perre, L. Challenges and enabling technologies for energy aware mobile radio networks. *IEEE Commun. Mag.* **2010**, *48*, 66–72. [CrossRef]
4. Han, C.; Harrold, T.; Armour, S.; Krikidis, I. Green radio: Radio techniques to enable energy-efficient wireless networks. *IEEE Commun. Mag.* **2011**, *49*, 46–54. [CrossRef]
5. Hasan, Z.; Boostanimehr, H.; Bhargava, V.K. Green cellular networks: A survey, some research issues and challenges. *IEEE Commun. Surveys Tutor.* **2011**, *13*, 524–540. [CrossRef]
6. Mancuso, V.; Alouf, S. Reducing costs and pollution in cellular networks. *IEEE Commun. Mag.* **2011**, *49*, 63–71. [CrossRef]
7. Wu, J.; Zhang, Y.; Zukerman, M.; Yung, E.K.-N. Energy-efficient base-stations sleep-mode techniques in green cellular networks: A survey. *IEEE Commun. Surveys Tutor.* **2015**, *17*, 803–826. [CrossRef]
8. Rodrigues, E.B.; Casadevall, F. Control of the trade-off between resource efficiency and user fairness in wireless networks using utility-based adaptive resource allocation. *IEEE Commun. Mag.* **2011**, *49*, 90–98. [CrossRef]
9. Wong, C.Y.; Cheng, R.S.; Lataief, K.B.; Murch, R.D. Multiuser OFDM with adaptive subcarrier, bit, and power allocation. *IEEE J. Select. Areas Commun.* **1999**, *17*, 1747–1758. [CrossRef]
10. Jeong, S.S.; Jeong, D.G.; Jeon, W.S. Cross-layer design of packet scheduling and resource allocation in OFDMA wireless multimedia networks. In Proceedings of the IEEE VTC 2006 Spring, Melbourne, Australia, 7–10 May 2006.
11. Kivanc, D.; Li, G.; Liu, H. Computationally efficient bandwidth allocation and power control for OFDMA. *IEEE Trans. Wirel. Commun.* **2003**, *2*, 1150–1158.

12. Madan, R.; Boyd, S.; Lall, S. Fast algorithms for resource allocation in wireless cellular networks. *IEEE/ACM Tans. Netw.* **2010**, *18*, 973–984. [CrossRef]
13. Ng, D.W.K.; Lo, E.S.; Schober, R. Energy-efficient resource allocation in OFDMA systems with large numbers of base station antennas. *IEEE Trans. Wirel. Commun.* **2012**, *11*, 3292–3304. [CrossRef]
14. Ng, D.W.K.; Lo, E.S.; Schober, R. Energy-efficient resource allocation in multi-cell OFDMA systems with limited backhaul capacity. *IEEE Trans. Wire. Commun.* **2012**, *11*, 3618–3631. [CrossRef]
15. He, S.; Huang, Y.; Jin, S.; Yang, L. Coordinated beamforming for energy efficient transmission in multicell multiuser systems. *IEEE Trans. Commun.* **2013**, *61*, 4961–4971. [CrossRef]
16. Piunti, P.; Cavdar, C.; Morosi, S.; Teka, K.E.; Del Re, E.; Zander, J. Energy efficient adaptive cellular network configuration with QoS guarantee. In Proceedings of the IEEE ICC, London, UK, 8–12 June 2015.
17. Morosi, S.; Piunti, P.; Del Re, E. Improving cellular network energy efficiency by joint management of sleep mode and transmission power. In Proceedings of the 24th TIWDC—Green ICT, Genova, Italia, 23–25 September 2013.
18. Chung, Y.-L. Energy-saving transmission for green macrocell-small cell systems: A system-level perspective. *IEEE Syst. J.* **2016**. [CrossRef]
19. Niu, Z.; Wu, Y.; Gong, J.; Yang, Z. Cell zooming for cost-efficient green cellular networks. *IEEE Commun. Mag.* **2010**, *48*, 74–79. [CrossRef]
20. 3GPP TR 32.826 V10.0.0. Telecommunication Management: Study on Energy Savings Management. Available online: http://www.qtc.jp/3GPP/Specs/32826-a00.pdf (accessed on 1 April 2016).
21. Micallef, G.; Mogensen, P.; Scheck, H.O. Dual-cell HSDPA for network energy saving. In Proceedings of the IEEE VTC Spring 2010, Taipei, Taiwan, 16–19 May 2010.
22. Lorincz, J.; Capone, A.; Begusic, D. Optimized network management for energy savings of wireless access networks. *Comput. Netw.* **2011**, *5*, 514–540. [CrossRef]
23. Chung, Y.-L.; Tsai, Z. A suboptimal power-saving transmission scheme in multiple component carrier networks. *IEICE Trans. Commun.* **2012**, *E95-B*, 2144–2147. [CrossRef]
24. Lorincz, J.; Capone, A.; Begusic, D. Impact of service rates and base station switching granularity on energy consumption of cellular networks. *EURASIP J. Wirel. Commun. Netw.* **2012**. [CrossRef]
25. Aleksic, S.; Deruyck, M.; Vereecken, W.; Joseph, W.; Pickavet, M.; Martens, L. Energy efficiency of femtocell deployment in combined wireless/optical access networks. *Comput. Netw.* **2013**, *57*, 1217–1233. [CrossRef]
26. Enokido, T.; Takizawa, M. An integrated power consumption model for distributed systems. *IEEE Trans. Ind. Electron.* **2013**, *60*, 824–836. [CrossRef]
27. Chung, Y.-L. Rate-and-power control based energy-saving transmissions in OFDMA-based multicarrier base station. *IEEE Syst. J.* **2015**, *9*, 578–584. [CrossRef]
28. Chung, Y.-L. Energy-efficient transmissions for green base stations with a novel carrier activation algorithm: A system-level perspective. *IEEE Syst. J.* **2015**, *9*, 1252–1263. [CrossRef]
29. Jain, R.; Chiu, D.-M.; Hawe, W. A Quantitative Measure of Fairness and Discrimination for Resource Allocation in Shared Computer Systems. Available online: http://www.cse.wustl.edu/~jain/papers/ftp/fairness.pdf (accessed on 1 April 2016).
30. Liu, S.; Wang, Z. On convergence analysis of iterative smoothing methods for a class of nonsmooth convex minimization problems. In Proceedings of the 7th International Joint Conference on Computational Sciences and Optimization (CSO 2014), Beijing, China, 4–6 July 2014.
31. Agrawal, D.P.; Zeng, Q.-A. *Introduction to Wireless & Mobile Systems*, 3rd ed.; CENGAGE Learning: Boston, MA, USA, 2011.
32. *COST Action 231: Digital Mobile Radio Towards Future Generation Systems: Final Report*; European Commission: Brussels, Belgium, 1999.
33. Andrews, J.G.; Buzzi, S.; Choi, W.; Hanly, S.V.; Lozano, A.; Soong, A.C.K.; Zhang, J.C. What will 5G be? *IEEE J. Sel. Areas Commun.* **2014**, *32*, 1065–1082. [CrossRef]
34. Bjornson, E.; Jorswieck, E.; Debbah, M.; Ottersten, B. Multi-objective signal processing optimization: The way to balance conflicting metrics in 5G systems. *IEEE Signal Process. Mag.* **2014**, *31*, 14–23. [CrossRef]

Article

Development of an Optimal Power Control Scheme for Wave-Offshore Hybrid Generation Systems

Seungmin Jung and Gilsoo Jang *

School of Electrical Engineering, Korea University, Seoul 136-713, Korea; muejuck@korea.ac.kr

* Author to whom correspondence should be addressed; gjang@korea.ac.kr; Tel.: +82-2-3290-3246; Fax: +82-2-3290-3692.

Academic Editor: William Holderbaum

Received: 28 May 2015; Accepted: 18 August 2015; Published: 25 August 2015

Abstract: Integration technology of various distribution systems for improving renewable energy utilization has been receiving attention in the power system industry. The wave-offshore hybrid generation system (HGS), which has a capacity of over 10 MW, was recently developed by adopting several voltage source converters (VSC), while a control method for adopted power conversion systems has not yet been configured in spite of the unique system characteristics of the designated structure. This paper deals with a reactive power assignment method for the developed hybrid system to improve the power transfer efficiency of the entire system. Through the development and application processes for an optimization algorithm utilizing the real-time active power profiles of each generator, a feasibility confirmation of power transmission loss reduction was implemented. To find the practical effect of the proposed control scheme, the real system information regarding the demonstration process was applied from case studies. Also, an evaluation for the loss of the improvement rate was calculated.

Keywords: hybrid generation system (HGS); reactive capability; reactive power assignment; power conversion system (PCS) control; voltage source converter (VSC)

1. Introduction

The growing interest in energy preservation in all industrial sectors has recently motivated the need to find sustainable technical solutions to reduce energy consumption. Nowadays, renewable energy sources are developed based on a geographically wide area and usually generate requirements for a management system to handle the entire system more appropriately [1]. Consequently, regarded industry areas have promoted the development of total control solutions, such as wind farm management systems, improving not only the mechanical conditions, but also the power control flexibility for the system operator [2,3]. In the case of the offshore generation industry, structural designs and integration studies have progressed by considering several different distribution systems including energy storage systems (ESSs) for increasing the reliability of the entire system's output profile [4,5]. Among these systems, an integrated system with various distribution sources, which is composed of an offshore cluster with some wind generators, has been receiving attention. These configurations have advantages in terms of efficiency as well as available energy quantity and, can lead to the reduction of construction costs by minimizing the related electrical systems.

In the case of the wave-offshore hybrid generation system (HGS), the configuration is suitable to increase the whole generation capacity with a number of distribution generators and the combined generation system can resolve the reliability issues of renewable sources [6]. In the current state of HGSs, several permanent magnetic synchronous generator (PMSG) wind turbines and permanent magnetic synchronous linear generator (PMSLG) wave generation systems have been built on a designed offshore platform, sharing various pieces of electrical equipment [7]. The electrical system's blueprint has almost been prepared but the optimized control logic has not yet been developed. General power control methods are available, but assigning an equivalent portion to several generation systems is counted as an inefficient solution. The HGS is intended to adopt a real time state monitoring system including power flow management, which can handle the power flow on a real time basis. The ongoing development of the central control algorithm would focus the utilization of the measured state of each system and find an optimal solution for the system's life cycle and operation efficiency. These countermeasures should include a wake effect analysis of wind turbines for considering active power efficiency and reliability, which are the main concerns of transmission and system operators (TSOs). Moreover, this type of large-scale generation system is responsible for reactive power support for the connected power grid, and integration of reactive/voltage management system to the conversion process is mandatory for utility grid. The most common distribution systems which have the reactive power supplying capability are the voltage source converter (VSC) interfaced generation system as mentioned in previous studies [8]. Active support by reactive power with VSC system has been continuously studied based on real/reactive power decoupled control considering own capability limitation [9,10]. An appropriate power flow management method by dealing with the above-mentioned issues is also considered as the final control form of HGS to offer a more appropriate required ancillary service for utility power grids.

In previous studies, only a real power assignment process has been considered due to the uncertainty regarding the environmental characteristics [11]. These considerations mean that the control logic of each generation system focuses only on the basic voltage reliability at each integration point (no responsibility for reactive power reserve) [12]. However, the HGS has unique characteristics in terms of the grid code because all requirements for the distribution system would be applied to the connection point of HGS as a single generation system. The previous active power assignment methods usually applied on wind farms are not suitable for HGS because the composition of the two systems somewhat differ. Therefore, novel power assignment methods should be developed to make not only the control topology meets the specially designated grid codes but also the entire system improves own power efficiency.

This paper deals with a power assignment plan based on the composed real time monitoring system. By considering the output profile of the HGS, the optimized reactive power assignment process will be built to reduce the entire system loss. The structure of this paper is as follows: Section 2 describes the principles of HGS and the related management system. Section 3 explains the proposed optimization algorithm for minimizing system loss. Section 4 gives a verification process with the composed EMTDC simulation and Section 5 shows the arrangement of the proposed method and application process.

2. Hybrid Generation System (HGS) Configuration

2.1. Wave-Offshore Generation System

The concept of HGS is developed to reduce the platform construction cost and improve the utilization of power transfer equipment [13,14]. The entire capacity of the system has increased over than 10 MW in terms of power capacity and the total area of the floating structure is expected to be over 40,000 m^2. All of the main generators are built on the floating structure and the required electricity devices also located inside the floating structure. Three MW PMSG wind turbines will be located at each vertex of the structure and 24 wave generators are erected at the corner of the structure to generate a 2.4 MW power profile. The total capacity of the recent HGS is 14.4 MW and the power conversion system (PCS) of the wave generator is shared by a number of generators. Figure 1 shows a conceptual image of the HGS. In the structure, several transmission cables are installed, and integration points occur at the center of each specific row. The AC cable is used for the wind power system and wave generators are installed at a single DC section based on the low voltage DC distribution system [15].

With the integration point as the center, a system can transfer generated energy to a center substation, which is responsible for following the reference signal by TSO.

The substation including DC/AC PCS is located in the central station to boost the primary voltage to the transmission level for integrating the HGS with the point of common coupling (PCC) through a high-voltage condition. Another main role of the central station is to act as a monitoring system for measuring the real-time condition of each system to manage the entire HGS for an optimized state. Especially, since all of the generation systems in the HGS utilize a VSC, studies on the appropriate voltage regulation method focusing on the PCC are being carried out. This paper deals with advanced reactive power control with an integrated monitoring system. The HGS first considers the wake effect of the wind system to achieve maximum power point tracking (MPPT) continuously and the proposed control scheme would make the reactive power assignment process efficient through developing an optimization process. In particular, the reactive power reserve would be maximized if we consider the output power of the entire system in a real time basis, and it will be the main strength of the proposed system.

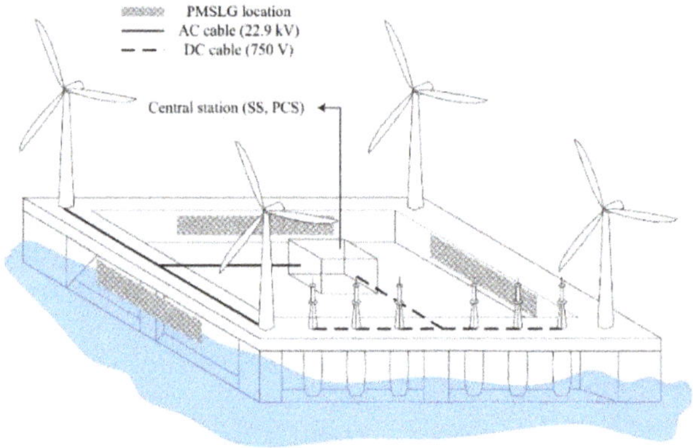

Figure 1. Concept of the wave-offshore hybrid generation system (HGS). PMSLG: synchronous linear generator; PCS: power conversion system.

2.2. *Integrated Monitoring System*

Since the power capacity of the composed HGS is significantly greater than that of the previous renewable sources, a further management solution should be considered for the system to improve reliability and the individual ancillary service. By focusing on the profile at the connection point of the HGS, the order of TSO for an active/reactive power signal should be controlled for matching the system requirement.

As a countermeasure to this issues, the HGS introduces a supervisory integrated monitoring and control system (IMCS) that can check the state of the system continuously, not only the mechanical load of each structure, but also the electrical conditions including voltage level and active/reactive output. The control signal for each generation system would be formulized based on the TSO orders and modified according to the output profile of the central transformer. In the case of active power, the wind turbines usually follow MPPT control and the profile exceeded above the reference could be limited thorough a supervisory control system with individual mechanical properties such as pitch control. The wave generation system does not adopt the power limitation method and is designed to follow the MPPT process continuously. On the other hand, reactive power could be controlled by applied full-type conversion systems of the wind system according to the operator's purpose, and this could enhance the HGS in term of a controllable reactive power capacity. Furthermore, the wave generation system also adopts a common converter system that includes a 2.4 MVA grid-side voltage source DC/AC inverter. These reactive sources can maximize individual reactive power capability by utilizing the output profile information on a real time basis. Because the available reactive power of the full-type converter depends on the profile of the active power of each generation system, it is worthwhile imposing these values in the power control scheme if the TSO demands more reactive power than the previous state. Above all, matching output profile with the active/reactive power order of the operator is obligated to these large-scaled power generation systems even for the systems with renewable sources [16], the importance of supervisory control will be growth, and the future power control scheme will focus to utilize the obtained information.

2.3. Wind System Characteristics

The wind systems in HGS were designed to obtain the real power from the wind resources of the prevailing wind direction, which is considered the main energy of HGS. However, the prevailing wind direction cannot be maintained continuously during the system operation, and the wake effects inside HGS should be analyzed appropriately to cope with the mentioned condition. The wake effect is not generally considered when the distance between each wind turbine is greater than the designated value [17]. However, for the HGS that includes a wind system having a relatively short distance among installed wind turbines, a significant wind speed reduction is expected when the changed wind condition is applied. Figure 2 presents the necessity of the proposed power control scheme. Almost all of the output profile of the wind system would be obtained from the prevailing wind direction, and the other wind direction (yawing control required) would normally not be considered; nevertheless, the actual wind energy reduction and related fluctuation occurs according to the previously analyzed effect when the wind passes through the front line of the wind system. If the HGS changes the status and confronts the mentioned situation, a significant gap of real power profile obtained from encountered wind energy occurs between the wind turbines of the front array and the wind turbines of the rear array.

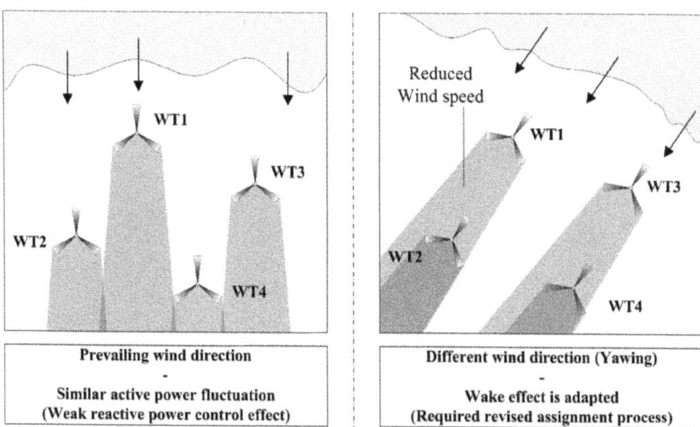

Figure 2. Target situation regarding wake effect.

Figure 3 shows the mentioned situation where the system confronts the entire wake effect. The applied wind speeds at the wind turbines of the rear array are significantly reduced. With this condition, the active power outputs of each wind turbine differ significantly.

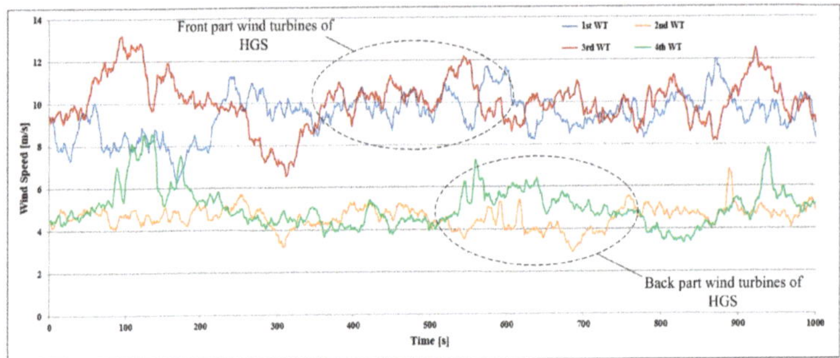

Figure 3. Different wind resources applied to HGS.

These differences could generate current and voltage different in the electrical system, which are related to the transfer efficiency. Hence, these current flow differences due to the active power demand further valancing control of the reactive power. Especially, since the duty of the reactive power supply of HGS is fully determined with the common coupling point, the current flow minimization ability of the inner system can result in a fine solution for the HGS (the requirements are achieved if the total reactive power output is equal to the designated value by TSO). To achieve this, the designed reactive power assignment process will use the real-time measured value obtained from the integrated management system and the purpose of this process is to minimize the current flow for the electric cable in HGS.

When the entire system utilizes the power control method and applies it in the PCS management system, the active power loss due to the active power difference could be mitigated continuously. The IMCS will transfer the required values to the management system by using individual measuring devices, softening the operation of the algorithm's interworking.

3. Power Control Algorithm

3.1. Proposed Method Description

The aim of the proposed algorithm is to assign reactive power requirements by focusing on the measured online active power profile of each wind turbine. The certified reactive power quantities are assigned through several stages by considering each device's current state. First, the unusual system structure of HGS is a major consideration for the assignment process because TSO does not consider each unit in the HGS as an individual controllable unit and the management system can control the available power according to the designed flow chart. The loss improvement can be achieved by considering the system layout because the components of the cable directly influence the loss occurrence. Additionally, the method should check the difference between the cable parameters of each section because the determination of cable specification depends on the expected current flow from each generation unit.

Figure 4 shows the electrical system structure that was mainly analyzed in the proposed process. Since the electric cable parameters of each section are different, several loss expectation formulas need to be included in the reactive assignment method to achieve the optimization process. Two classified sections, named "array", will be interconnected with the center substation through a thicker electric cable than the individual electric cable of the wind turbine. The proposed method first focuses on establishing a proportional equation at each section in relation to reactive power by using designated cable components. The cable parameters used in the HGS are shown in Table 1. This values will be imposed to controller before operation and utilized in loss expectation process. The wind turbines on the connection point (marked at the Figure 4) are classified with two different turbine "A" and "B". In Section 3.2, assignment equations are formulized with this classification. The configured formulas could be applied to different connection points through the divided calculation mode with an additional consideration such as voltage fluctuations.

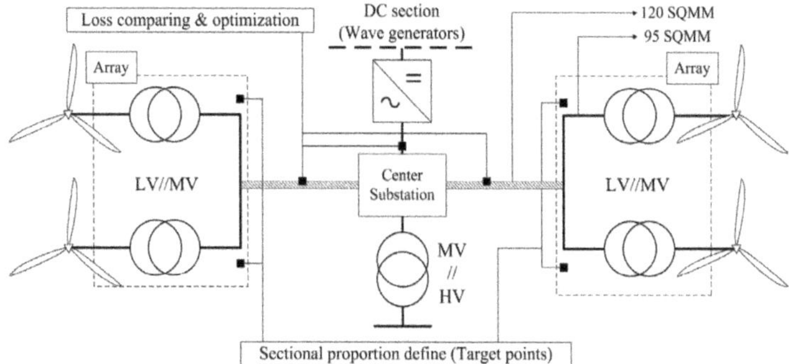

Figure 4. Electrical system structure of HGS and control purpose description.

Table 1. Numerical cable parameter in HGS.

Voltage (kV)	Size (mm^2)	Allowable current (A)	Conductor resistance (Ω/km)	Inductance (mH/km)	Capacitance (μF/km)
0.75 (DC)	35	228	0.565	0.277	0.08
	50	289	0.393	0.266	0.09
22.9	95	291	0.193	0.42	0.17
	120	330	0.153	0.41	0.18

The following section will discuss the loss equation. The general assignment process that can be applied to each point will then be introduced. The proposed decision process for the proportion focuses on the current balance of a certain section according to the current output of wind turbines. When the proportion of the reactive power is designated with the measured active power, the main system can calculate the expected loss including the wave generation system. By comparing the expected loss according to the assigned quantity, the algorithm can set the reference signal of the reactive power. Through the designed process, loss expectation and minimization could be performed by balancing the current flow of each section.

3.2. System Loss Equation

To obtain the loss reduction process, a real power loss equation should be built in each section. Figure 5 shows the system structure of HGS by dividing it into several sections to illustrate the mentioned equations. Each generation system demands an electric cable that would be located on the outer deck of HGS. Basically, every outer deck will include a DC cable for the wave generation system. Additionally, AC cables for integrating wind turbines would be included in some of the outer decks.

The wave generation system utilizes a linear generation system and the generated power would be transferred to the center power system in a DC electric form by using an individual AC/DC converter [18]. Because the DC current would change in the AC form at the center of the power station, the represented DC section in the figure is operated as a type of low voltage distribution system. In this paper, because we focus on the active/reactive power flow on the AC system, the AC cable structure and related cable information should mainly be discussed. There are two arrays (grouping two wind turbines as one array) in the HGS and both arrays are interconnected with center substation to transfer the controlled output value to the utility grid.

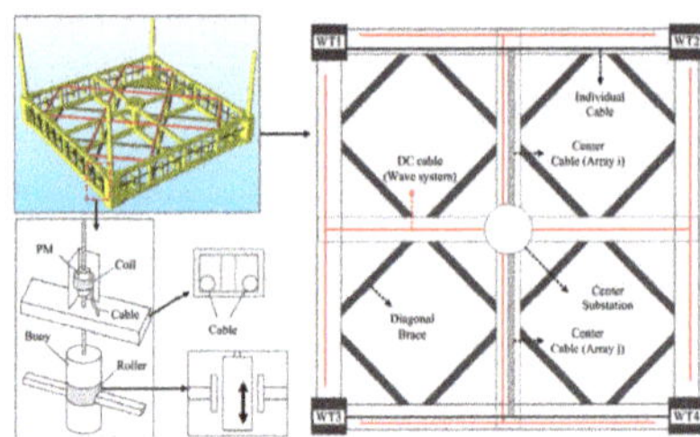

Figure 5. HGS structure analysis-electric cable location.

Basically, the distance between connected wind turbine and the center cable is the same; therefore, the loss occurring in the two cables fully depends on the power flow states. Since the power control station is directly interconnected with a dedicated line, the entire profile of the real system could be defined as Equation (1):

$$P_{\text{HGS}} = P_{\text{wave}} + P_{\text{wind}} \tag{1}$$

where P_{HGS} is the real output power of the entire HGS, P_{wave} is the real output power of wave generators, and P_{wind} is the real output power of wind turbines.

The output profile of total wind system could be divided into the each turbine's output value and each section's loss term. Equation (2) shows the divided quantity, and the loss of a certain array could also be divided as shown in Equation (3) by using the terms defined above:

$$P_{\text{wind}} = \sum P_{\text{wt}} - \sum P_{\text{L}} \tag{2}$$

$$P_{\text{L}} = P_{\text{L}\cdot\text{CNT}} + \sum P_{\text{L}\cdot\text{IND}} \tag{3}$$

where P_{wt} is the measured output power of the wind turbine, P_{L} is the system loss at a certain array, $P_{\text{L}\cdot\text{CNT}}$ is the system loss by the center cable, and $P_{\text{L}\cdot\text{IND}}$ is the system loss by the individual cable.

If we assume that a wind turbine can precisely generate reactive power according to the reference signal, the loss equation can be created directly. The real loss at a certain array can be composed of two individual loss equations as follows:

$$P_{\text{L}\cdot\text{CNT}} = r_{\text{CNT}} \cdot \frac{(\sum P_{\text{wt}})^2 + (\sum Q_{\text{ref}})^2}{V^2} \tag{4}$$

$$P_{\text{L}\cdot\text{IND}} = r_{\text{IND}} \cdot \frac{{P_{\text{wt}}}^2 + {Q_{\text{ref}}}^2}{V^2} \tag{5}$$

where r_{CNT} is the resistance of the central cable, r_{IND} is the resistance of the individual cable, and Q_{ref} is the reactive power reference of a certain wind turbine.

According to the above equation, the sectional loss is dependent on not only the active power, but also the reactive power. Therefore, the current control can be achieved with a reactive assignment process by considering the active power variation. The voltage variation at the connection point can normally be neglected due to the continuous regulation by the full-type converters in the wind turbines. However, owing to the significant power fluctuation or system fault, the voltage gap can be too high to ignore in some cases. The proposed method considers both conditions and divides the calculation process into two classified assignment processes.

3.3. Reactive Power Assignment Method

As mentioned above, the total reactive power order for HGS is specified by TSO as a single distribution source. Therefore, the inner control system could designate the appropriate value to each system to match the output profile with the order. The total value of the reactive power reference is shown in Equation (6) and the reference signal for wind system can also be divided into each array as shown in Equations (7) and (8):

$$Q_{\text{order}} = Q_{\text{wave}} + \sum Q_{\text{array}} \tag{6}$$

$$Q_i = Q_{\text{ref1}} + Q_{\text{ref2}} \tag{7}$$

$$Q_j = Q_{\text{ref3}} + Q_{\text{ref4}} \tag{8}$$

where Q_{order} is the reactive power order designated by TSO, Q_{wave} is the reactive power order for wave generators, Q_{array} is the reactive power order for array, Q_i is the total reactive power reference of array i, and Q_j is the total reactive power reference of array j.

The established formulas will be applied at a connection point for assigning the designated reactive power quantity to the two different turbines named A and B in the above section. Considering the voltage variation level, the assignment processes are classified according to the two following processes.

3.3.1. Low Voltage Variation

Normally, the AC voltage levels at all sections in the HGS are almost the same because the scale of the entire system is small to cause a large voltage difference. In this case, the voltage level in Equations (4) and (5) could be neglected in the loss comparison process (V = 1.0 p.u.). Furthermore, since both resistance values in all assignment points are equal due to the structural characteristic, matching the two expected losses could be represented in Equation (9):

$$P_{\text{A}}{}^2 + Q_{\text{A}}{}^2 = P_{\text{B}}{}^2 + Q_{\text{B}}{}^2 \tag{9}$$

where P_{A} is the measured active power output of turbine A, P_{B} is the measured active power output of turbine B, Q_{A} is the required reactive power quantity of turbine A, and Q_{B} is the required reactive power quantity of turbine B.

As the active power value is continuously checked and utilized as a constant value, the reactive power quantity of each wind turbine could be calculated directly by using Equation (10). With this value, the reactive power flow will be modified according to the active power flow by wind turbines. To prevent a negative reference signal due to the low reference quantity by the upper process, the modification processes were established in Equations (11) and (12) as follows:

$$Q_{\text{A}} = \frac{(Q_{\text{array}})^2 + P_{\text{B}}^2 - P_{\text{A}}^2}{2Q_{\text{array}}}, \quad Q_{\text{B}} = \frac{(Q_{\text{array}})^2 + P_{\text{A}}^2 - P_{\text{B}}^2}{2Q_{\text{array}}} \tag{10}$$

$$Q_{\text{A}} = Q_{\text{A}} - |Q_{\text{B}}|, \quad (Q_{\text{B}} < 0) \tag{11}$$

$$Q_{\text{B}} = Q_{\text{B}} - |Q_{\text{A}}|, \quad (Q_{\text{A}} < 0) \tag{12}$$

These equations are applied to find a solution to the inner-array assignment process. The current flow of each cable in the HGS could be more balanced than with the proportional distribution method, increasing the power transfer efficiency. The voltage variation for selecting the assignment mode would be checked for every calculation state, and if the variation is higher than the designated value, the following assignment process will be performed.

3.3.2. Considering Voltage Variation

If the voltage variation is increased due to the abnormal condition, the voltage level of each point should be considered. The voltage condition is checked by IMCS and transferred to the main control algorithm to determine whether or not the variation is considerable. If the value is high and can generate error in the calculation process, the voltage value will be imposed at the current balancing process. As the above section's assignment process, these calculation processes also consider two different points as an assignment required section. By adapting the above equation including voltage values, the reference of a certain point could be designated in Equation (13):

$$Q_A^2 = \left(\frac{V_A}{V_B}\right)^2 \cdot Q_B + \left(\frac{V_A}{V_B}\right)^2 \cdot P_B^2 - P_A^2 \tag{13}$$

where V_A is the measured voltage of turbine A and V_B is the measured voltage of turbine B.

Since the sum of the required reactive power of each point should match the reactive power order, the designated reactive power quantity of the certain connection point can be represented as follows:

$$\left[\left(\frac{V_A}{V_B}\right)^2 - 1\right] Q_A^2 - 2\left(\frac{V_A}{V_B}\right)^2 Q_A Q_{array} + \left(\frac{V_A}{V_B}\right)^2 P_B^2 - P_A^2 + \left(\frac{V_A}{V_B}\right)^2 Q_{array} \tag{14}$$

By solving Equation (14), two reference values for each turbine can be calculated. The modification processes in Equations (11) and (12) are equally applied in this process. With the allocation process, the loss expectation of HGS could be directly determined.

3.3.3. Incremental Loss Comparison

To achieve the loss minimization process, an incremental loss calculation process can be performed as follows:

$$\frac{dLoss}{dQ_i} = \frac{dLoss}{dQ_j} = \frac{dLoss}{dQ_{wave}} = \lambda \tag{15}$$

The minimized power loss by reactive power can be calculated with the incremental loss. When the incremental losses of each part are equal, the references quantity of the wind and wave system is designated and it will be available in the control processes. Then, the reference signal for each generator could be designated as follow above equation. Figure 6 shows the flowchart of the entire assignment process.

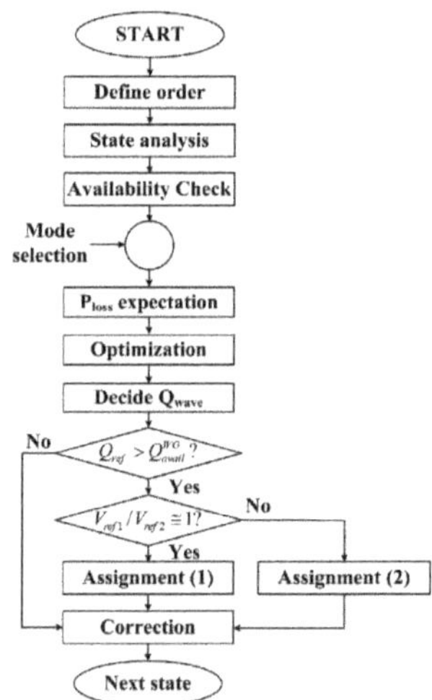

Figure 6. Introduced entire allocation process.

The IMCS first checks the current status and system requirements to confirm the availability of the reactive power control. The capability curve for checking the reactive power reserve depends on the composed conversion system specification and the designated power factor (PF) as Equation (16).

The operator could determine the available reactive power quantity using the real power output with the power conversion capacity as defined in Equation (17):

$$|Q_{max}| = \sqrt{1 - PF^2} S_{cap} \tag{16}$$

$$|Q_{ava}| = \sqrt{S_{cap}^2 - P^2} \tag{17}$$

where Q_{max} is the maximum reactive power capability and S_{cap} is the power conversion capacity of the converter (MVA) Q_{ava} is the available reactive power.

The reactive power control would then be performed according to the designated control mode to reflect another system requirement. In the next stage, the power loss due to the reactive power reference for each section (array 1, array 2, wave generation system) will be calculated with the above assignment process. The loss expectation can be easily carried out with the assigned formulas because the reference signal for each system will be designated with the mentioned linear components of HGS. With the measured active power profile by IMCS, expecting loss, and finding the optimized value for incremental values can be accomplished. Finally, the reference signal for the PCS of the wave generation system will be designated and an optimized solution can also be found for the wind system.

4. Simulation

4.1. System Design

To verify the proposed power control algorithm, the HGS was configured with EMTDC simulation. A 3.3 MVA PCS is utilized to integrate three MW PMSG wind turbines, and the wave power generation system adapts 2.4 MVA PCS to change the individual electrical form of the output profile. The whole PCS was configured with full switching modules. Figure 7 presents the configured PCS in the HGS. Four wind turbines were connected to the grid through a 3-level neutral point clamped (NPC) voltage source inverter [19]. The 4 rotor-side convertors follow MPPT control independently, and the reactive reference currents are also controlled by the individual grid-side convertors utilizing the system states pulled by the phase locked loop and order of the system operator. In order to capture the maximum power from encountered wind, P-ω relation applied look-up table which predefines the points of maximum aerodynamic efficiency is contained. The wind system will generate optimized power according to the maximum power coefficient (c_{opt}) during the simulation. Taking into account the system specification, the wave generation system has adopted two level PCS modules. Also, the VSC for wave generators have previously been configured with individual generator-side converters and a single common grid-side convertor, but in this paper, both PCSs were applied to a single PCS by combining 24 wave generators to reduce simulation complexity. Except for generating the switching signal, both power control modules utilize the measured grid information and follow almost the same topologies.

The converter control is divided into generator side control and grid side control. Since the reactive power supply to the grid is independent from the reactive profile of generator, reactive power orders for the grid side converter including system limits are mainly treated in this paper and the wind power system including mechanical values will follow the referred previous studies. In the current reference designation process, the capability limitations are applied to impose both system's configuration. The generated signal is used to generate a set of three-phase reference voltage to control pulse width modulation (PWM) converter.

The EMTDC simulation for analyzing the proposed algorithm was configured by utilizing real wind data that applied the full wake effect. In the simulation, the wind speed was applied to the turbine with a time-table form. The back part wind turbines would encounter a reduced wind speed compared to the previous state. The reactive power signal for system electrical efficiency could be verified through the designed system.

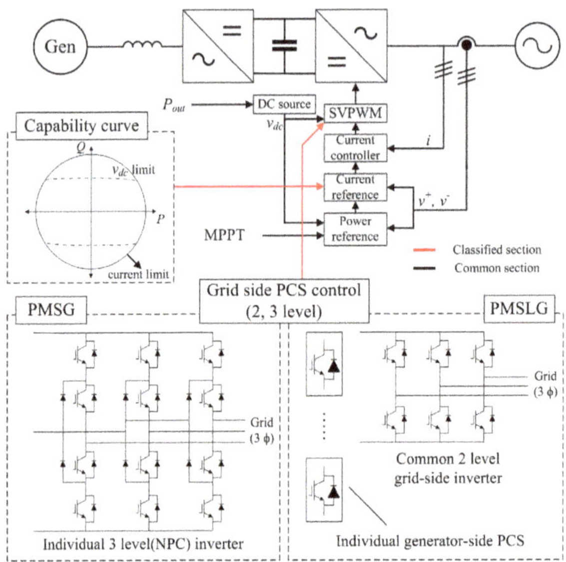

Figure 7. VSC control concept of HGS in the simulation.

The case studies could be verified by designing a feed-back loop to match the reference signal for the reactive power with a real output profile. For appropriate handling the voltage/reactive power according to the real-time variation value, the system profile should be included in the control process (a PI control scheme was added to the reactive power control and performed for each case study).

4.2. Simulation

In order to estimate the proposed method in terms of system efficiency, PSCAD/EMTDC simulation was performed with the realistic power fluctuation condition. The wake analysis result was implemented through the PSCAD/EMTDC simulation for demonstration. To confirm the suitability of the proposed method, not only the efficiency regarding power loss but also the absorbed reactive power flow at PCC need to be verified. Especially, the grid connection requirements should be satisfied regardless of the dynamic power fluctuation according to the wind resources. Figures 8 and 9 show the active power profile of the HGS by dividing the wind generation system and the wave generation system. The entire simulation time is 25 s and the front wind turbine is designated to WT1 and WT3.

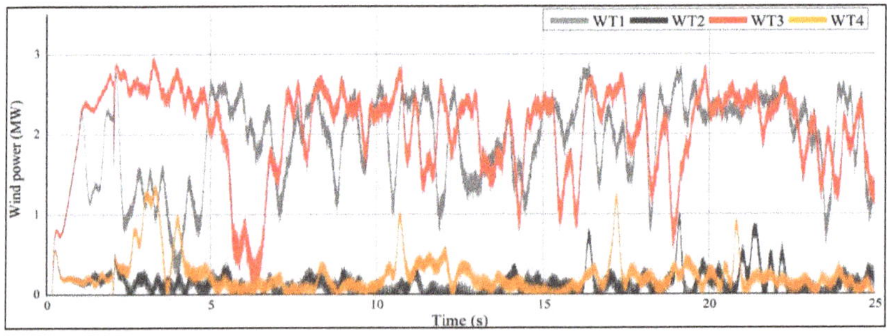

Figure 8. Wind power fluctuation in the designed case studies.

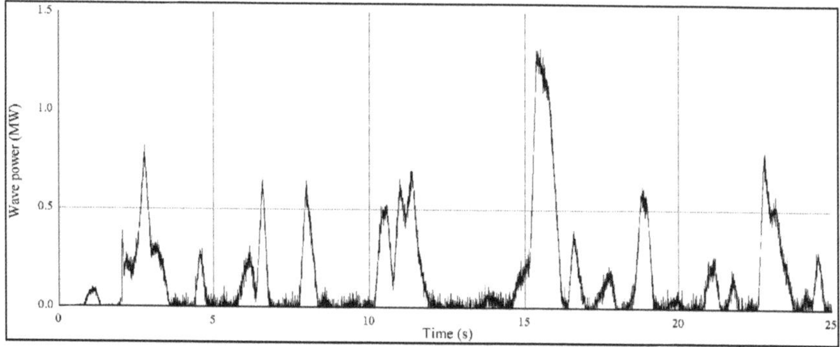

Figure 9. Wave power fluctuation in the designed case studies.

As shown in the mentioned structure (Figure 5), the wind turbines in the simulation were electrically integrated with the central power system through designated cable components. The cable data in Table 1 are utilized in the simulation.

In order to check the control effect and the following state, a caparison between normal and adapted control was equally carried out for the above power generation condition. Table 2 represents the simulation parameters of the designed system in this paper. As mentioned above, the wave generators are comprised of a DC system and are connected with the single PCS in the center of the station. The equivalent source is incorporated by considering the short circuit ratio (SCR), which is used to estimate the system's robustness. Order changes during the simulation are represented in Table 3. In the simulation, the required reactive power changes from 0 MVar to 4.73 MVar (0.95 lagging power factor of entire capacity of HGS). The cases that applied the proposed algorithm are divided into two different simulations for confirming the feasibility of the capacity limitation which is represented in Equation (16). The representative simulation was designed without considering reactive power capability limitation. The simulation of designating maximum reactive power reserve with 0.9 PF was also progressed in the below section. After a short initializing section, the proposed control scheme is applied to the system.

Table 2. Numerical data of the performed simulation.

Number (n)		Rate power (MW)			Grid data		WT-WT distance (m)	Simulation time (s)
WT	WG	WT	WG	Total	SCR	X/R		
4	24	3	0.1	14.4	15	15	100	25

Table 3. Reference signal for case studies.

Case	Initialize section (s)	Normal control (s)	Proposed control (s)	Q order (MVar)
Non-adapted	0–2	2–25	-	4.73
Adapted	0–2	2–3	3–25	4.73

Figure 10 shows the original reactive power curves of the wind and wave generation system in the case study. Every generation system is given the same reference signal by the system operator and generates reactive power equally. Although slight differences exist between each generator due to the electrical condition, the overall output characteristics are constant during the simulation. In the adapted simulation case that represented in Figure 11, however, the reactive power output continuously changes during the simulation. After applying the algorithm at three second, the conversion system automatically imposes calculated reference value; hence the graph indicates the new power curve with the proposed algorithm for the same time period and operational condition. The reactive power curves show opposite characteristics which is contrary to the generated real power as shown in Figure 8.

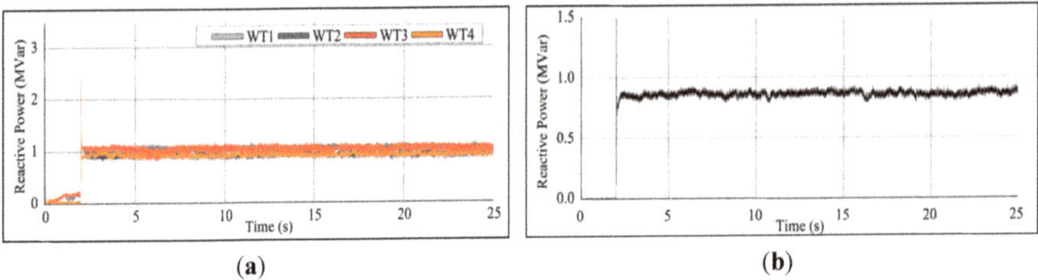

Figure 10. Reactieve power output with normal proportional control: (**a**) wind turbines; and (**b**) wave generation system.

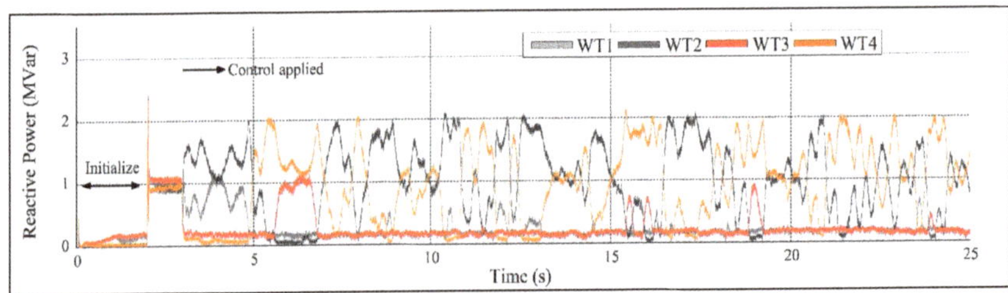

Figure 11. Reactive power output with proposed control-without capability limit (wind turbines).

As shown in the reactive profile curve of Figure 12, the proposed algorithm was performed ordinarily with capability limitation designated by system operator. Although the opposite characteristics about real power is not imposed rather than Figure 11, the fluctuation still follow the current minimization process. The reactive power profile of each generator is adjusted to reduce the apparent power flow in the cable. Not only the wind power but also the wave generation system participates in the control scheme.

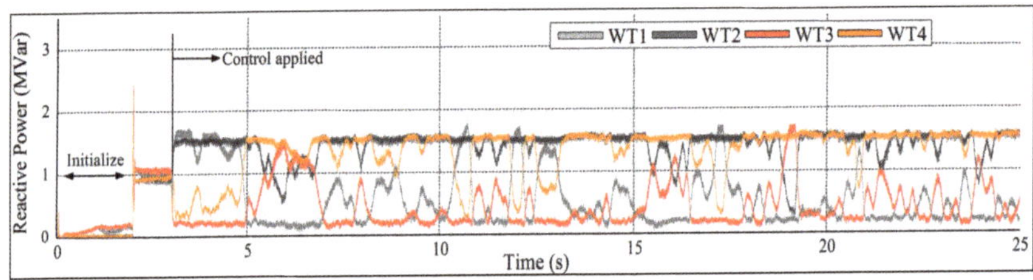

Figure 12. Reactive power output with proposed control-with capability limit (wind turbines).

Figures 13 and 14 show the reactive power fluctuation of the wave generation system with proposed control method. Conversion systems were performed according to the calculated signal by the mentioned formulas, and no measured errors were observed in the simulation using a full scale switching model. Without reactive power capability limit, the converter for wave generation system is fully utilized and the active power change directly affect the reactive power fluctuation. In case of limit-imposed simulation, however, pre-calculated maximum quantity of reactive power regulate the utilization of the applied conversion system.

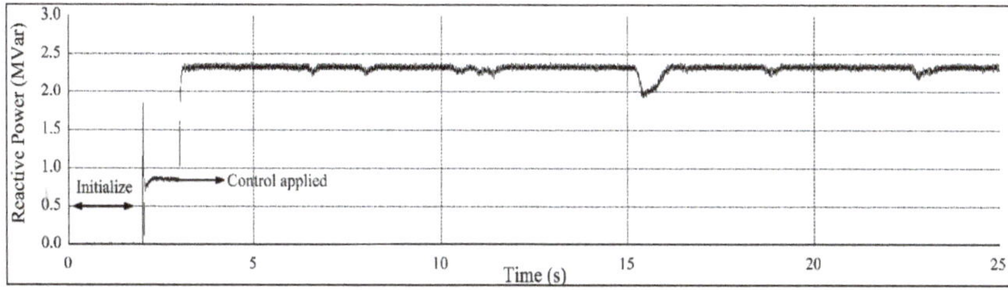

Figure 13. Reactive power output with proposed control-without capability limit (wave generation system).

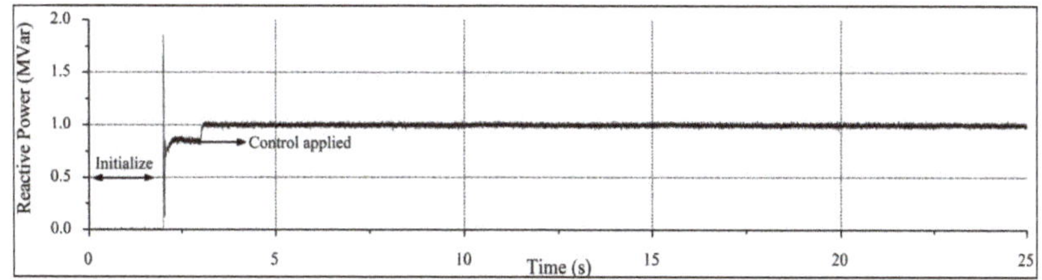

Figure 14. Reactive power output with proposed control-with capability limit (wave generation system).

In order to check the system impact and improvement in terms of system loss, two graphs are shown in Figure 15. The system has continuous reactive power fluctuation during the control process, but the absorbed reactive power at PCC is not affected. At PCC, the reactive power absorption between the two control schemes does not differ, as shown in Figure 15, even if the control method changes. The proposed controls show similar output characteristics in terms of reactive power supply. As shown in the Figure 15b, both curve stably supply reactive power and the averaged quantity is same with the normal operation. Mitigation of the energy loss in the system is depicted in Figure 16, due to the reduction of current flow. The averaged system loss of both methods, the energy loss in the simulation, and the absorbed reactive power at PCC are presented in Table 4. The loss reduction percentages are slightly over 6% and the improved power profile quantity is larger than 0.1 kW/s. Since the energy loss was measured during the entire simulation, the percent improvement value slightly differs with that of power loss. Even if the improved quantity of power loss is not significant, the annual production improvement (assumed to 919.8 kWh) can be a considerable benefit to the system owner. As the reactive support request from utility grid is growing, the impact could be significant than expected state. Moreover, the measuring and integrating processes of precise reactive power reserve will improve the active support plan regarding voltage/reactive power management methods.

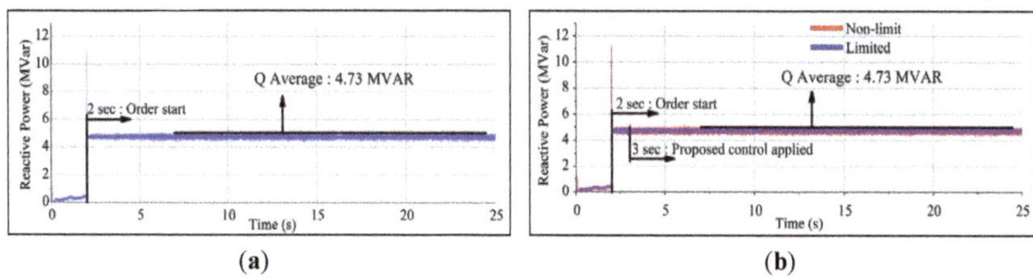

Figure 15. Measured reactive power at point of common coupling (PCC): (**a**) normal; and (**b**) proposed.

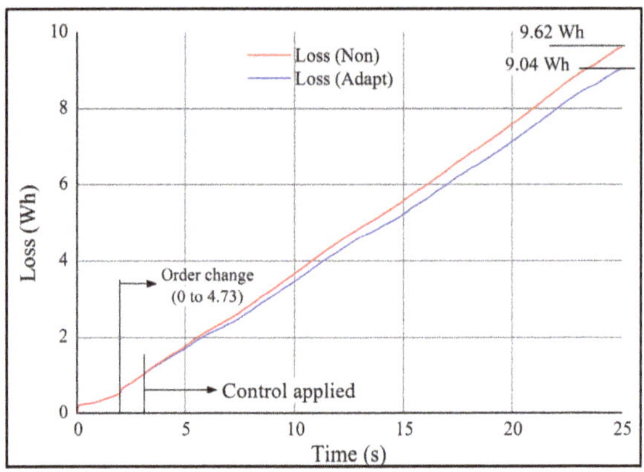

Figure 16. Power loss comparison between both controls.

Table 4. Reference signal for case studies.

Case	Average power loss	Energy loss (25 s)	Absorbed Q at PCC
Non-adapted	1.411 kW	9.62 Wh	4.73 MVar
Adapted	1.316 kW	9.04 Wh	4.73 MVar
Improvement	6.7%	6.02%	-

5. Conclusions

This paper suggests a new power control algorithm for the HGS to achieve an optimization process for the inner system's power flow. Through case studies, it is verified that the proposed algorithm contributes to the system efficiency while satisfying the reactive power reliability demands. Owing to the continuous change of active power flow by generation systems, the control scheme changes the reference signal to ensure the current flow balance. The impact of the algorithm would be significant when considering the entire life cycle of the general renewable energy source. As the application of a real time monitoring system, these current flow management methods can generate additional benefits to the operator, and further reactive power reserves can be utilized for the integrated power system.

Acknowledgments: This work was supported by the MOF grant (Development of the design technologies for a 10 MW class wave and offshore wind hybrid power generation system and establishment of the sea test infra-structure) and by Human Resources Development of KETEP grant (No. 20134030200340) funded by the Korea government Ministry of Knowledge Economy.

Conflicts of Interest: The authors declare no conflict of interest.

References

1. Morales, A.; Robe, X.; Sala, M.; Prats, P.; Aguerri, C.; Torres, E. Advanced grid requirements for the integration of wind farms into the Spanish transmission system. *IET Renew. Power Gener.* **2008**, *2*, 47–57. [CrossRef]
2. Muyeen, S.M.; Takahashi, R.; Murata, T.; Tamura, J. A Variable speed wind turbine control strategy to meet wind farm grid code requirements. *IEEE Trans. Power Syst.* **2010**, *25*, 331–340. [CrossRef]
3. Yun-Hyuk, C.; Sang-Gyun, K.; Byongjun, L. Coordinated Voltage-Reactive Power Control Schemes Based on PMU Measurement at Automated Substations. *J. Electr. Eng. Technol.* **2015**, *10*, 1400–1407.
4. Hartmann, B.; Dan, A. Cooperation of a grid-connected wind farm and an energy storage unit—Demonstration of a simulation tool. *IEEE Trans. Sustain. Energy* **2012**, *3*, 49–56. [CrossRef]

5. Dicorato, M.; Forte, G.; Trovato, M. Voltage compensation for wind integration in power systems. In Proceedings of the 2012 3rd IEEE International Symposium on Power Electronics for Distributed Generation Systems (PEDG), Aalborg, Danmark, 25–28 June 2012; pp. 464–469.
6. Seungmin, J.; Hyun-Wook, K.; Gilsoo, J. Adaptive power control method considering reactive power reserve for wave-offshore hybrid power generator system. *Energy Procedia* **2014**, *61*, 1703–1706.
7. Pérez-Collazo, C.; Greaves, D.; Iglesias, G. A review of combined wave and offshore wind energy. *Renew. Sustain. Energy Rev.* **2015**, *42*, 141–153. [CrossRef]
8. Majumder, R. Aspect of voltage stability and reactive power support in active distribution. *Gener. Transm. Distrib. IET* **2014**, *8*, 442–450. [CrossRef]
9. Zhao, C.; Guo, C. Complete-independent control strategy of active and reactive power for VSC based HVDC system. In Proceedings of the IEEE Power & Energy Society General Meeting, Calgary, AB, Canada, 26–30 July 2009; pp. 1–6.
10. Senturk, O.S.; Helle, L.; Rodriguez, P.; Teodorescu, R. Power capability investigation based on electrothermal models of press-pack IGBT three-level NPC and ANPC VSCs for multimegawatt wind turbines. *IEEE Trans. Power Electron.* **2012**, *27*, 3195–3206. [CrossRef]
11. Tian, J.; Su, C.; Soltani, M.; Chen, Z. Active power dispatch method for a wind farm central controller considering wake effect. In Proceedings of the IEEE Conference on Industrial Electronics Society, Dallas, TX, USA, 29 October–1 November 2014; pp. 5450–5456.
12. Rosyadi, M.; Muyeen, S.M.; Takahashi, R.; Tamura, J. Voltage stability control of wind farm using PMSG based variable speed wind turbine. In Proceedings of the 2012 International Conference on Electrical Machines (ICEM), Marseille, France, 2–5 September 2012; pp. 2192–2197.
13. Beerens, J. Offshore hybrid wind-wave energy converter system. Master's Thesis, Delft University of Technology, Delft, The Netherlands, 26 February 2010.
14. Chunhua, L.; Chau, K.T.; Lee, C.H.; Fei, L. An efficient offshore wind-wave hybrid generation system using direct-drive multitoothed rotating and linear machines. In Proceedings of the 2014 17th International Conference on Electrical Machines and Systems (ICEMS), Hangzhou, Zhejiang, China, 22–25 October 2014; pp. 273–279.
15. Rekola, J.; Tuusa, H. Comparison of line and load converter topologies in a bipolar LVDC distribution. In Proceedings of the 2011 14th European Conference on Power Electronics and Applications, Birmingham, UK, 30 August–1 September 2011; pp. 1–10.
16. Ullah, N.R.; Bhattacharya, K.; Thiringer, T. Wind farms as reactive power ancillary service providers—Technical and economic issues. *IEEE Trans. Energy Convers.* **2009**, *24*, 661–672. [CrossRef]
17. Barthelmie, R.J.; Hansen, K.S.; Pryor, S.C. Meteorological controls on wind turbine wakes. *IEEE Proc.* **2013**, *101*, 1010–1019. [CrossRef]
18. Hyeon-Jae, S.; Jang-Young, C.; Yu-Seop, P.; Min-Mo, K.; Seok-Myeong, J.; Hyungsuk, H. Electromagnetic vibration analysis and measurements of double-sided axial-flux permanent magnet generator with slotless stator. *IEEE Trans. Magn.* **2014**, *50*, 1–4.
19. Bunjongjit, K.; Kumsuwan, Y.; Sriuthaisiriwong, Y. An implementation of three-level BTB NPC voltage source converter based-PMSG wind energy conversion system. In Proceedings of the TENCON 2014-2014 IEEE Region 10 Conference, Bangkok, Thailand, 22–25 October 2014; pp. 1–5.

MDPI AG
St. Alban-Anlage 66
4052 Basel, Switzerland
Tel. +41 61 683 77 34
Fax +41 61 302 89 18
http://www.mdpi.com

Energies Editorial Office
E-mail: energies@mdpi.com
http://www.mdpi.com/journal/energies

www.ingramcontent.com/pod-product-compliance
Lightning Source LLC
Chambersburg PA
CBHW051715210326
41597CB00032B/5487

* 9 7 8 3 0 3 8 4 2 4 9 4 9 *